传世经典书丛
Eternal Classics

架构之美

行业思想领袖揭秘软件设计之美
（评注版）

Beautiful Architecture
Leading Thinkers Reveal the Hidden Beauty in Software Design

【美】Diomidis Spinellis 编
Georgios Gousios

张逸　评注

电子工业出版社
Publishing House of Electronics Industry
北京·BEIJING

内 容 简 介

本书是荟萃了软件架构领域各位思想领袖真知灼见的经典之作,内容覆盖了软件架构的方方面面,包括架构理论、企业架构、系统架构、应用架构等。这些架构大师们用简洁的文本、真实的案例向读者勾勒出美丽架构的模样,并由此提出设计美丽架构的原则、实践与演进过程。全书传递的架构知识既有高屋建瓴的系统描述,又有深入系统的全面剖析,全面体现了架构设计中的简洁之美、清晰之美、风格之美、灵活之美和演进之美。

本书适合期待提高架构能力,学会欣赏架构之美的开发人员与架构师。

© 2009 by O'Reilly Media,Inc.

English Adaptation Edition, jointly published by O'Reilly Media, Inc. and Publishing House of Electronics Industry, 2018. Authorized translation of the English edition, 2009 O'Reilly Media, Inc., the owner of all rights to publish and sell the same. All rights reserved including the rights of reproduction in whole or in part in any form.

本书英文影印改编版专有出版权由 O'Reilly Media, Inc.授予电子工业出版社。未经许可,不得以任何方式复制或抄袭本书的任何部分。专有出版权受法律保护。

版权贸易合同登记号　图字：01-2013-4027

图书在版编目（CIP）数据

架构之美：行业思想领袖揭秘软件设计之美：评注版 /（美）迪奥米德斯·斯宾耐立思（Diomidis Spinellis），（美）乔治斯·郭西奥斯（Georgios Gousios）编；张逸评注. —北京：电子工业出版社，2018.6

（传世经典书丛）

书名原文：Beautiful Architecture: Leading Thinkers Reveal the Hidden Beauty in Software Design

ISBN 978-7-121-33807-6

Ⅰ. ①架… Ⅱ. ①迪… ②乔… ③张… Ⅲ. ①软件设计 Ⅳ. ①TP311.5

中国版本图书馆 CIP 数据核字（2018）第 042441 号

策划编辑：符隆美
责任编辑：付　睿

印　　刷：北京虎彩文化传播有限公司
装　　订：北京虎彩文化传播有限公司
出版发行：电子工业出版社
　　　　　北京市海淀区万寿路 173 信箱　邮编：100036
开　　本：850×1168　1/16　印张：26.25　字数：550 千字
版　　次：2018 年 6 月第 1 版
印　　次：2021 年 5 月第 9 次印刷
定　　价：89.00 元

凡所购买电子工业出版社图书有缺损问题，请向购买书店调换。若书店售缺，请与本社发行部联系，联系及邮购电话：(010) 88254888, 88258888。

质量投诉请发邮件至 zlts@phei.com.cn，盗版侵权举报请发邮件至 dbqq@phei.com.cn。

本书咨询联系方式：(010) 51260888-819，faq@phei.com.cn。

出版说明

悦读上品 得乎益友

孔子云:"取乎其上,得乎其中;取乎其中,得乎其下;取乎其下,则无所得矣"。

对于读书求知而言,这句古训教我们去读好书,最好是好书中的上品——经典书。其中,科技人员要读的技术书,因为直接关乎客观是非与生产效率,阅读选材本更应慎重。然而,随着技术图书品种的日益丰富,发现经典书越来越难,尤其对于涉世尚浅的新读者,更为不易,而他们又往往是最需要阅读、提升的重要群体。

所谓经典书,或说上品,是指选材精良、内容精练、讲述生动、外延丰盈、表现手法体贴入微的读品,它们会成为读者的知识和经验库中的重要组成部分,并且拥有从不断重读中汲取养分的空间。因此,选择阅读上品的问题便成了有效阅读的首要问题。当然,这不只是效率问题,上品促成的既是对某一种技术、思想的真正理解和掌握,同时又是一种感悟或享受,是一种愉悦。

与技术本身类似,经典 IT 技术书多来自国外。深厚的积累、良好的写作氛围,使一批大师为全球技术学习者留下了璀璨的智慧瑰宝。就在那个年代即将远去之时,无须回眸,也能感受到这一部部厚重而深邃的经典著作,在造福无数读者后发出从未蒙尘的熠熠光辉。而这些凝结众多当今国内技术中坚美妙记忆与绝佳体验的技术图书,虽然尚在国外图书市场上大放异彩,却已逐渐淡出国人的视线。最为遗憾的是,迟迟未有可以填补空缺的新书问世。而无可替代,不正是经典书被奉为圭臬的原因?

为了不让国内读者,尤其是即将步入技术生涯的新一代读者,就此错失这些滋养过先行者们的好书,以出版 IT 精品图书,满足技术人群需求为己任的我们,愿意承担这一使命。本次机遇惠顾了我们,让我们有机会精心推出"传世经典书丛"。

在我们眼中,"传世经典"的价值首先在于——既适合喜爱科技图书的读者,也符合专家们挑剔的标准。幸运的是,我们的确找到了这些堪称上品的佳作。丛书带给我们的幸运颇多,细数一下吧。

得以引荐大师著作

有恐思虑不周,我们大量参考了国外权威机构和网站的评选结果,又进一步对符合标准之图书的国内外口碑与销售情况进行细致分析，也听取了国内技术专家的宝贵建议，才有幸选出对国内读者最富有技术养分的大师上品。

■ 向深邃的技术内涵致敬

中外技术环境存在差异，很多享誉国外的好书未必适用于国内读者；且技术与应用瞬息万变，很容易让人心生迷惘或疲于奔命。本丛书的图书遴选，注重打好思考方法与技术理念的根基，旨在帮助读者修炼内功，提升境界，将技术真正融入个人知识体系，从而可以一通百通，从容面对随时涌现的技术变化。

■ 翻译与评注的双项选择

引进优秀外版著作，将其翻译为中文供国内读者阅读，较为有效与常见。但另有一些外语水平较高、喜好阅读原版的读者，苦于对技术理解不足，不能充分体会原文表述的精妙，需要有人指导与点拨。而一批本土技术精英经过长期经典熏陶及实践锤炼，已足以胜任这一工作。有鉴于此，本丛书在翻译版的同时推出融合英文原著与中文点评、注释的评注版，供不同志趣的读者自由选择。

■ 承蒙国内一流译(注)者的扶持

优秀的英文原著最终转化为真正的上品，尚需跨越翻译鸿沟，外版图书的翻译质量一直屡遭国内读者诟病。评注版的增值与含金量，同样依赖于评注者的高卓才具。好在，本丛书得到了久经考验的权威译(注)者的认可和支持，首肯我们选用其佳作，或亲自参与评注工作。正是他们的参与保证了经典的品质，既再次为我们的选材把关，更提供了一流的中文表述。

■ 期望带给读者良好的阅读体验

一本好书带给人的愉悦不止于知识收获，良好的阅读感受同样不可缺少，且对学业不无助益。为让读者收获与上品相称的体验，我们在图书装帧设计与选材用料上同样不敢轻率，惟愿送到读者手中的除了珠玑章句，还有舒适与熨帖的视觉感受。

所有参与丛书出版的人员，尽管能力有限，却无不心怀严谨之心与完美愿望。如果读者朋友能从潜心阅读这些上品中偶有获益，不啻为对我们工作的最佳褒奖。若有阅读感悟，敬请拨冗告知，以鼓励我们继续在这一道路上贡献绵薄之力。如有不周之处，也请不吝指教。

电子工业出版社博文视点

All royalties from this book will be donated to Doctors Without Borders.

评注者序

软件架构终归属于工程学的范畴，不能一概以"只可意会不可言传"来搪塞，因为架构知识是可以传递的，架构文档是可以共享的，最重要的是，架构自身是可以评审、验证与实现的。

Stephen J. Mellor 在"*Beautiful Architecture*"一书的序中，画龙点睛地勾勒出美丽架构的模样，即必须遵循的一些普遍原则，分别为：

- One fact In one place（一处一事实）
- Automatic propagation（自动传播）
- Architecture includes construction（架构包含构建）
- Minimize mechanisms（最小化机制）
- Constuct engines（构建引擎）
- O(G), the order of growth（O(G)，增长的阶）
- Resist entropy（抵制熵）

这些原则，其实就是架构师的智慧，没有足够深刻的理解与深入实践，是不可能给出如此言简意赅的架构建议的。按照我的理解，这些普适性原则其实就是在说明所谓美丽的架构，**就是简单、一致、适应变化并能去除重复的架构**。就如 Mellor 所言——**美丽的架构能用更少的机制做更多的工作**。这就是"*Beautiful Architecture*"一书不凡的开篇。

若是一本平庸的书，必然会惧怕这样精彩绝伦的序，因为它愈发的美，就愈发能照映出正文的丑；它愈发的言之有物，又愈发会衬托出正文的空洞无味。然而，若是内容是超乎寻常的精彩绝伦，这样的序就无异于锦上添花，珠联璧合了。通透点儿，就是齐活！这就好比一首歌曲的领唱者，倘若一开始就飙出高音，声入云霄。后续跟着唱的人要是没有点儿本事，恐怕就难以为继了；可要都是高手呢？那就真是一场音乐的盛宴了。

"*Beautiful Architecture*"荟萃了全球最顶级的架构师和意见领袖，他们在这本书中唱出了架构思想、实践与原则的最强音。全书共分为以下 5 个主题。

- On Architeture
- Enterprise Application Architecture
- Systems Architecture
- End-User Application Architectures
- Languages and Architecture

这些主题几乎覆盖了软件架构的方方面面，精选的每篇文章可谓字字珠玑，充满了写作者的真知灼见。开卷阅读，如与大师对话，聆听者必须凝神应对，稍不留神就可能遗漏那些重要而正确的意见，影响到对整篇文章的理解。整本书正文不足 400 页，然而每次阅读皆有新意，书的内容仿佛博尔赫斯笔下小径分叉的花园，花园虽小，景色却变幻多姿，路途虽短，距离却无穷无尽，咫尺天涯。

　　因此，作为本书的评注者，真可以说是战战兢兢、如履薄冰。我的每句点评都尽力追求达到个人最大努力的完美，不求锦上添花，只求不得"狗尾续貂"之嫌。安全地说，这些评注不过是我写在这本大书边上的感悟罢了。这些感悟，或是阅读到精彩段落的击节赞叹，或是不明其义而反复研读之后的醍醐灌顶，或是触类旁通体会到架构本质因而不揣冒昧地给出自己的心得体会。在评注过程中，我恪守"扬长避短"的原则，不懂就不装懂，默不作声，当一位沉默的看客；一旦涉猎到自己擅长的部分，却也不妨洋洋洒洒高谈阔论，坦承自己的观点。

　　对于这些架构领袖们，我怀揣敬意，却也不愿以一种卑微的心态被动接受。我需要做一个具有自己人格和高度的评注者。至于对否，就交给读者诸君对评注再做一次"评注"吧！

张　逸

个人博客：http://zhangyi.xyz

目录

序   xix
    Stephen J. Mellor
前言   xxiii

第 1 部分   论架构

第 1 章   何谓架构   3
    John Klein 与 David Weiss
简介   3
创建软件架构   10
架构结构   14
好的架构   19
美丽架构   20
致谢   23
参考文献   23

第 2 章   两个系统的故事：摩登时代的软件神话   25
    Pete Goodliffe
混乱大都市   26
设计之城   33
然后呢   41
轮到你了   41
参考文献   42

第 2 部分   企业应用架构

第 3 章   可伸缩架构   45
    Jim Waldo
简介   45
上下文   47
架构   51

	对架构的思考	57
第 4 章	记忆留存	63
	Michael Nygard	
	功能与约束	64
	工作流	65
	架构要素	66
	用户反应	87
	结论	88
	参考文献	88
第 5 章	面向资源架构：在 Web 之中	89
	Brian Sletten	
	简介	89
	传统的 Web 服务	90
	Web	92
	面向资源架构	98
	数据驱动应用	102
	运用面向资源架构	103
	结论	109
第 6 章	数据增长：Facebook 平台的架构	111
	Dave Fetterman	
	简介	111
	创建社交 Web 服务	117
	创建社交数据查询服务	124
	创建社交 Web 门户：FBML	133
	系统的支持功能	146
	总结	151

第 3 部分　系统架构

第 7 章	XEN 与虚拟化之美	155
	Derek Murray 与 Keir Fraser	
	简介	155
	Xenoservers	156
	虚拟化的挑战	159
	半虚拟化	159

	Xen 的变化	163
	变化中的硬件，变化中的 Xen	169
	前车之鉴	172
	延伸阅读	173

第 8 章　Guardian：一个容错操作系统环境　175
Greg Lehey

	Tandem/16：未来所有计算机都将像这样构建	176
	硬件	176
	机械设计	178
	处理器架构	179
	处理器间总线	184
	输入/输出	184
	进程结构	185
	消息系统	186
	文件系统	190
	逸闻趣事	195
	弊端	195
	后继者	197
	延伸阅读	198

第 9 章　JPC：纯 Java 的 X86 PC 模拟器　199
Rhys Newman 与 Christopher Dennis

	简介	200
	概念验证	202
	PC 架构	205
	Java 性能技巧	206
	4GB 放入 4GB：这不会奏效	207
	保护模式的危险	210
	屡败屡战	214
	劫持 JVM	217
	终极灵活性	229
	终极安全性	231
	吃一堑长一智	232

第 10 章　元循环虚拟机的力量：Jikes RVM　235
Ian Rogers 与 Dave Grove

	背景	236

运行时环境之谜	237
Jikes RVM 简史	240
自部署运行时的自举	241
运行时组件	246
前车之鉴	259
参考文献	259

第 4 部分　终端用户应用架构

第 11 章　GNU Emacs：滋生的特性为其优势　263
Jim Blandy

Emacs 在使用	264
Emacs 的架构	266
滋生的特性	272
另外两个架构	275

第 12 章　当集市开始构建教堂　279
Till Adam 与 Mirko Boehm

简介	279
KDE 项目的历史与架构	282
Akonadi	287
ThreadWeaver	303

第 5 部分　语言与架构

第 13 章　软件架构：面向对象 vs. 面向函数　315
Bertrand Meyer

概览	315
函数式示例	318
评估函数式解决方案的模块化	321
面向对象视图	330
评估和改进面向对象的模块化	336
代理：将操作包裹到对象中	341
致谢	345
参考文献	346

第 14 章　重读经典　349
Panagiotis Louridas

万物皆对象	353
类型被隐式定义	361
问题	367
土木建筑架构	372
参考文献	380

跋　　　　　　　　　　　　　　　　　　　　383
William J. Mitchell

贡献者　　　　　　　　　　　　　　　　　　387

CONTENTS

	FOREWORD *by Stephen J. Mellor*	xix
	PREFACE	xxiii

Part One   ON ARCHITECTURE

1	WHAT IS ARCHITECTURE? *by John Klein and David Weiss*	3
	Introduction	3
	Creating a Software Architecture	10
	Architectural Structures	14
	Good Architectures	19
	Beautiful Architectures	20
	Acknowledgments	23
	References	23
2	A TALE OF TWO SYSTEMS: A MODERN-DAY SOFTWARE FABLE *by Pete Goodliffe*	25
	The Messy Metropolis	26
	Design Town	33
	So What?	41
	Your Turn	41
	References	42

Part Two   ENTERPRISE APPLICATION ARCHITECTURE

3	ARCHITECTING FOR SCALE *by Jim Waldo*	45
	Introduction	45
	Context	47
	The Architecture	51
	Thoughts on the Architecture	57
4	MAKING MEMORIES *by Michael Nygard*	63
	Capabilities and Constraints	64
	Workflow	65
	Architecture Facets	66
	User Response	87

	Conclusion	88
	References	88
5	**RESOURCE-ORIENTED ARCHITECTURES: BEING "IN THE WEB"**	**89**
	by Brian Sletten	
	Introduction	89
	Conventional Web Services	90
	The Web	92
	Resource-Oriented Architectures	98
	Data-Driven Applications	102
	Applied Resource-Oriented Architecture	103
	Conclusion	109
6	**DATA GROWS UP: THE ARCHITECTURE OF THE FACEBOOK PLATFORM**	**111**
	by Dave Fetterman	
	Introduction	111
	Creating a Social Web Service	117
	Creating a Social Data Query Service	124
	Creating a Social Web Portal: FBML	133
	Supporting Functionality for the System	146
	Summation	151

Part Three   SYSTEMS ARCHITECTURE

7	**XEN AND THE BEAUTY OF VIRTUALIZATION**	**155**
	by Derek Murray and Keir Fraser	
	Introduction	155
	Xenoservers	156
	The Challenges of Virtualization	159
	Paravirtualization	159
	The Changing Shape of Xen	163
	Changing Hardware, Changing Xen	169
	Lessons Learned	172
	Further Reading	173
8	**GUARDIAN: A FAULT-TOLERANT OPERATING SYSTEM ENVIRONMENT**	**175**
	by Greg Lehey	
	Tandem/16: Some Day All Computers Will Be Built Like This	176
	Hardware	176
	Mechanical Layout	178
	Processor Architecture	179
	The Interprocessor Bus	184
	Input/Output	184
	Process Structure	185
	Message System	186
	File System	190
	Folklore	195
	The Downside	195

	Posterity	197
	Further Reading	198
9	JPC: AN X86 PC EMULATOR IN PURE JAVA *by Rhys Newman and Christopher Dennis*	199
	Introduction	200
	Proof of Concept	202
	The PC Architecture	205
	Java Performance Tips	206
	Four in Four: It Just Won't Go	207
	The Perils of Protected Mode	210
	Fighting A Losing Battle	214
	Hijacking the JVM	217
	Ultimate Flexibility	229
	Ultimate Security	231
	It Feels Better the Second Time Around	232
10	THE STRENGTH OF METACIRCULAR VIRTUAL MACHINES: JIKES RVM *by Ian Rogers and Dave Grove*	235
	Background	236
	Myths Surrounding Runtime Environments	237
	A Brief History of Jikes RVM	240
	Bootstrapping a Self-Hosting Runtime	241
	Runtime Components	246
	Lessons Learned	259
	References	259

Part Four   END-USER APPLICATION ARCHITECTURES

11	GNU EMACS: CREEPING FEATURISM IS A STRENGTH *by Jim Blandy*	263
	Emacs in Use	264
	Emacs's Architecture	266
	Creeping Featurism	272
	Two Other Architectures	275
12	WHEN THE BAZAAR SETS OUT TO BUILD CATHEDRALS *by Till Adam and Mirko Boehm*	279
	Introduction	279
	History and Structure of the KDE Project	282
	Akonadi	287
	ThreadWeaver	303

Part Five   LANGUAGES AND ARCHITECTURE

13	SOFTWARE ARCHITECTURE: OBJECT-ORIENTED VERSUS FUNCTIONAL *by Bertrand Meyer*	315
	Overview	315

	The Functional Examples	318
	Assessing the Modularity of Functional Solutions	321
	An Object-Oriented View	330
	Assessing and Improving OO Modularity	336
	Agents: Wrapping Operations into Objects	341
	Acknowledgments	345
	References	346
14	**REREADING THE CLASSICS**	349
	by Panagiotis Louridas	
	Everything Is an Object	353
	Types Are Defined Implicitly	361
	Problems	367
	Brick and Mortar Architecture	372
	References	380
	AFTERWORD	383
	by William J. Mitchell	
	CONTRIBUTORS	387

Foreword

Stephen J. Mellor

THE CHALLENGES OF DEVELOPING HIGH-PERFORMANCE, HIGH-RELIABILITY, and high-quality software systems are too much for ad hoc and informal engineering techniques that might have worked in the past on less demanding systems. The complexity of our systems has risen to the point where we can no longer cope without developing and maintaining a single overarching architecture that ties the system into a coherent whole and avoids piecemeal implementation, which causes testing and integration failures.

But building an architecture is a complex task. Examples are hard to come by, due to either proprietary concerns or the opposite, a need to "sell" a particular architectural style into a wide range of environments, some of which are inappropriate. And architectures are big, which makes them difficult to capture and describe without overwhelming the reader.

Yet beautiful architectures exhibit a few universal principles, some of which I outline here:

> One fact in one place
>> Duplication leads to error, so it should be avoided. Each fact must be a single, nondecomposable unit, and each fact must be independent of all other facts. When change occurs, as it inevitably does, only one place need be modified. This principle is well known to database designers, and it has been formalized under the name of *normalization*. The principle also applies less formally to behavior, under the name *factoring*, such that common functionality is factored out into separate modules.

▶ One fact in one place即"DRY (Don't Repeat Yourself)"原则。重复是糟糕架构的典型体现。针对一个问题域，提供多个大同小异的解决方案，会最终导致"解决方案蔓延"。一旦问题域发生变化，就需要多处进行修改。若某个解决方案忘记修改，则可能引入潜在的缺陷。

（未完见下页）

（接上页）

要避免重复，其手段可以选择抽象（Abstraction）或分解（Partition）。利用抽象可寻求其共性特征，进行共性与可变性分析，既精简了模型，又可重用共性部分。利用分解，可缩小逻辑单元，使其变得更加可重用。然而，此两种方式都是有代价的。抽象会导致间接层次增多，从而影响性能；分解则会带来大量细粒度单元，使得系统的实体（模块、类、方法）数量增加，增加了系统的复杂性。

因此，面对重复，我们需要权衡。设计的重点就在于权衡。

Beautiful architectures find ways to localize information and behavior. At runtime, this manifests as *layering*, the notion that a system may be factored into layers, each representing a *layer of abstraction* or *domain*.

Automatic propagation

One fact in one place sounds good, but for efficiency's sake, some data or behavior is often duplicated. To maintain consistency and correctness, propagation of these facts must be carried out automatically at construction time.

Beautiful architectures are supported by construction tools that effect *meta-programming*, propagating one fact in one place into many places where they may be used efficiently.

Architecture includes construction

An architecture must include not only the runtime system, but also how it is constructed. A focus solely on the runtime code is a recipe for deterioration of the architecture over time.

Beautiful architectures are *reflective*. Not only are they beautiful at runtime, but they are also beautiful at construction time, using the same data, functions, and techniques to build the system as those that are used at runtime.

Minimize mechanisms

The best way to implement a given function varies case by case, but a beautiful architecture will not strive for "the best." There are, for example, many ways of storing data and searching it, but if the system can meet its performance requirements using one mechanism, there is less code to write, verify, maintain, and occupy memory.

Beautiful architectures employ a minimal set of mechanisms that satisfy the requirements of the whole. Finding "the best" in each case leads to proliferation of error-prone mechanisms, whereas adding mechanisms parsimoniously leads to smaller, faster, and more robust systems.

Construct engines

If you wish to build brittle systems, follow Ivar Jacobson's advice and base your architecture on use cases and one function at a time (i.e., use "controller" objects). Extensible systems, on the other hand, rely on the construction of virtual machines—engines that are "programmed" by data provided by higher layers, and that implement multiple application functions at a time.

This principle appears in many guises. "Layering" of virtual machines goes back to Edsger Dijkstra. "Data-driven systems" provide engines that rely on coding invariants in the system, letting the data define the specific functionality in a particular case. These engines are highly reusable—and beautiful.

O(G), the order of growth

Back in the day, we thought about the "order" of algorithms, analyzing the performance of sorting, say, in terms of the time it takes to sort a set of a certain number of elements. Whole books have been written on the subject.

The same applies for architecture. Polling, for example, works well for a small number of elements, but is a response-time disaster as the number of items increases. Organizing everything around interrupts or events works well until they all go off at once. Beautiful architectures consider the direction of likely growth and account for it.

Resist entropy

Beautiful architectures establish a path of least resistance for maintenance that preserves the architecture over time and so slows the effects of the Law of System Entropy, which states that systems become more disorganized over time. Maintainers must internalize the architecture so that changes will be consistent with it and not increase system entropy.

One approach is the Agile concept of the *Metaphor*, which is a simple way to represent what the architecture is "like." Another is extensive documentation and threats of unemployment, though that seldom works for long. Usually, however, it generally means tools, especially for generating the system. A beautiful architecture must remain beautiful.

These principles are highly interrelated. One fact in one place can work only if you have automatic propagation, which in turn is effective when the architecture takes construction into account. Similarly, constructing engines and minimizing mechanisms support one fact in one place. Resisting entropy is a requirement for maintaining an architecture over time, and it relies on the architecture including construction and support for propagation. Moreover, a failure to consider the way in which a system will likely grow will cause the architecture to become unstable, and eventually fail under extreme but predictable circumstances. And combining minimal mechanisms with the notion of constructing engines means that beautiful architectures usually feature a limited set of patterns that enable construction of arbitrary system extensions, a kind of "expansion by pattern."

In short, beautiful architectures do more with less.

As you read this book, ably assembled and introduced by Diomidis Spinellis and Georgios Gousios, you might look for these principles and consider their implications, using the specific examples presented in each chapter. You might also look for violations of these principles and ask whether the architecture is thus ugly or whether some higher principle is involved.

During the development of this Foreword, your authors asked me if I might say a few words about how someone becomes a good architect. I laughed. If we only knew that…. But then I recalled from my own experience that there is a powerful, if nonanalytic, way of becoming a

▶ 当今的软件系统，规模越来越大，数据越来越海量，空间复杂度与时间复杂度会成为系统设计的主要关注点。这带来的影响是，无论架构还是编码，都需要提前考虑系统可能面对的极限环境。例如内存是否会溢出，分布式环境下网络连接是否中断，请求数过多是否会造成应用服务的阻塞……

这些皆可视为"风险"。在进行软件架构时，需要提前识别这些风险，并对风险进行优先级排列；然后从这些风险出发去寻找合理的解决方案。应该对解决方案进行为期较短的技术预研（Spike），并搭建与真实环境相当的虚拟测试环境。

beautiful architect. That way* is never to believe that the last system you built is the only way to build systems, and to seek out many examples of different ways of solving the same type of problem. The example beautiful architectures presented in this book are a step forward in helping you meet that goal.

* Or exercise more and eat less.

Preface

THE IDEA FOR THE BOOK YOU'RE READING WAS CONCEIVED IN 2007 as a successor to the award-winning, best-selling *Beautiful Code*: a collection of essays about innovative and sometimes surprising solutions to programming problems. In *Beautiful Architecture*, the scope and purpose is different, but similarly focused: to get leading software designers and architects to describe a software architecture of their choice, peeling back the layers of their creations to show how they developed software that is functional, reliable, usable, efficient, maintainable, portable, and, yes, elegant.

To put together this book, we contacted leading architects of well-known or less-well-known but highly innovative software projects. Many of them replied promptly and came back to us with thought-provoking ideas. Some of the contributors even caught us by surprise by proposing not to write about a specific system, but instead investigating the depth and the extent of architectural aspects in software engineering.

All chapter authors were glad to hear that the work they put in their chapters is also helping a good cause, as the royalties of this book are donated to *Medécins Sans Frontières* (Doctors Without Borders), an international humanitarian aid organization that provides emergency medical assistance to suffering people.

How This Book Is Organized

We have organized the contents of this book around five thematic areas: overviews, enterprise applications, systems, end-user applications, and programming languages. There is an obvious, but not deliberate, lack of chapters on desktop software architectures. Having approached more than 50 software architects, this result was another surprise for us. Are there really no shining examples of beautiful desktop software architectures? Or are talented architects shying away from an area often driven by a quest to continuously pile ever more features on an application? We are really looking forward to hearing from you on these issues.

Part I: On Architecture

Part I of this book examines the breadth and scope of software architecture and its implications for software development and evolution.

Chapter 1, *What Is Architecture?*, by John Klein and David Weiss, defines software architecture by examining the subject through the perspectives of quality concerns and architectural structures.

Chapter 2, *A Tale of Two Systems: A Modern-Day Software Fable*, by Pete Goodliffe, provides an allegory on how software architectures can affect system evolution and developer engagement with a project.

Part II: Enterprise Application Architecture

Enterprise systems, the IT backbone of many organizations, are large and often tailor-made conglomerates of software usually built from diverse components. They serve large, transactional workloads and must scale along with the enterprise they support, readily adapting to changing business realities. Scalability, correctness, stability, and extensibility are the most important concerns when architecting such systems. Part II of this book includes some exemplar cases of enterprise software architectures.

Chapter 3, *Architecting for Scale*, by Jim Waldo, demonstrates the architectural prowess required to build servers for massive multiplayer online games.

Chapter 4, *Making Memories*, by Michael Nygard, goes through the architecture of a multistage, multisite data processing system and presents the compromises that must be made to make it work.

Chapter 5, *Resource-Oriented Architectures: Being "In the Web"*, by Brian Sletten, discusses the power of resource mapping when constructing data-driven applications and provides an elegant example of a purely resource-oriented architecture.

Chapter 6, *Data Grows Up: The Architecture of the Facebook Platform*, by Dave Fetterman, advocates data-centric systems, explaining how a good architecture can create and support an application ecosystem.

Part III: Systems Architecture

Systems software is arguably the most demanding type of software to design, partly because efficient use of hardware is a black art mastered by a selected few, and partly because many consider systems software as infrastructure that is "simply there." Seldom are great systems architectures designed on a blank sheet; most systems that we use today are based on ideas first conceived in the 1960s. The chapters in Part III walk you through four innovative systems software architectures, discussing the complexities behind the architectural decisions that made them beautiful.

Chapter 7, *Xen and the Beauty of Virtualization*, by Derek Murray and Keir Fraser, gives an example of how a well-thought-out architecture can change the way operating systems evolve.

Chapter 8, *Guardian: A Fault-Tolerant Operating System Environment*, by Greg Lehey, presents a retrospective on the architectural choices and building blocks (both software and hardware) that made Tandem the platform of choice in high-availability environments for nearly two decades.

Chapter 9, *JPC: An x86 PC Emulator in Pure Java*, by Rhys Newman and Christopher Dennis, describes how carefully designed software and a good understanding of domain requirements can overcome the perceived deficiencies of a programming system.

Chapter 10, *The Strength of Metacircular Virtual Machines: Jikes RVM*, by Ian Rogers and Dave Grove, walks us through the architectural choices required for creating a self-optimizable, self-hosting runtime for a high-level language.

Part IV: End-User Application Architectures

End-user applications are those that we interact with in our everyday computing lives, and the software that our CPUs burn the most cycles to execute. This kind of software normally does not need to carefully manage resources or serve large transaction volumes. However, it does need to be usable, secure, customizable, and extensible. These properties can lead to popularity and widespread use and, in the case of free and open source software, to an army of volunteers willing to improve it. In Part IV, the authors dissect the architectures and the community processes required to evolve two very popular desktop software packages.

Chapter 11, *GNU Emacs: Creeping Featurism Is a Strength*, by Jim Blandy, explains how a set of very simple components and an extension language can turn the humble text editor into ~~an operating system~~* the Swiss army knife of a programmer's toolchest.

---

* As some die-hard users say, "Emacs is my operating system; Linux just provides the device drivers."

Chapter 12, *When the Bazaar Sets Out to Build Cathedrals*, by Till Adam and Mirko Boehm, demonstrates how community processes such as sprints and peer-reviews can help software architectures evolve from rough sketches into beautiful systems.

Part V: Languages and Architecture

As many people have pointed out in their works, the programming language we use affects the way we solve a problem. But can a programming language also affect a system's architecture and, if so, how? In the architecture of buildings, new materials and the adoption of CAD systems allowed the expression of more sophisticated and sometimes strikingly beautiful designs; does the same also apply to computer programs? Part V, which contains the last two chapters, investigates the relationship between the tools we use and the designs we produce.

Chapter 13, *Software Architecture: Object-Oriented Versus Functional*, by Bertrand Meyer, compares the affordances of object-oriented and functional architectural styles.

Chapter 14, *Rereading the Classics*, by Panagiotis Louridas, surveys the architectural choices behind the building blocks of modern and classical object-oriented software languages.

Finally, in the thought-provoking Afterword, William J. Mitchell, an MIT Professor of Architecture and Media Arts and Sciences, ties the concept of beauty between the building architectures we encounter in the real world and the software architectures residing on silicon.

Principles, Properties, and Structures

Late in this book's review process, one of the reviewers asked us to provide our personal opinion, in the form of commentary, on what a reader could learn from each chapter. The idea was intriguing, but we did not like the fact that we would have to second-guess the chapter authors. Asking the authors themselves to provide a meta-analysis of their writings would lead to a Babel tower of definitions, terms, and architectural constructs guaranteed to confuse readers. What was needed was a common vocabulary of architectural terms; thankfully, we realized we already had that in our hands.

In the Foreword, Stephen Mellor discusses seven principles upon which all beautiful architectures are based. In Chapter 1, John Klein and David Weiss present four architecture building blocks and six properties that beautiful architectures exhibit. A careful reader will notice that Mellor's principles and Klein's and Weiss's properties are not independent of each other. In fact, they mostly coincide; this happens because great minds think alike. All three, being very experienced architects, have seen many times in action the importance of the concepts they describe.

We merged Mellor's architectural principles with the definitions of Klein and Weiss into two lists: one containing principles and properties (Table P-1), and one containing structures (Table P-2). We then asked the chapter authors to mark the terms they thought applied to their chapters, and produced a corresponding legend for each chapter. In these tables, you can see the definition of each principle, property, or architectural construct that appears in the chapter legend. We hope the legends will guide your reading of this book by giving you a clean overview of the contents of each chapter, but we urge you to delve into a chapter's text rather than simply stay with the legend.

TABLE P-1. Architectural principles and properties

Principle or property	The ability of an architecture to...
Versatility	...offer "good enough" mechanisms to address a variety of problems with an economy of expression.
Conceptual integrity	...offer a single, optimal, nonredundant way for expressing the solution of a set of similar problems.
Independently changeable	...keep its elements isolated so as to minimize the number of changes required to accommodate changes.
Automatic propagation	...maintain consistency and correctness, by propagating changes in data or behavior across modules.
Buildability	...guide the software's consistent and correct construction.
Growth accommodation	...cater for likely growth.
Entropy resistance	...maintain order by accommodating, constraining, and isolating the effects of changes.

TABLE P-2. Architectural structures

Structure	A structure that...
Module	...hides design or implementation decisions behind a stable interface.
Dependency	...organizes components along the way where one uses functionality of another.
Process	...encapsulates and isolates the runtime state of a module.
Data access	...compartmentalizes data, setting access rights to it.

Conventions Used in This Book

The following typographical conventions are used in this book:

Italic
: Indicates new terms, URLs, email addresses, filenames, and file extensions.

Constant width
> Used for program listings, as well as within paragraphs to refer to program elements such as variable or function names, databases, data types, environment variables, statements, and keywords.

Constant width bold
> Shows commands or other text that should be typed literally by the user.

Constant width italic
> Shows text that should be replaced with user-supplied values or by values determined by context.

Using Code Examples

This book is here to help you get your job done. In general, you may use the code in this book in your programs and documentation. You do not need to contact us for permission unless you're reproducing a significant portion of the code. For example, writing a program that uses several chunks of code from this book does not require permission. Selling or distributing a CD-ROM of examples from O'Reilly books does require permission. Answering a question by citing this book and quoting example code does not require permission. Incorporating a significant amount of example code from this book into your product's documentation does require permission.

We appreciate, but do not require, attribution. An attribution usually includes the title, author, publisher, and ISBN. For example: "*Beautiful Architecture*, edited by Diomidis Spinellis and Georgios Gousios. Copyright 2009 O'Reilly Media, Inc., 978-0-596-51798-4."

If you feel your use of code examples falls outside fair use or the permission given here, feel free to contact us at *permissions@oreilly.com*.

Safari® Books Online

When you see a Safari® Books Online icon on the cover of your favorite technology book, that means the book is available online through the O'Reilly Network Safari Bookshelf.

Safari offers a solution that's better than e-books. It's a virtual library that lets you easily search thousands of top tech books, cut and paste code samples, download chapters, and find quick answers when you need the most accurate, current information. Try it for free at *http://safari.oreilly.com*

PART I

第 1 部分

On Architecture

论架构

Chapter 1　What Is Architecture?
第 1 章　何谓架构

Chapter 2　A Tale of Two Systems: A Modern-Day Software Fable
第 2 章　两个系统的故事：摩登时代的软件神话

CHAPTER ONE

What Is Architecture?

John Klein
David Weiss

▶ 这段话基本上概况了美丽架构的特征：良好的用户体验、及时的反馈与响应、维护性好，没有重大错误、可靠、易于安装、以标准方式与其他系统通信。我参与的一些系统达到了这些要求。通过引入现场客户和快速迭代随时掌握客户需求，快速响应；通过引入持续集成和自动部署规避重大错误，提供一键式安装；通过面向契约的服务设计对实现封装，提供更好的重用单元，并满足服务的独立演化；通过REST保障系统之间通信方式的标准化，使得系统更简单、更容易理解。

Introduction

BUILDERS, MUSICIANS, WRITERS, COMPUTER DESIGNERS, NETWORK DESIGNERS, and software developers all use the term architecture, as do others (ever hear of a food architect?), yet each produces different results. A building is very different from a symphony, but both have architectures. Further, all architects talk about beauty in their work and its results. A building architect might say that a building should provide an environment suitable for working or living, and that it should be beautiful to behold; a musician that the music should be playable, with a discernible theme, and that it should be beautiful to the ear; a software architect that the system should be friendly and responsive to the user, maintainable, free of critical errors, easy to install, reliable, that it should communicate in standard ways with other systems, and that it, too, should be beautiful.

This book provides you with detailed examples of beautiful architectures drawn from the fields of computerized systems, a relatively young discipline. Because we are young, we have fewer examples to emulate than fields such as building, music, or writing, and therefore we need them even more. This book intends to help fill that need.

Before you proceed to the examples, we would like you to consider what an architecture is and what the attributes of a beautiful architecture might be. As you will see from the different definitions of architecture in this chapter, each discipline has its own definition, so we will first explore what is common among architectures in different disciplines and what problems one tries to solve with an architecture. Particularly, an architecture can help assure that the system satisfies the concerns of its stakeholders, and it can help deal with the complexity of conceiving, planning, building, and maintaining the system.

We then proceed to a definition of architecture and show how we can apply that definition to software architecture, since software is central to many of the later examples. Key to the definition is that an architecture consists of a set of structures designed to let the architects, builders, and other stakeholders see how their concerns are satisfied.

We end this chapter with a discussion of the attributes of beautiful architectures and cite a few examples. Central to beauty is conceptual integrity—that is, a set of abstractions and the rules for using them throughout the system as simply as possible.

In our discussion we will use "architecture" as a noun to denote a set of artifacts, including documentation such as blueprints and building specifications that describe the object to be built, wherein the object is viewed as a set of structures. The term is also used by some as a verb to describe the process of creating the artifacts, including the resulting work. As Jim Waldo and others have pointed out, however, there is no process that you can learn that guarantees you will produce a good system architecture, let alone a beautiful one (Waldo 2006), so we will focus more on artifacts than process.

> **Architecture: "The art or science of building; esp. the art or practice of designing and building edifices for human use, taking both aesthetic and practical factors into account."**
>
> —The Shorter Oxford English Dictionary, *Fifth Edition, 2002*

In all disciplines, architecture provides a means for solving a common problem: assuring that a building, or bridge, or composition, or book, or computer, or network, or system has certain properties and behaviors when it has been built. Put another way, the architecture is both a plan for the system so that the result can have the desired properties and a description of the built system. Wikipedia says: "According to the earliest surviving work on the subject, Vitruvius' 'On Architecture,' good building should have Beauty (Venustas), Firmness (Firmitas), and Utility (Utilitas); architecture can be said to be a balance and coordination among these three elements, with no one overpowering the others."

> **We speak of the "architecture" of a symphony, and call architecture, in its turn, "frozen music."**
>
> —Deryck Cooke, *The Language of Music*

A good system architecture exhibits conceptual integrity; that is, it comes equipped with a set of design rules that aid in reducing complexity and that can be used as guidance in detailed design and in system verification. Design rules may incorporate certain abstractions that are always used in the same way, such as virtual devices. The rules may be represented as a pattern, such as pipes and filters. In the best case there are verifiable rules, such as "any virtual device of the same type may replace any other virtual device of the same type in the event of device failure," or "all processes contending for the same resource must have the same scheduling priority."

A contemporary architect might say that the object or system under construction must have the following characteristics.

- It has the functionality required by the customer.
- It is safely buildable on the required schedule.
- It performs adequately.
- It is reliable.
- It is usable and safe to use.
- It is secure.
- It is affordable.
- It conforms to legal standards.
- It will outlast its predecessors and its competitors.

> The architecture of a computer system we define as *the minimal set of properties that determine what programs will run and what results they will produce.*
>
> —Gerrit Blaauw & Frederick Brooks, Computer Architecture

We've never seen a complex system that perfectly satisfies all of the preceding characteristics. Architecture is a game of trade-offs—a decision that improves one of these characteristics often diminishes another. The architect must determine what is sufficient to satisfy, by discovering the important concerns for a particular system and the conditions for satisfying them sufficiently.

Common among the notions of architecture is the idea of structures, each defined by components of various sorts and their relations: how they fit together, invoke each other, communicate, synchronize, and otherwise interact. Components could be support beams or internal rooms in a building, individual instruments or melodies in a symphony, book chapters or characters in a story, CPUs and memory chips in a computer, layers in a communications stack or processors connected to a network, cooperating sequential processes, objects, collections of compile-time macros, or build-time scripts. Each discipline has its own sets of components and its own relationships among them.

▶ 架构是一种 trade-off，或者说架构师的职责就是做出设计权衡与决策。

在架构设计的早期，我喜欢引入名为 Value Sliders 的活动，详见P42。这一活动主要用于帮助团队进行项目因素或质量因素的权衡。通常情况下，应该是架构师、项目经理及客户一起参与讨论。

Value Sliders 图的左侧low代表"可以商量"，右侧high代表"不能商量"。在这两者之间可以用即时贴表示该因素是否可以协调和商量。纵向地看，在一条纵线上只能存在一个即时贴，制约着各个因素必须分出不同的优先级，从而做出合理权衡。

> **In wider use, the term "architecture" always means "unchanging deep structure."**
>
> —*Stewart Brand*, How Buildings Learn

In the face of increasing complexity of systems and their interactions, both internally and with each other, an architecture comprising a set of structures provides the primary means for dealing with complexity in order to ensure that the resulting system has the required properties. Structures provide ways to understand the system as sets of interacting components.

Each structure is intended to help the architect understand how to satisfy particular concerns, such as changeability or performance. The job of demonstrating that particular concerns are satisfied may fall to others, but the architect must be able to demonstrate that *all* concerns have been met.

> **Network architecture: the communication equipment, protocols, and transmission links that constitute a network, and the methods by which they are arranged.**
>
> —http://www.wtcs.org/snmp4tpc/jton.htm

The Role of Architect

When buildings are designed, constructed, or renovated, we designate key designers as "architects" and give them a broad range of responsibilities. An architect prepares initial sketches of the building, showing both external appearance and internal layout, and discusses these sketches with clients until all concerned have agreed that what is shown is what they want. The sketches are abstractions: they focus attention on the pertinent details of a particular aspect of the building, omitting other concerns.

After the clients and architects agree on these abstractions, the architects prepare, or supervise the preparation of, much more detailed drawings, as well as associated textual specifications. These drawings and specifications describe many "nitty-gritty" details of a building, such as plumbing, siding materials, window glazing, and electrical wiring.

On rare occasions, an architect simply hands the detailed plans to a builder who completes the project in accordance with the plans. For more important projects, the architect remains involved, regularly inspects the work, and may propose changes or accept suggestions for change from both the builder and customer. When the architect supervises the project, it is not considered complete until he certifies that it is in substantial compliance with the plans and specifications.

We employ an architect to assure that the design (1) meets the needs of the client, including the characteristics previously noted; (2) has conceptual integrity by using the same design rules throughout; and (3) meets legal and safety requirements. An important part of the architect's role is to ensure that the design concepts are consistently realized during the implementation.

Sometimes the architect also acts as a mediator between builder and client. There is often some disagreement about which decisions are in the realm of the architect and which are left to others, but it is always clear that the architect makes the major decisions, including all that can affect the usability, safety, and maintainability of the structure.

MUSIC COMPOSITION AND SOFTWARE ARCHITECTURE

Whereas building architecture is often used as an analogy for software architecture, music composition may be a better analogy. A building architect creates a static description (blueprints and other drawings) of a relatively static structure (the architecture must account for movement of people and services within the building as well as the load-bearing structure). In music composition and software design, the composer (software architect) creates a static description of a piece of music (architecture description and code) that is later performed (executed) many times. In both music and software the design can account for many components interacting to produce the desired result, and the result varies depending on the performers, the environment in which it is performed, and the interpretation imposed by the performers.

The Role of the Software Architect

Software development projects need people who play the same role for software construction that traditional architects play when buildings are constructed or renovated. For software systems, however, it has never been clear exactly which decisions are the purview of the architect and which can be left to the implementers. The definition of what an architect does in a software project is more difficult than the analogous definition for building architects because of three factors: lack of tradition, the intangible nature of the product, and the complexity of the system. (See Grinter [1999] for a portrayal of how a software architect carries out her role within a large software development organization.)

In particular:

- Building architects can look back at thousands of years of history to see what architects have done in the past; they can visit and study buildings that have been standing for hundreds, and sometimes a thousand years or more, and that are still in use. In software we have only a few decades of history and our designs are often not public. Furthermore, building architects have and use standards for describing the drawings and specifications that the architects produce, allowing present architects to take advantage of the recorded history of architecture.
- Buildings are physical products; there is a clear distinction between the plans produced by the architects and the building produced by the workers.

▶ 本章并没有明确给出架构师究竟应该在团队中扮演什么样的角色。这实际上取决于架构的定义。

▶ 架构有诸多定义，但从某个视角来讲，我们可以认为架构是一种知识的分享，是沟通，每个团队成员都必须充分理解整个系统的架构。在进行架构设计时，架构师不应该闭门造车，奢求通过自己掌握的知识与技能独立完成系统的架构，而应该成为团队架构活动的促进者，带领和引导整个团队成员一起参与架构活动，在保障满足合理的架构原则前提下，促进各个成员对整个系统的认识，形成知识分享。

ARCHITECTURAL REUSE

The Hagia Sophia (top), built in Istanbul in the sixth century, pioneered the use of structures called pendentives to support its enormous dome, and is an example of beauty in Byzantine architecture. Christopher Wren, 1,100 years later, used the same design for the dome of St. Paul's cathedral (bottom), a London landmark. Both still stand and are used today.

On major software projects, there are often many architects. Some architects are quite specialized in disciplines, such as databases and networks, and usually work as part of a team, but for now we will write as if there were only one.

What Constitutes a Software Architecture?

It is a mistake to think of "an architecture" as if it were a simple entity that could be described by a single document or drawing. Architects must make many design decisions. To be useful, these decisions must be documented so that they can be reviewed, discussed, modified, and approved, and then serve to constrain subsequent decision making and construction. For software systems, these design decisions are behavioral and structural.

External behavioral descriptions show how the product will interface with its users, other systems, and external devices, and should take the form of requirements. Structural descriptions show how the product is divided into parts and the relations between those parts. Internal behavioral descriptions are needed to describe the interfaces between components. Structural descriptions often show several distinct views of the same part because it is impossible to put all the information in one drawing or document in a meaningful way. A component in one view may be a part of a component in another.

Software architectures are often presented as layered hierarchies that tend to commingle several different structures in one diagram. In the 1970s Parnas pointed out that the term "hierarchy" had become a buzzword, and then precisely defined the term and gave several different examples of structures used for different purposes in the design of different systems (Parnas 1974). Describing the structures of an architecture as a set of *views*, each of which addresses different concerns, is now accepted as a standard architecture practice (Clements et al. 2003; IEEE 2000). We will use the word "architecture" to refer to a set of annotated diagrams and functional descriptions that specify the structures used to design and construct a system. In the software development community there are many different forms used, and proposed, for such diagrams and descriptions. See Hoffman and Weiss (2000, chaps. 14 and 16) for some examples.

> **The software architecture of a program or computing system is the structure or structures of the system, which comprise software elements, the externally visible properties of those elements, and the relationships among them.**
>
> **"Externally visible" properties are those assumptions other elements can make of an element, such as its provided services, performance characteristics, fault handling, shared resource usage, and so on.**
>
> —Len Bass, Paul Clements, and Rick Kazman, Software Architecture in Practice, *Second Edition*

Architecture Versus Design

Architecture is a part of the design of the system; it highlights some details by abstracting away from others. Architecture is thus a subset of design. A developer focused on implementing a component of the system may not be very aware of how all the components fit together, but rather is primarily concerned with the design and development of a small number of component(s), including the architectural constraints that they must obey and the rules they can use. As such, the developer is working on a different aspect of the system design than the architect.

If architecture is concerned with the relationships among components and the externally visible properties of system components, then design will additionally be concerned with the internal structure of those components. For example, if one set of components consists of information-hiding modules, then the externally visible properties form the interfaces to those components, and the internal structure is concerned with the data structures and flow of control within a module (Hoffman and Weiss 2000, chaps. 7 and 16).

Creating a Software Architecture

So far, we have considered architecture in general and looked at how software architecture is both similar to and different from architecture in other domains. We now turn our attention to the "how" of software architecture. Where should the architect focus her attention when she is creating the architecture for a software system?

The first concern of a software architect is not the functionality of the system.

That's right—the first concern of a software architect is not the functionality of the system.

For example, if we offer to hire you to develop the architecture for a "web-based application," would you start by asking us about page layouts and navigation trees, or would you ask us questions such as:

- Who will host it? Are there technology restrictions in the hosting environment?
- Do you want to run on a Windows Server or on a LAMP stack?
- How many simultaneous users do you want to support?
- How secure does the application need to be? Is there data that we need to protect? Will the application be used on the public Internet or a private intranet?
- Can you prioritize these answers for me? For example, is number of users more important than response time?

Depending on our answers to these and a few other questions, you can begin sketching out an architecture for the system. And we still haven't talked about the functionality of the application.

Now, admittedly, we cheated a bit here because we asked for a "web-based application," which is a well-understood domain, so you already knew what decisions would have the most influence on your architecture. Similarly, if we had asked for a telecommunications system or an avionics system, an architect experienced in one of those domains would have some notion of required functionality in mind. But still, you were able to begin creating the architecture without worrying too much about the functionality. You did this by focusing on *quality concerns* that needed to be satisfied.

Quality concerns specify the way in which the functionality must be delivered in order to be acceptable to the system's stakeholders, the people with a vested interest in the outcome of the system. Stakeholders have certain concerns that the architect must address. Later, we will discuss concerns that are typically raised when trying to assure that the system has the required qualities. As we said earlier, one role of the architect is to ensure that the design of the system will meet the needs of the client, and we use quality concerns to help us understand those needs.

This example highlights two key practices of successful architects: stakeholder involvement and a focus on *both* quality concerns and functionality. As the architect, you began by asking us what we wanted from the system, and in what priority. In a real project, you would have sought out other stakeholders. Typical stakeholders and their concerns include:

- Funders, who want to know if the project can be completed within resource and schedule constraints
- Architects, developers, and testers, who are first concerned with initial construction and later with maintenance and evolution
- Project managers, who need to organize teams and plan iterations
- Marketers, who may want to use quality concerns to differentiate the system from competitors
- Users, including end users, system administrators, and the people who do installation, deployment, provisioning, and configuration
- Technical support staff, who are concerned with the number and complexity of Help Desk calls

Every system has its own set of quality concerns. Some, such as performance, security, and scalability, may be well-specified, but other, often equally important concerns, such as changeability, maintainability, and usability, may not be defined with enough detail to be useful. Odd, isn't it, that stakeholders want to put functions in software and not hardware so that they can be easily and quickly modified, and then often give short shrift to changeability when stating their quality concerns? Architecture decisions will have an impact on what kinds of changes can be done easily and quickly and what changes will take time and be hard to do. So shouldn't an architect understand his stakeholders' expectations for qualities such as "changeability" as well as he understands the functional requirements?

Once the architect understands the stakeholders' quality concerns, what does she do next? Consider the trade-offs. For example, encrypting messages improves security but hurts performance. Using configuration files may increase changeability but could decrease usability unless we can verify that the configuration is valid. Should we use a standard representation for these files, such as XML, or invent our own? Creating the architecture for a system involves making many such difficult trade-offs.

The first task of the architect, then, is to work with stakeholders to understand and prioritize quality concerns and constraints. Why not start with functional requirements? Because there are usually many possible system decompositions. For example, starting with a data model would lead to one architecture, whereas starting with a business process model might lead to a different architecture. In the extreme case, there is no decomposition, and the system is developed as a monolithic block of software. This might satisfy all functional requirements, but it probably will not satisfy quality concerns such as changeability, maintainability, or scalability. Architects often must do architecture-level refactoring of a system, for example to move from simplex to distributed deployment, or from single-threaded to multithreaded in order to meet scalability or performance requirements, or hardcoded parameters to external configuration files because parameters that were *never* going to change now need to be modified.

Although there are many architectures that can meet functional requirements, only a subset of these will also satisfy quality requirements. Let's go back to the web application example. Think of the many ways to serve up web pages—Apache with static pages, CGI, servlets, JSP, JSF, PHP, Ruby on Rails, or ASP.NET, to name just a few. Choosing one of these technologies is an architecture decision that will have significant impact on your ability to meet certain quality requirements. For example, an approach such as Ruby on Rails might provide the fast time-to-market benefit, but could be harder to maintain as both the Ruby language and the Rails framework continue to evolve rapidly. Or perhaps our application is a web-based telephone and we need to make the phone "ring." If you need to send true asynchronous events from the server to the web page to satisfy performance requirements, an architecture based on servlets might be more testable and modifiable.

In real-world projects, satisfying stakeholder concerns requires many more decisions than simply selecting a web framework. Do you really need an "architecture," and do you need an "architect" to make the decisions? Who should make them? Is it the coder, who may make many of them unintentionally and implicitly, or is it the architect, who makes them explicitly with a view in mind of the entire system, its stakeholders, and its evolution? Either way, you will have an architecture. Should it be explicitly developed and documented, or should it be implicit and require reading of the code to discover?

Often, of course, the choice is not so stark. As the size of the system, its complexity, and the number of people who work on it increase, however, those early decisions and the way that they are documented will have greater and greater impact.

We hope you understand by now that architecture decisions are important if your system is going to meet its quality requirements, and that you want to pay attention to the architecture and make these decisions intentionally rather than just "letting the architecture emerge."

What happens when the system is very large? One of the reasons that we apply architecture principles such as "divide and conquer" is to reduce complexity and enable work to proceed in parallel. This allows us to create larger and larger systems. Can the architecture itself be decomposed into parts, and those parts worked on by different people in parallel? In considering computer architecture, Gerrit Blaauw and Fred Brooks asserted:

> ...if, after all techniques to make the task manageable by a single mind have been applied, the architectural task is still so large and complex that it cannot be done in that way, the product

conceived is too complex to be usable and should not be built. In other words, the mind of a single user must comprehend a computer architecture. If a planned architecture cannot be designed by a single mind, it cannot be comprehended by one. (1997)

Do you need to understand all aspects of an architecture in order to use it? An architecture separates concerns so, for the most part, the developer or tester using the architecture to build or maintain a system does not need to deal with the entire architecture at once, but can interact with only the necessary parts to perform a given function. This allows us to create systems larger than a single mind can comprehend. But, before we completely ignore the advice of the people who built the IBM System/360, one of the longest-lived computer architectures, let's look at what prompted them to make this statement.

Fred Brooks said that conceptual integrity is the most important attribute of an architecture: "It is better to have a system...reflect one set of design ideas, than to have one that contains many good but independent and uncoordinated ideas" (1995). It is this conceptual integrity that allows a developer who already knows about one part of a system to quickly understand another part. Conceptual integrity comes from consistency in things such as decomposition criteria, application of design patterns, and data formats. This allows a developer to apply experience gained working in one part of the system to developing and maintaining other parts of the system. The same rules apply throughout the system. As we move from system to "system-of-systems," the conceptual integrity must also be maintained in the architecture that integrates the systems, for example by selecting an architecture style such as *publish/subscribe message bus* and then applying this style uniformly to all system integrations in the system-of-systems.

The challenge for an architecture team is to maintain a single-mindedness and a single philosophy as they go about creating the architecture. Keep the team as small as possible, work in a highly collaborative environment with frequent communication, and have one or two "chiefs" act as benevolent dictators with the final say on all decisions. This organizational pattern is commonly seen in successful systems, whether corporate or open source, and results in the conceptual integrity that is one of the attributes of a beautiful architecture.

Good architects are often formed by having better architects mentor them (Waldo 2006). One reason may be that there are certain concerns that are common to nearly all projects. We have already alluded to some of them, but here is a more complete list, with each concern phrased as a question that the architect may need to consider during the course of a project. Of course, individual systems will have additional critical concerns.

Functionality
    What functionality does the product offer to its users?

Changeability
    What changes may be needed in the software in the future, and what changes are unlikely and need not be especially easy to make in the future?

Performance

What will the performance of the product be?

Capacity

How many users will use the system simultaneously? How much data will the system need to store for its users?

Ecosystem

What interactions will the system have with other systems in the ecosystem in which it will be deployed?

Modularity

How is the task of writing the software organized into work assignments (modules), particularly modules that can be developed independently and that suit each other's needs precisely and easily?

Buildability

How can the software be built as a set of components that can be independently implemented and verified? What components should be reused from other products and which should be acquired from external suppliers?

Producibility

If the product will exist in several variations, how can it be developed as a product line, taking advantage of the commonality among the versions, and what are the steps by which the products in the product line can be developed (Weiss and Lai 1999)? What investment should be made in creating a software product line? What is the expected return from creating the options to develop different members of the product line?

In particular, is it possible to develop the smallest minimally useful product first and then develop additional members of the product line by adding (and subtracting) components without having to change the code that was written previously?

Security

If the product requires authorization for its use or must restrict access to data, how can security of data be ensured? How can "denial of service" and other attacks be withstood?

Finally, a good architect realizes that the architecture affects the organization. Conway noted that the structure of a system reflects the structure of the organization that built it (1968). The architect may realize that Conway's Law can be used in reverse. In other words, a good architecture may influence an organization to change so as to be more efficient in building systems derived from the architecture.

Architectural Structures

How, then, does a good architect deal with these concerns? We have already mentioned the need to organize the system into structures, each defining specific relationships among certain types of components. The architect's chief focus is to organize the system so that each structure helps answer the defining questions for one of the concerns. Key structural decisions divide the product into components and define the relationships among those components (Bass, Clements, and Kazman 2003; Booch, Rumbaugh, and Jacobson 1999; IEEE 2000; Garlan and Perry 1995). For any given product, there are many structures that need to be designed. Each must be designed separately so that it is viewed as a separate concern. In the next few sections we discuss some structures that you can use to address the concerns on our list. For example, the Information Hiding Structures show how the system is organized into work assignments. They can also be used as a roadmap for change, showing for proposed changes which modules accommodate those changes. For each structure we describe the components and the relations among them that define the structure. Given the concerns on our list, we consider the following structures to be of primary importance.

The Information Hiding Structures

COMPONENTS AND RELATIONS: The primary components are Information Hiding Modules, where each module is a work assignment for a group of developers, and each module embodies a design decision. We say that a design decision is the secret of a module if the decision can be changed without affecting any other module (Hoffman and Weiss 2000, chaps. 7 and 16). The most basic relation between the modules is "part of." Information Hiding Module A is part of Information Hiding Module B if A's secret is a part of B's secret. Note that it must be possible to change A's secret without changing any other part of B; otherwise, A is not a submodule according to our definition. For example, many architectures have virtual device modules, whose secret is how to communicate with certain physical devices. If virtual devices are organized into types, then each type might form a submodule of the virtual device module, where the secret of each virtual device type would be how to communicate with devices of that type.

Each module is a work assignment that includes a set of programs to be written. Depending on language, platform, and environment, a "program" could be a method, a procedure, a function, a subroutine, a script, a macro, or other sequence of instructions that can be made to execute on a computer. A second Information Hiding Module Structure is based on the relation "contained in" between programs and modules. A program P is contained in a module M if part of the work assignment M is to write P. Note that every program is contained in a module because every program must be part of some developer's work assignment.

Some of these programs are accessible on the module's interface, whereas others are internal. Modules may also be related through interfaces. A module's interface is a set of assumptions that programs outside of the module may make about the module and the set of assumptions that the module's programs make about programs and data structures of other modules. A is said to "depend on" B's interface if a change to B's interface might require a change in A.

The "part of" structure is a hierarchy. At the leaf nodes of the hierarchy are modules that contain no identified submodules. The "contained in" structure is also a hierarchy, since each program is contained in only one module. The "depends on" relation does not necessarily define a hierarchy, as two modules may depend on each other either directly or through a longer loop in the "depends on" relation. Note that "depends on" should not be confused with "uses" as defined in a later section.

Information Hiding Structures are the foundation of the object-oriented design paradigm. If an Information Hiding Module is implemented as a class, the public methods of the class belong to the interface for the module.

CONCERNS SATISFIED: The Information Hiding Structures should be designed so that they satisfy changeability, modularity, and buildability.

The Uses Structures

COMPONENTS AND RELATION: As defined previously, Information Hiding Modules contain one or more programs (as defined in the previous section). Two programs are included in the same module if and only if they share a secret. The components of the Uses Structure are programs that may be independently invoked. Note that programs may be invoked by each other or by the hardware (for example, by an interrupt routine), and the invocation may come from a program in a different namespace, such as an operating system routine or a remote procedure. Furthermore, the time at which an invocation may occur could be any time from compile time through runtime.

We will consider forming a Uses Structure only among programs that operate at the same binding time. It is probably easiest first just to think about programs that operate at runtime. Later, we may also think about the uses relation among programs that operate at compile time or load time.

▶一旦在Uses Relation中出现loop，就说明Components之间即使具有逻辑的分离关系，在物理上仍然是不可分离的。在部署视图中，这些存在循环Uses Relation 的Components就不能做到分布式部署，存在物理上的耦合Coupling关系。

We say that program A uses program B if B must be present and satisfy its specification for A to satisfy its specification. In other words, B must be present and operate correctly for A to operate correctly. The Uses Relation is sometimes known as "requires the presence of a correct version of." For a further explanation and example, see Chapter 14 of Hoffman and Weiss (2000).

> The Uses Structure determines what working subsets can be built and tested. A desirable property in the Uses Relation for a software system is that it defines a hierarchy, meaning that there are no loops in it. When there is a loop in the Uses Relation, all programs in the loop must be present and working in the system for any of them to work. Since it may not be possible to construct a completely loop-free Uses Relation, an architect may treat all of the programs in a Uses loop as a single program for the purpose of creating subsets. A subset must include either the whole program or none of it.

When there are no loops in the Uses Relation, a levels structure is imposed on the software. At the bottom level, level 0, are all programs that use no other programs. Level n consists of all programs that use programs in level $n-1$ or below. The levels are often depicted as a series of layers, with each layer representing one or several levels in the Uses Relation. Grouping adjacent levels in Uses helps to simplify the representation and allows for cases where there are small loops in the relation. One guideline in performing such a grouping is that programs at one layer should execute approximately 10 times as quickly and 10 times as often as programs in the next layer above it (Courtois 1977).

A system that has a hierarchical Uses Structure can be built one or a few layers at a time. These layers are sometimes known as "levels of abstraction," but this is a misnomer. Because the components are individual programs, not whole modules, they do not necessarily abstract from (hide) anything.

Often a large software system has too many programs to make the description of the Uses Relation among programs easily understandable. In such cases, the Uses Relation may be formed on aggregations of programs, such as modules, classes, or packages. Such aggregated descriptions lose important information but help to present the "big picture." For example, one can sometimes form a Uses Relation on Information Hiding Modules, but unless all programs in a module are on the same level of the programmatic Uses hierarchy, important information is lost.

In some projects, the Uses Relation for a system is not fully determined until the system is implemented, because the developers determine what programs they will use as the implementation proceeds. The architects of the system may, however, create an "Allowed-to-Use" Relation at design time that constrains the developers' choices. Henceforth, we will not distinguish between "Uses" and "Allowed-to-Use."

A well-defined Uses Structure will create proper subsets of the system and can be used to drive iterative or incremental development cycles.

CONCERNS SATISFIED: Producibility and ecosystem.

The Process Structures

COMPONENTS AND RELATION: The Information Hiding Module Structures and the Uses Structures are static structures that exist at design and code time. We now turn to a runtime structure. The components that participate in the Process Structure are Processes. Processes are runtime sequences of events that are controlled by programs (Dijkstra 1968). Each program executes as part of one or many Processes. The sequence of events in one Process proceed independently of the sequence of events in another Process, except where the Processes synchronize with each other, such as when one Process waits for a signal or a message from the other. Processes are allocated resources, including memory and processor time, by support systems. A system may contain a fixed number of Processes, or it may create and destroy Processes while running. Note that *threads* implemented in operating systems such as Linux and Windows fall under this definition of Processes. Processes are the components of several distinct relations. Some examples follow.

Process gives work to

One Process may create work that must be completed by other Processes. This structure is essential in determining whether a system can get into a deadlock.

CONCERNS SATISFIED: Performance and capacity.

Process gets resources from

In systems with dynamic resource allocation, one Process may control the resources used by another, where the second must request and return those resources. Because a requesting Process may request resources from several controllers, each resource may have a distinct controlling Process.

CONCERNS SATISFIED: Performance and capacity.

Process shares resources with

Two Processes may share resources such as printers, memory, or ports. If two Processes share a resource, synchronization is necessary to prevent usage conflicts. There may be distinct relations for each resource.

CONCERNS SATISFIED: Performance and capacity.

Process contained in module

Every Process is controlled by a program and, as noted earlier, every program is contained in a module. Consequently, we can consider each Process to be contained in a module.

CONCERNS SATISFIED: Changeability.

Access Structures

The data in a system may be divided into segments with the property so that if a program has access to any data in a segment, it has access to all data in that segment. Note that to simplify the description, the decomposition should use maximally sized segments by adding the condition that if two segments are accessed by the same set of programs, those two segments should be combined. The data access structure has two kinds of components, programs and segments. This relation is entitled "has access to," and is a relation between programs and segments. A system is thought to be more secure if this structure minimizes the access rights of programs and is tightly enforced.

CONCERNS SATISFIED: Security.

Summary of Structures

Table 1-1 summarizes the preceding software structures, how they are defined, and the concerns that they satisfy.

TABLE 1-1. Structure summary

Structure	Components	Relations	Concerns
Information Hiding	Information Hiding Modules	Is a part of Is contained in	Changeability Modularity Buildability
Uses	Programs	Uses	Producibility Ecosystem
Process	Processes (tasks, threads)	Gives work to Gets resources from Shares resources with Contained in ...	Performance Changeability Capacity
Data Access	Programs and Segments	Has access to	Security Ecosystem

Good Architectures

Recall that architects play a game of trade-offs. For a given set of functional and quality requirements, there is no single correct architecture and no single "right answer." We know from experience that we should evaluate an architecture to determine whether it will meet its requirements before spending money to build, test, and deploy the system. Evaluation attempts to answer one or more of the concerns discussed in previous sections, or concerns specific to a particular system.

There are two common approaches to architecture evaluation (Clements, Kazman, and Klein 2002). The first class of evaluation methods determines properties of the architecture, often by modeling or simulation of one or more aspects of the system. For example, performance modeling is carried out to assess throughput and scalability, and fault tree models can be used to estimate reliability and availability. Other types of models include using complexity and coupling metrics to assess changeability and maintainability.

The second, and broadest, class of evaluation methods is based on questioning the architects to assess the architecture. There are many structured questioning methods. For example, the

Software Architecture Review Board (SARB) process developed at Bell Labs uses experts from within the organization and leverages their deep domain expertise in telecommunications and related applications (Maranzano et al. 2005).

Another variation of the questioning approach is the Architecture Trade-off Analysis Method (ATAM) (Clements, Kazman, and Klein 2002), which looks for risks that the architecture will not satisfy quality concerns. ATAM uses scenarios, each describing a particular stakeholder's quality concern for the system. The architects then explain how the architecture supports each of the scenarios.

Active reviews are another type of questioning approach that turns the process on its head, requiring the architects to provide the reviewers with the questions that the architects think are important to answer (Hoffman and Weiss 2000, chap. 17). The reviewers then use the existing architecture documents and descriptions to answer the questions. Finally, searching the Web for "software architecture review checklist" returns dozens of checklists, some very general and some specific to an application domain or technology framework.

Beautiful Architectures

All of the preceding methods help to evaluate whether an architecture is "good enough"—that is, whether it is likely to guide the developer and testers to produce a system that will satisfy the functional and quality concerns of the system's stakeholders. There are many good architectures in systems that we use every day.

But what about architectures that are more than good enough? What if there were a "Software Architecture Hall of Fame"? Which architectures would line the walls of that gallery? The idea is not as far-fetched as you might think—in the field of software product lines, just such a Hall of Fame exists.* The criteria for induction into the Software Product Line Hall of Fame include commercial success, influence on other product line architectures (others have "borrowed, copied, or stolen" from the architecture), and sufficient documentation that others can understand the architecture "without resorting to hearsay."

What criteria would we add to these for nominees for a more general "Architecture Hall of Fame," or perhaps a "Gallery of Beautiful Architectures"?

First, we should recognize that this is a gallery of software systems, not art, and our systems are built to be used. So, perhaps we should begin by looking at the Utility of the architecture: it should be used every day by many people.

But before an architecture can be used, it must be built, and so we should look at the Buildability of the architecture. We would look for architectures with a well-defined Uses Structure that would support incremental construction, so that at each iteration of construction we would have a useful, testable system. We would also look for architectures that have

---

* See *http://www.sei.cmu.edu/productlines/plp_hof.html*.

well-defined module interfaces and that are inherently testable, so that the construction progress is transparent and visible.

Next, we want architectures that demonstrate Persistence—that is, architectures that have stood the test of time. We work in an era when the technical environment is changing at an ever-increasing rate. A beautiful architecture should anticipate the need for change, and allow expected changes to be made easily and efficiently. We want to find architectures that have avoided the "aging horizon" (Klein 2005) beyond which maintenance becomes prohibitively expensive.

Finally, we would want to include architectures that have features that delight the developers and testers who use the architecture and build it and maintain it, as well as the users of the system(s) built from it. Why delight developers? It makes their job easier and is more likely to result in a high-quality system. Why delight testers? They are the ones who have to attempt to emulate what the users will do as part of the testing process. If they are delighted, it is likely that the users will be, too. Think of the chef who is unhappy with his culinary creations. His customers, who consume those creations, are likely to be unhappy, too.

Different systems and application domains offer opportunities for architectures to exhibit specific delightful features, but Conceptual Integrity is a feature that cuts across all domains and that always delights. A consistent architecture is easier and faster to learn, and once you know a little, you can begin to predict the rest. Without the need to remember and handle special cases, code is cleaner and test sets are smaller. A consistent architecture does not offer two (or more) ways to do the same thing, forcing the user to waste time choosing. As Ludwig Mies van der Rohe said of good design, "Less is more," and Albert Einstein might say that beautiful architectures are as simple as possible, but no simpler.

Given these criteria, we propose some initial candidates for our "Gallery of Beautiful Architectures."

The first entry is the architecture for the A-7E Onboard Flight Processor (OFP), developed at the Naval Research Laboratory (NRL) in the late 1970s, and described in Bass, Clements, and Kazman (2003). Although this particular system never went into production, it meets every other criterion for inclusion. This architecture has had tremendous influence on the practice of software architecture by demonstrating in a real-world system the separation of a design-time Information Hiding Module and Uses structures from the runtime Process Structures. It showed that information hiding could be used as a primary decomposition principle for a complex system. Since the U.S. government funded and developed the architecture, all project documentation is available in the public domain.† The architecture had a well-defined Uses structure that facilitated incremental construction of the system. Finally, the Information Hiding Module structure provided clear and consistent criteria for decomposing the system,

---

† See, for example, Chapters 6, 15, and 16 in Hoffman and Weiss (2000), or conduct a search for "A-7E" in the NRL Digital Archives (*http://torpedo.nrl.navy.mil/tu/ps*).

resulting in strong Conceptual Integrity. As an exemplar of embedded system software architecture, the A-7E OFP certainly belongs in our gallery.

Another architecture that we would want to include in our gallery is the software architecture for the Lucent 5ESS telephone switch (Carney et al. 1985). The 5ESS has been a global commercial success, providing core telephone network switching for networks in countries around the world. It has set the standard for performance and availability, with each unit capable of handling over one million call connections per hour with less than 10 seconds of unplanned downtime per year (Alcatel-Lucent 1999). The architecture's unifying concepts, such as the "half call model" for managing telephone connections, have become standard patterns in the domains of telephony and network protocols (Hanmer 2001). In addition to keeping the number of call types that must be handled to $2n$, where n is the number of call protocols, the half call pattern links the operating system concept of process to the telephony concept of call type, thereby providing a simple design rule and introducing a beautiful Conceptual Integrity. A development team of up to 3,000 people has evolved and enhanced the system over the past 25 years. Based on success, persistence, and influence, the 5ESS architecture is a fine addition to our gallery.

Another system to consider for inclusion in our Gallery of Beautiful Architectures is the architecture of the World Wide Web (WWW), created by Tim Berners-Lee at CERN, and described in Bass, Clements, and Kazman (2003). The WWW has certainly been commercially successful, transforming the way that people use the Internet. The architecture has remained intact, even as new applications are created and new capabilities introduced. The overall simplicity of the architecture contributes to its Conceptual Integrity, but decisions such as using a single library for both clients and servers and creating a layered architecture to separate concerns have ensured that the integrity of the architecture remains intact. The persistence of the core WWW architecture and its ability to continue to support new extensions and features certainly qualify it for inclusion in our gallery.

Our last example is the Unix system, which exhibits conceptual integrity, is widely used, and has had great influence. The pipe and filters design is a lovely abstraction that permits rapid construction of new applications.

WHAT'S AN ARCHITECT?

A stranger is traveling down a road on a hot summer day. As he progresses, he comes upon a man working by the side of the road breaking rocks.

"What are you doing?" he asks the man.

The man looks up at him. "I'm breaking rocks. What does it look like I'm doing? Now get out of my way and let me get back to it."

The stranger continues down the road and soon comes upon a second man breaking rocks in the hot sun. The man is working hard and sweating freely.

"What are you doing?" asks the stranger.

The man looks up and smiles.

"I'm working for a living," he says. "But it's hard work. Maybe you have a better job for me?"

The stranger shakes his head and moves on. Pretty soon he comes on a third man breaking rocks. The sun is at its zenith now, the man is straining, and sweat is pouring off him.

"What are you doing?" asks the stranger. The man pauses, takes a drink of water, smiles, and raises his arms to the sky.

"I'm building a cathedral," he breathes.

The stranger looks at him for a moment and says, "We're starting a new company. How would you like to be our chief architect?"

We have gone to considerable length to describe architectures, the role of architects, and considerations that go into creating architectures, and we have offered several brief examples of beautiful architectures. We invite you now to read more detailed examples from accomplished architects in the following chapters as they describe the beautiful architectures that they have created and used.

Acknowledgments

David Parnas defined many of the structures we described in several papers, including his "Buzzword" paper (Parnas 1974). Jon Bentley was an inspiration in this work and he, Deborah Hill, and Mark Klein made many useful suggestions on earlier drafts.

References

Alcatel-Lucent. 1999. "Lucent's record-breaking reliability continues to lead the industry according to latest quality report." *Alcatel-Lucent Press Releases.* June 2. *http://www.alcatel-lucent.com/wps/portal/NewsReleases/DetailLucent?LMSG_CABINET=Docs_and_Resource_Ctr&LMSG_CONTENT_FILE=News_Releases_LU_1999/LU_News_Article_007318.xml* (accessed May 15, 2008).

Bass, L., P. Clements, and R. Kazman. 2003. *Software Architecture in Practice,* Second Edition. Boston, MA: Addison-Wesley.

Blaauw, G., and F. Brooks. 1997. *Computer Architecture: Concepts and Evolution.* Boston, MA: Addison-Wesley.

Booch, G., J. Rumbaugh, and I. Jacobson. 1999. *The UML Modeling Language User Guide.* Boston, MA: Addison-Wesley.

Brooks, F. 1995. *The Mythical Man-Month*. Boston, MA: Addison-Wesley.

Carney, D. L., et al. 1985. "The 5ESS switching system: Architectural overview." *AT&T Technical Journal*, vol. 64, no. 6, p. 1339.

Clements, P., et al. 2003. *Documenting Software Architectures: Views and Beyond*. Boston, MA: Addison-Wesley.

Clements, P., R. Kazman, and M. Klein. 2002. *Evaluating Software Architectures*. Boston: Addison-Wesley.

Conway, M. 1968. "How do committees invent." *Datamation*, vol. 14, no. 4.

Courtois, P. J. 1977. *Decomposability: Queuing and Computer Systems*. New York, NY: Academic Press.

Dijkstra, E. W. 1968. "Co-operating sequential processes." *Programming Languages*. Ed. F. Genuys. New York, NY: Academic Press.

Garlan, D., and D. Perry. 1995. "Introduction to the special issue on software architecture." *IEEE Transactions on Software Engineering*, vol. 21, no. 4.

Grinter, R. E. 1999. "Systems architecture: Product designing and social engineering." *Proceedings of ACM Conference on Work Activities Coordination and Collaboration (WACC '99)*. 11–18. San Francisco, CA.

Hanmer, R. 2001. "Call processing." *Pattern Languages of Programming (PLoP)*. Monticello, IL. http://hillside.net/plop/plop2001/accepted_submissions/PLoP2001/rhanmer0/PLoP2001_rhanmer0_1.pdf.

Hoffman, D., and D. Weiss. 2000. *Software Fundamentals: Collected Papers by David L. Parnas*. Boston, MA: Addison-Wesley.

IEEE. 2000. "Recommended practice for architectural description of software intensive systems." Std 1471. Los Alamitos, CA: IEEE.

Klein, John. 2005. "How does the architect's role change as the software ages?" *Proceedings of the 5th Working IEEE/IFIP Conference on Software Architecture (WICSA)*. Washington, DC: IEEE Computer Society.

Maranzano, J., et al. 2005. "Architecture reviews: Practice and experience." *IEEE Software*, March/April 2005.

Parnas, David L. 1974. "On a buzzword: Hierarchical structure." *Proceedings of IFIP Congress*. Amsterdam, North Holland. [Reprinted as Chapter 9 in Hoffman and Weiss (2000).]

Waldo, J. 2006. "On system design." *OOPLSA '06*. October 22–26. Portland, OR.

Weiss, D., and C. T. R. Lai. 1999. *Software Product Line Engineering*. Boston, MA: Addison-Wesley.

CHAPTER TWO

A Tale of Two Systems: A Modern-Day Software Fable

Pete Goodliffe

Architecture is the art of how to waste space.

—*Philip Johnson*

A software system is like a city—an intricate network of highways and hostelries, of back roads and buildings. There's a lot going on in a busy city; flows of control are continually being born, weaving their life through it, and dying. A wealth of data is amassed, stored, and destroyed. There are a range of buildings: some tall and beautiful, some squat and functional, others dilapidated and falling into disrepair. As data flows around them there are traffic jams and tailbacks, rush hours and road works. The quality of your software city is directly related to how much town planning went into it.

Some software systems are lucky, created through thoughtful design from experienced architects. They are structured with a sense of elegance and balance. They are well-mapped and easy to navigate. Others are not so lucky, and are essentially software settlements that grew up around the accidental gathering of some code. The transport infrastructure is inadequate, and the buildings are drab and uninspiring. Placed in the middle of it, you'd get completely lost trying to find a route out.

Where would your code rather live? What kind of software city would you rather construct?

In this chapter, I tell the story of two such software cities. It's a true story and, like all good stories, this one has a moral at the end. They say *experience is a great teacher,* but other people's experience is even better—if you can learn from these projects' mistakes and successes, you might save yourself (and your software) a lot of pain.

The two systems in this chapter are particularly interesting because they turned out very differently, despite being superficially very similar:

- They were of similar size (around 500,000 lines of code).
- They were both "embedded" consumer audio appliances.
- Each software ecosystem was mature and had gone through many product releases.
- Both solutions were Linux-based.
- The code was written in C++.
- They were both developed by "experienced" programmers (who, in some cases, *should* have known better).
- The programmers themselves were the architects.

In this story, names have been changed to protect the innocent (and the guilty).

The Messy Metropolis

> **Build up, build up, prepare the road! Remove the obstacles out of the way of my people.**
>
> —*Isaiah 57:14*

The first software system we'll look at is known as the Messy Metropolis. It's one I look back on fondly—not because it was good or because it was enjoyable to work with, but because it taught me a valuable lesson about software development when I first came across it.

My first contact with the Messy Metropolis was when I joined the company that created it. It initially looked like a promising job. I was to join a team working on a Linux-based, "modern" C++ codebase that had been in development for a number of years. Exciting stuff, if you have the same peculiar fetishes as me.

The work wasn't smooth sailing at first, but you never expect an easy ride when you start to work in a new team on a new codebase. However, it didn't get any better as the days (and weeks) rolled by. The code took a *fantastically* long time to learn, and there were no obvious routes into the system. That was a warning sign. At the microlevel, looking at individual lines, methods, and components, the code was messy and badly put together. There was no consistency, no style, and no unifying concepts drawing the separate parts together. That was another warning sign. Control flew around the system in unfathomable and unpredictable ways. That was yet another warning sign. There were so many bad "code smells" (Fowler 1999) that the codebase was not just putrid, it was a pungent landfill site on a hot summer's day. A clear warning sign. The data was rarely kept near where it was used. Often extra baroque caching layers were introduced to try to persuade it to hang around in more convenient places. Again, a warning sign.

As I tried to build a mental picture of the Metropolis, no one was able to explain the structure; no one knew all of its layers, tendrils, and dark, secluded corners. In fact, no one actually knew how any of it really worked (it was actually by a combination of luck and heroic maintenance programmers). People knew the small areas they had worked on, but no one had an overall comprehension of the system. And, naturally, there was no documentation. That was a warning sign. What I needed was a map.

This was the sad story I had become a part of: the Metropolis was a town planning disaster. Before you can improve a mess, you need to understand that mess, so with much effort and perseverance we pulled together a map of the "architecture." We charted every highway, all the arterial roads, the uncharted back roads, and all of the dimly lit side passages, and placed them on one master diagram. For the first time we could see what the software looked like. Not a pretty sight. It was a tangle of blobs and lines. In an effort to make it more comprehensible, we color-coded the control paths to signify their type. Then we stood back.

It was stunning. It was psychedelic. It was as if a drunk spider had stumbled into a few pots of poster paint and then spun a chromatic web across a piece of paper. It looked something like Figure 2-1 (it's a simplified version, with details changed to protect the guilty). Then it became clear. We had all but drawn a map of the London Underground. It even had the circle line.

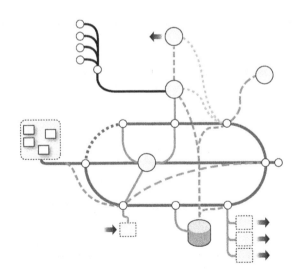

FIGURE 2-1. *The Messy Metropolis "architecture"*

This was the kind of system that would vex a traveling salesman. In fact, the architectural similarity to the London Underground was remarkable: there were many routes to get from one end of the system to the other, and it was rarely obvious how best to do so. Often a destination was geographically nearby but not accessible, and you wished you could bore a new tunnel between two points. Sometimes it would have actually have been better to get out and take a bus. Or walk.

That's not a "good" architecture by any metric. The Metropolis's problems went beyond the design, right up to the development process and company culture. These problems had actually caused a lot of the architectural rot. The code had grown "organically" over a period of years, which is a polite way to say that no one had performed any architectural design of note, and that various bits had been bolted on over time without much thought. No one had ever stopped to impose a sane structure on the code. It had grown by accretion, and was a classic example of a system that had received absolutely no architectural design. But a codebase never has *no* architecture. This just had a very poor one.

The Metropolis's state of affairs was understandable (but not condonable) when you looked at the history of the company that built it: it was a startup with heavy pressure to get many new releases out rapidly. Delays were not tolerable—they would spell financial ruin. The software engineers were driven to get code shipping as quickly as humanly possible (if not sooner). And so the code had been thrown together in a series of mad dashes.

> **NOTE**
> Poor company structure and unhealthy development processes will be reflected in a poor software architecture.

Down the Tubes

The Metropolis's lack of town planning had many consequences, which we'll see here. These ramifications were severe and went far beyond what you might naïvely expect of a bad design. The underground train had turned into a roller coaster, headed rapidly downward.

Incomprehensibility

As you can already see, the Metropolis's architecture and its lack of imposed structure had led to a software system that was remarkably tricky to comprehend, and practically impossible to modify. New recruits coming into the project (like myself) were stunned by the complexity and unable to come to grips with what was going on.

The bad design actually encouraged further bad design to be bolted onto it—in fact, it literally forced you to do so—as there was no way to extend the design in a sane way. The path of least resistance for the job in hand was always taken; there was no obvious way to fix the structural problems, and so new functionality was thrown in wherever it would cause less hassle.

> **NOTE**
> It's important to maintain the quality of a software design. Bad architectural design leads to further bad architectural design.

Lack of cohesion

The system's components were not at all cohesive. Where each one should have had a single, well-defined role, instead each component contained a grab bag of functionality that wasn't necessarily related. This made it hard to determine why a component existed at all, and hard to work out where a particular piece of functionality had been implemented in the system.

Naturally, this made bug fixing a nightmare, which seriously affected the quality and reliability of the software.

Both functionality and data were located in the wrong place in the system. Many things you'd consider "core services" were not implemented in the hub of the system, but were simulated by the outlying modules (at great pain and expense).

Further software archaeology showed why: there had been personality struggles in the original team, and so a few key programmers had begun to build their own little software empires. They'd grab the functionality they thought was cool and plonk it into their module, even if it didn't belong there. To deal with this, they would then make ever more baroque communication mechanisms to stitch the control back to the correct place.

> **NOTE**
> The health of the working relationships in your development team will feed directly into the software design. Unhealthy relationships and inflated egos lead to unhealthy software.

COHESION AND COUPLING

Key qualities of software design are *cohesion* and *coupling*. These are not newfangled "object-oriented" concepts; developers have been talking about them for many years, since the emergence of structured design in the early 1970s. We aim to design systems with components that have:

Strong cohesion
Cohesion is a measure of how related functionality is gathered together and how well the parts *inside* a module work as a whole. Cohesion is the glue holding a module together.

Weakly cohesive modules are a sign of bad decomposition. Each module must have a clearly defined role, and not be a grab bag of unrelated functionality.

Low coupling
Coupling is a measure of the interdependency *between* modules—the amount of wiring to and from them. In the simplest designs, modules have little coupling and so are less reliant on one

▶ 违背高内聚、松耦合原则是导致混乱结构的错误之源。站在架构的角度，我们可以参考Robert Martin提出的Clean Architecture思想。

Clean Architecture提出的模型是一个可测试的模型，无须依赖于任何基础设施就可以对它进行测试，只需通过边界对象发送和接收对应的数据结构即可。它们都遵循稳定依赖原则，不对变化或易于变化的事物形成依赖。为实现这一原则，Clean Architecture模型让外部易变的部分依赖于更加稳定的部分，如Domain Model，而非形成相反的依赖关系。这样还可使实现变得更易于变化；多变的部分依赖于稳定的部分。

好架构就要能轻松地改变那些易变的决定。

another. Obviously, modules can't be totally decoupled, or they wouldn't be working together at all!

Modules interconnect in many ways, some direct, some indirect. A module can call functions on other modules or be called by other modules. It may use web services or facilities published by another module. It may use another module's data types or share some data (perhaps variables or files).

▶ Clean Architecture图见P42。

Good software design limits the lines of communication to only those that are absolutely necessary. These communication lines are part of what determines the architecture.

Unnecessary coupling

The Metropolis had no clear layering. Dependencies between modules were not unidirectional, and coupling was often bidirectional. Component A would hackily reach into the innards of component B to get its work done for one task. Elsewhere, component B had hardcoded calls onto component A. There was no bottom layer or central hub to the system. It was one monolithic blob of software.

This meant that the individual parts of the system were so tightly coupled that you couldn't bring up a skeletal system without creating every single component. Any change in a single component rippled out, requiring changes in many dependent components. The code components did not make sense in isolation.

This made low-level testing impossible. Not only were code-level unit tests impossible to write, but component-level integration tests could not be constructed, as every component depended on almost every other component. Of course, testing had never been a particularly high priority in the company (we didn't have anywhere near enough time to do that), so this "wasn't a problem." Needless to say, the software was not very reliable.

> **NOTE**
> Good design takes into account connection mechanisms and the number (and nature) of inter-component connections. The individual parts of a system should be able to stand alone. Tight coupling leads to untestable code.

Code problems

The problems with bad top-level design had wormed their way down to the code level. Problems beget problems (see the discussion of broken windows in Hunt and Davis [1999]). Since there was no common design and no overall project "style," no one bothered with common coding standards, using common libraries, or employing common idioms. There were no naming conventions for components, classes, or files. There was not even a common build system; duct tape, shell scripts, and Perl glue nestled alongside makefiles and Visual Studio project files. Compiling this monster was considered a rite of passage!

One of the most subtle yet serious Metropolis problems was duplication. Without a clear design and a clear place for functionality to live, wheels had been reinvented across the entire codebase. Simple things like common algorithms and data structures were repeated across many modules, each implementation with its own set of obscure bugs and quirky behavioral traits. Larger-scale concerns such as external communication and data caching were also implemented multiple times.

More software archaeology showed why: the Metropolis started out as a series of separate prototypes that got tacked together when they should have been thrown away. The Metropolis was actually an accidental conurbation. When stitched together, the code components had never really fit together properly. Over time, the careless stitches began to tear, so the components pulled against one another and caused friction in the codebase, rather than working in harmony.

> **NOTE**
> A lax and fuzzy architecture leads to individual code components that are badly written and don't fit well together. It also leads to duplication of code and effort.

Problems outside the code

The problems within the Metropolis spilled out from the codebase to cause havoc elsewhere in the company. There were problems in the development team, but the architectural rot also affected the people supporting and using the product.

The development team
New recruits coming into the project (like myself) were stunned by the complexity and were unable to come to grips with what was going on. This partially explains why very few new recruits stayed at the company for any length of time—staff turnover was very high.

Those who remained had to work very hard, and stress levels on the project were high. Planning new features instilled a dread fear.

Slow development cycle
Since maintaining the Metropolis was a frightful task, even simple changes or "small" bug fixes took an unpredictable length of time. Managing the software development cycle was difficult, timescales were hard to plan, and the release cycle was cumbersome and slow. Customers were left waiting for important features, and management got increasingly frustrated at the development team's inability to meet business requirements.

Support engineers
The product support engineers had an awful time trying to support a flaky product while working out the intricate behavioral differences between relatively minor software releases.

Third-party support

An external control protocol had been developed, enabling other devices to control the Metropolis remotely. Since it was a thin veneer over the guts of the software, it reflected the Metropolis's architecture, which means that it was baroque, hard to understand, prone to fail randomly, and impossible to use. Third-party engineers' lives were also made miserable by the poor structure of the Metropolis.

Intra-company politics

The development problems led to friction between different "tribes" in the company. The development team had strained relations with the marketing and sales guys, and the manufacturing department was permanently stressed every time a release loomed on the horizon. The managers despaired.

> **NOTE**
> The consequence of a bad architecture is not constrained within the code. It spills outside to affect people, teams, processes, and timescales.

Clear requirements

Software archaeology highlighted an important reason that the Messy Metropolis turned out so messy: at the very beginning of the project *the team did not know what it was building*.

The parent startup company had an idea of which market it wanted to capture, but didn't know which kind of product to capture it with. So they hedged their bets and asked for a software platform that could do *many* things. *Oh, and we wanted it yesterday*. So the programmers rushed to create a hopelessly general infrastructure that could potentially do many things (badly), rather than craft an architecture that supported one thing well and could be extended to do more in the future.

> **NOTE**
> It's important to know what you're designing before you start designing it. If you don't know what it is and what it's supposed to do, *don't design it yet*. Only design what you know you need.

At the earliest stages of Metropolis planning there were far too many architects. With woolly requirements, they all took a disjoint piece of the puzzle and tried to work on it individually. They didn't keep the entire project in sight as they worked, so when they tried to put the puzzle pieces back together, they simply didn't fit. Without time to work on the architecture further, the parts of the software design were left overlapping slightly, and thus began the Metropolis town planning disaster.

▶ 在进行架构之前，必须要明确架构的目标。例如要设计一个高性能的产品，就需要在架构设计中充分考虑哪些因素可能成为性能瓶颈，然后针对这些问题寻找针对性的方案，例如引入缓存、并行处理、合理的资源分配等方案。确定了架构目标，就意味着它是你做出架构决策所要遵循的原则，是设计时明确的约束，要通过约束驱动架构。

Where Is It Now?

The Metropolis's design was almost completely irredeemable—believe me, over time we tried to fix it. The amount of effort required to rework, refactor, and correct the problems with the code structure had become prohibitive. A rewrite wasn't a cheap option, as support for the old, baroque control protocol was a requirement.

As you can see, the consequence of the Metropolis's "design" was a diabolical situation that was inexorably getting worse. It was so hard to add new features that people were just applying more kludges, Band-Aids, and calculated fudges. No one enjoyed working with the code, and the project was heading in a downward spiral. The lack of design had led to bad code, which led to bad team morale and increasingly lengthy development cycles. This eventually led to severe financial problems for the company.

Eventually, management acknowledged that the Messy Metropolis had become uneconomical, and it was thrown away. This is a brave step for any organization, especially one that is constantly running 10 paces ahead of itself while trying to tread water. With all of the C++ and Linux experience the team had gained form the previous version, the system was rewritten in C# on Windows. Go figure.

A Postcard from the Metropolis

So what have we learned? Bad architecture can have a profound effect and severe repercussions. The lack of foresight and architectural design in the Messy Metropolis led to:

- A low-quality product with infrequent releases
- An inflexible system that couldn't accommodate change or the addition of new functionality
- Pervasive code problems
- Staffing problems (stress, low morale, turnover, etc.)
- A lot of messy internal company politics
- Lack of success for the company
- Many painful headaches and late nights working on the code

Design Town

Form ever follows function.

—*Louis Henry Sullivan*

The Design Town software project was superficially very similar to the Messy Metropolis. It too was a consumer audio product written in C++, running on a Linux operating system.

However, it was built in a very different way, and so the internal structure worked out very differently.

I was involved with the Design Town project from the very start. A brand-new team of capable developers had been assembled to build it from scratch. The team was small (initially four programmers) and, like the Metropolis, the team structure was flat. Fortunately, there was none of the interpersonal rivalry apparent in the Metropolis project, or any vying for positions of power in the team. The members didn't know each other well beforehand and didn't know how well we'd work together, but we were all enthused about the project and relished the challenge.

So far, so good.

Linux and C++ were early decisions for the project, and that shaped the team that had been assembled. From the outset the project had clearly defined goals: a particular first product and a roadmap of future functionality that the codebase had to accommodate. This was to be a general-purpose codebase that would be applied in a number of product configurations.

The development process employed was eXtreme Programming (or XP) (Beck and Andres 2004), which many believe eschews design: *code from the hip, and don't think too far ahead*. In fact, some observers were shocked at our choice and predicted that it would all end in tears, just like the Metropolis. But this is a common misconception. XP does not discourage design; it discourages work that isn't necessary (this is the YAGNI, or *You Aren't Going To Need It*, principle). However, where upfront design is required, XP requires you to do that. It also encourages rapid prototypes (known as *spikes*) to flesh out and prove the validity of designs. Both of these were very useful and contributed greatly to the final software design.

First Steps into Design Town

Early in the design process, we established the main areas of functionality (these included the core audio path, content management, and user control/interface). We considered where they each fit in the system, and an initial architecture was fleshed out, including the core threading models that were necessary to achieve performance requirements.

The relative positions of the separate parts of the system was established in a conventional layer diagram, a simplified part of which is shown in Figure 2-2. Notice that this was *not* a big upfront design. It was an intentionally simple conceptual model of the Design Town: just some blobs on a diagram, a basic system design that could grow easily as pieces of functionality were added. Although basic, this initial architecture proved a solid basis for growth. Whereas the Metropolis had no overall picture and saw functionality grafted (or bodged) in wherever was "convenient," this system had a clear model of what belonged where.

Extra design time was spent on the heart of the system: the audio path. It was essentially an internal subarchitecture of the entire system. To define this, we considered the flow of data through a series of components and arrived at a filter-and-pipeline audio architecture, similar to Figure 2-3. The products involved a number of these pipelines, depending on their physical configuration. Again, at first this pipeline was nothing more than a concept—more blobs on a diagram. We hadn't decided how it would all be stitched together.

FIGURE 2-2. *The Design Town initial architecture*

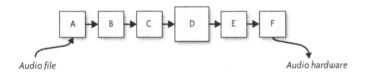

FIGURE 2-3. *The Design Town audio pipeline*

We also made an early choice of supporting libraries the project would employ (for example, the Boost C++ libraries available at *http://www.boost.org* and a set of database libraries). Decisions about some of the basic concerns were made at this point to ensure that the code would grow easily and cohesively, including:

- The top-level file structure
- How we would name things
- A "house" presentation style
- Common coding idioms
- The choice of unit test framework
- The supporting infrastructure (e.g., source control, a suitable build system, and continuous integration)

These "fine detail" factors were very important: they allied closely with the software architecture and, in turn, influenced many later design decisions.

The Story Unfolds

Once the initial design had been established by the team, the Design Town project proceeded following the XP process. Design and code construction was either done in pairs or carefully reviewed to ensure that work was correct.

The design and the code developed and matured over time, and as the story of Design Town unfolded, there were the following consequences.

Locating functionality

With a clear overview of the system structure in place from the very beginning, new units of functionality were consistently added to the correct functional areas of the codebase. There was never a question about where code belonged. It was also easy to find the implementation of existing functionality in order to extend it or to fix problems.

Now, sometimes putting new code in the "right" place was harder than simply bodging it into a more convenient, but less tasteful, place. So the existence of an architectural plan sometimes made the developers work harder. The payoff for this extra effort was a *much* easier life later on, when maintaining or extending the system—there was very little cruft to trip over.

> **NOTE**
> An architecture helps you to locate functionality: to add it, to modify it, or to fix it. It provides a template for you to slot work into and a map to navigate the system.

Consistency

The entire system was consistent. Every decision at every level was taken in the context of the whole design. The developers did this intentionally from the outset so all the code produced matched the design fully, and matched all the other code written.

Over the project's history, despite many changes ranging across the entire scope of the codebase—from individual lines of code to the system structure—everything followed the original design template.

> **NOTE**
> A clear architectural design leads to a consistent system. All decisions should be made in the context of the architectural design.

The good taste and elegance of the top-level design naturally fed down to the lower levels. Even at the lowest levels, the code was uniform and neat. A clearly defined software design ensured that there was no duplication, that familiar design patterns were used throughout, familiar interface idioms were adopted, and that there were no unusual object lifetimes or odd resource management issues. Lines of code were written in the context of the town plan.

NOTE
Clear architecture helps reduce duplication of functionality.

Growing the architecture

Some entirely new functional areas appeared in the "big picture" design—storage management and an external control facility, for example. In the Metropolis project, this was a crushing blow and incredibly hard to do. But in Design Town, things worked differently.

The system design, like the code, was considered malleable and refactorable. One of the development team's core principles was to stay nimble—that nothing should be set in stone—and so the architecture could be changed when necessary. This encouraged us to keep our designs simple and easy to change. Consequently, the code could grow rapidly and maintain a good internal structure. Accommodating new functional blocks was not a problem.

NOTE
Software architecture is not set in stone. Change it if you need to. To be changeable, the architecture must remain simple. Resist changes that compromise simplicity.

▶ 评注见下页。

Deferring design decisions

One of the XP principles that really enhanced the quality of the Design Town architecture was YAGNI (don't do anything if *you aren't going to need it*). It encouraged us to design only the important stuff early on, and to defer all remaining decisions until later, when we had a clearer picture of the actual requirements and how best to fit them into the system. This is an immensely powerful design approach, and quite liberating.

- One of the worst things you can do is design something you don't yet understand. YAGNI forces you to wait until you know what the problem really is and how it should be accommodated by the design. It eliminates guesswork and ensures the design will be correct.
- It is dangerous to add everything you *might* need (including the kitchen sink) to a software design when you first create it. Most of your design work will be wasted effort, and produce extra baggage that you'll need to support over the entire changing life of the software. It costs more at first, and continues to cost over the life of the project.

NOTE
Defer design decisions until you *have* to make them. Don't make architectural decisions when you don't know the requirements yet. Don't guess.

Maintaining quality

From the outset, the Design Town project put a number of quality control processes in place:

- Pair programming

第 2 章 两个系统的故事：摩登时代的软件神话 37

- Code/design reviews for anything not pair-programmed
- Unit tests for every piece of code

These processes ensured that the system never had an incorrect, badly fitting change applied. Anything that didn't mesh with the software design was rejected. This might sound draconian, but they were processes that the developers bought into.

This buy-in highlights an important attitude: the developers believed in the design, and considered it important enough to protect. They took ownership of, and personal responsibility for, the design.

> **NOTE**
> Architectural quality must be maintained. This can happen only when the developers are given and take responsibility for it.

Managing technical debt

Despite these quality control measures, Design Town development was fairly pragmatic. As deadlines approached, a number of corners were cut to allow projects to ship on time. Small code "sins" or design warts were allowed to enter the codebase, either to get functionality working quickly or to avoid high-risk changes near a release.

However, unlike the Messy Metropolis project, these fudges were marked as *technical debt* and scheduled for later revision. These warts stood out clearly, and the developers were not happy about them until they were dealt with. Again, we see the developers taking responsibility for the quality of the design.

Unit tests shape design

One of the core decisions about the codebase (which is also mandated by XP development) was that everything should be unit tested. Unit testing brings many advantages, one of which is the ability to change sections of the software without worrying about destroying everything else in the process. Some areas of the Design Town internal structure received quite radical rework, and the unit tests gave us confidence that the rest of the system had not been broken. For example, the thread model and interconnection interface of the audio pipeline was changed fundamentally. This was a serious design change relatively late in the development of that subsystem, but the rest of the code interfacing with the audio path continued executing perfectly. The unit tests enabled us to change the design.

This kind of "major" design change slowed down as Design Town matured. After an amount of design rework, things settled down, and subsequently there were only minor design changes. The system developed quickly, in an iterative manner, with each step improving the design, until it reached a relatively stable plateau.

▶ Managing technical debt与Deferring design decisions一脉相承。正是因为我们不能在需求不明确的情况下对未来做出决策，才可能导致在整个开发过程中源源不断地涌现出技术债。技术债是不可避免的，关键是需要我们积极地去应对它们。随时随地识别这些技术债，并将其可视化出来，例如按照不同象限以及重要程度标记在雷达图中，张贴在团队触目可及的地方。同时，要做好规划，将处理技术债的任务合理地安排在各个迭代中，与开发任务一视同仁。

> **NOTE**
> Having a good set of automated tests for your system allows you to make fundamental architectural changes with minimal risk. It gives you space in which to work.

Another major benefit of the unit tests was their remarkable shaping of the code design: they practically enforced good structure. Each small code component was crafted as a well-defined entity that could stand alone, as it had to be constructible in a unit test without requiring the rest of the system to be built up around it. Writing unit tests ensured that each module of code was internally cohesive and loosely coupled from the rest of the system. The unit tests forced careful thought about each unit's interface, and ensured that the unit's API was meaningful and internally consistent.

> **NOTE**
> Unit testing your code leads to better software designs, so design for testability.

Time for design

One of the contributing factors to Design Town's success was the allotted development timescale, which was neither too long nor too short (just like Goldilocks's porridge). A project needs a conducive environment in which to thrive.

Given too much time, programmers often want to create their magnum opus (the kind of thing that will always be *almost* ready, but never quite materializes). A little pressure is a wonderful thing, and a sense of urgency helps to get things done. However, given too little time, it simply isn't possible to achieve any worthwhile design, and you'll get only a half-baked solution rushed out—just like the Metropolis.

> **NOTE**
> Good project planning leads to superior designs. Allot sufficient time to create an architectural masterpiece—they don't appear instantly.

Working with the design

Although the codebase was large, it was coherent and easily understood. New programmers could pick it up and work with it relatively easily. There were no unnecessarily complex interconnections to understand, or weird legacy code to work around.

Since the code has generated relatively few problems and is still enjoyable to work with, there has been very, very low turnover of team members. This is due in part to the developers taking ownership of the design and continually wanting to improve it.

It was interesting to observe how the development team dynamics followed the architecture. Design Town project principles mandated that no one "owned" any area of the design, meaning that any developer could work anywhere in the system. Everyone was expected to write

▶ 最近几年，随着云计算（或虚拟化）的发展，软件测试搭建环境的成本变得越来越低，且更易于重用和维护，这使得集成测试的自动化门槛降低。由于单元测试不可避免地存在诸多问题，例如单元测试的脆弱性、维护成本、Mock引入的问题，业内人士也开始对自动化测试开始了反思，观点也在发生变化——从过去较多地重视单元测试覆盖率改为对集成测试的重视。而微服务提倡的小代码库，使得单元测试重构的重要性也不如过去。

high-quality code. Whereas the Metropolis was a sprawling mess created by many uncoordinated, fighting programmers, Design Town was clean and cohesive, closely cooperating software components created by closely cooperating colleagues. In many ways, Conway's Law* worked in reverse, and the team gelled together as the software did.

> **NOTE**
> A team's organization has an inevitable affect on the code it produces. Over time, the architecture also affects how well the team works together. When teams separate, the code interacts clumsily. When they work together, the architecture integrates well.

Where Is It Now?

After some time, the Design Town architecture looked like Figure 2-4. That is, it was remarkably similar to the original design, with a few notable changes—and a lot more experience to prove the design was right. A healthy development process, a smaller, more thoughtful development team, and an appropriate focus on ensuring consistency led to an incredibly simple, clear, and consistent design. This simplicity worked to the advantage of the Design Town, leading to malleable code and rapidly developed products.

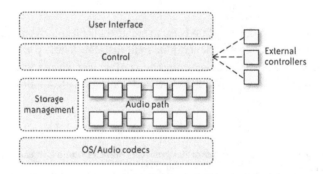

FIGURE 2-4. The Design Town final architecture

At the time of this writing, the Design Town project has been alive for three years. The codebase is still in production use and has spawned a number of successful products. It is still being developed, still growing, still being extended, and still being changed daily. Its design next month might be quite different from how it looks this month, but it probably won't.

Let me make this clear: the code is by no means perfect. It has areas of technical debt that need work, but they stick out against the backdrop of neatness and will be addressed in the future. Nothing is set in stone, and thanks to the adaptable architecture and flexible code structure,

---

* Conway's Law states that code structure follows team structure. Simply stated, it says, "If you have four groups working on a compiler, you'll get a four-pass compiler."

these things can be fixed. Almost everything is in the right place, because the architecture is sound.

So What?

> **When perfection comes, the imperfect disappears.**
> —1 Corinthians 13:10

This simple story about two software systems is certainly not an exhaustive treatise on software architecture, but I have shown how architecture profoundly affects a software project. An architecture influences almost everything that comes into contact with it, determining the health of the codebase and also the health of the surrounding areas. Just as a thriving city can bring prosperity and renown to its local area, a good software architecture will help its project to flourish and bring success to those depending on it.

Good architecture is the product of many factors, including (but not limited to):

- Actually doing intentional upfront design. (Many projects fail in this way before they even start.)
- The quality and experience of the designers. (It helps to have made a few mistakes beforehand to point you in the right direction next time! The Metropolis project certainly taught me a thing or two.)
- Keeping the design clearly in view as development progresses.
- The team being given and taking responsibility for the overall design of the software.
- Never being afraid of changing the design: nothing is set in stone.
- Having the right people on the team, including designers, programmers, and managers, and ensuring the development team is the right size. Ensure they have healthy working relationships, as these relationship will inevitably feed into the structure of the code.
- Making design decisions at the appropriate time, when you know all the information necessary to make them. Defer design decisions you cannot yet make.
- Good project management, with the right kind of deadlines.

Your Turn

> **Never lose a holy curiosity.**
> —Albert Einstein

You are reading this book right now because you *care* about software architecture, and you care about improving your own software. So here's an excellent opportunity. Consider these simple questions about your software experience to date:

1. What's the best system architecture you've ever seen?
   - How did you recognize it as good?
   - What were the consequences of this architecture, both inside and outside the codebase?
   - What have you learned from it?
2. What's the worst architecture system you've ever seen?
   - How did you recognize it as bad?
   - What were the consequences of this architecture, both inside and outside the codebase?
   - What have you learned from it?

References

Beck, Kent, with Cynthia Andres. 2004. *Extreme Programming Explained*, Second Edition. Boston, MA: Addison-Wesley Professional.

Fowler, Martin. 1999. *Refactoring: Improving the Design of Existing Code*. Boston, MA: Addison-Wesley Professional.

Hunt, Andrew, and David Thomas. 1999. *The Pragmatic Programmer*. Boston, MA: Addison-Wesley Professional.

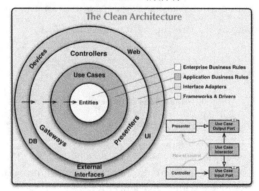

Value Sliders 的活动

Clean Architecture 图

PART II

第 2 部分

Enterprise Application Architecture
企业应用架构

Chapter 3　Architecting for Scale
第 3 章　可伸缩架构

Chapter 4　Making Memories
第 4 章　记忆留存

Chapter 5　Resource-Oriented Architectures: Being "In the Web"
第 5 章　面向资源架构：在 Web 之中

Chapter 6　Data Grows Up: The Architecture of the Facebook Platform
第 6 章　数据增长：Facebook 平台的架构

	Principles and properties		Structures
✓	Versatility		Module
✓	Conceptual integrity	✓	Dependency
	Independently changeable		Process
	Automatic propagation	✓	Data access
	Buildability		
✓	Growth accommodation		
	Entropy resistance		

CHAPTER THREE

Architecting for Scale

Jim Waldo

Introduction

ONE OF THE MORE INTERESTING PROBLEMS IN DESIGNING AN ARCHITECTURE for a system is ensuring flexibility in the scale of that system. Scaling is becoming increasingly important, as more of our systems are run on networks or are available on the Web. For such systems, the idea of capacity planning is absurd if you want a margin of error that is under a couple of orders of magnitude. If you put up a site and it becomes popular, you might suddenly find that there are millions of users accessing your site. Just as easily (and just as much of a disaster), you can put up a site and find that no one is particularly interested, and all of the equipment in which you invested now lies idle, soaking up money in energy costs and administrative effort. In the networked world, a site can transition from one of these states to the other in a matter of minutes.

The scaling problem is faced by anyone who attaches a system to a network, but it is particularly interesting in the case of massively multiplayer online games (MMOs) and virtual worlds. These systems must be capable of scaling to large numbers of users. Unlike web servers, however, where the users are requesting fairly static information and are not interacting with each other, players in an MMO or residents in a virtual world are there to interact with both the world (changing the underlying information in the world) and each other. These interplays complicate the scaling of the infrastructures for such systems, as the user interactions with the system are mostly independent (except when they aren't) and don't change much state in the world. Given any two participants in such a world, the likelihood that they are interacting at any given time is vanishingly small. But nearly every player will be interacting with someone nearly all the time. The result is a kind of system that is embarrassingly parallel but interdependent in a small number of interactions.

Scaling of MMOs and virtual worlds is further complicated by the culture that has grown up around these systems. Both MMOs and virtual worlds trace their descent from the production of video games. This is a design culture that grew up in the PC and console game tradition, a tradition in which the programmer could assume that the game ran on a standalone machine or game console. In such an environment, all of the resources of the machine are at the command of the game program, and problems with the program are confined to the single user playing the game (and, in fact, bugs or odd behavior could often be taken as part of the logic of the game itself).

These games, and the companies that write, produce, and enhance them, are part of the entertainment industry. Teams writing a game are led by a producer, and there are scripts and back stories. The goal of a game is to be immersive, persuasive, and most of all, fun. Reliability is nice, but hardly required. Extensibility is a property of the game, allowing new plot lines and themes to be released as upgrades to the game, rather than a property of the code that allows the code to be used in new and different ways.

The rise of online games and virtual worlds brings this culture into an environment where the requirements are much more like those that are faced by the enterprise developer. With multiple players interacting on a server over the network, the crash of a server brought about by the unexpected actions of a player will affect many other players. As these worlds develop economies (some of which interact with the economy of the real world), the stability and consistency of the online world becomes more than just a game. And as the number of players or inhabitants in these worlds reaches the millions, the ability to scale becomes a primary requirement of any architecture.

Project Darkstar (referred to in the rest of this chapter as simply Darkstar) is a response to these changing needs of the builders of games and virtual worlds. The project, undertaken by a research group inside of Sun Microsystems Laboratories, is an ongoing exploration in the architecture of scale. What makes the project particularly interesting is that it is targeted to the MMO and virtual-world builder, a group of programmers who have very different needs from those that we (as system designers) have been used to. The resulting architecture has much that seems familiar until you look at it closely, at which point you can see why it differs from what your experience told you it must be. The result is an architecture with its own sort of beauty, and an object lesson in how different requirements can change the way you have to think about building a system.

Context

Like the physical architecture of a building or a city, the architecture of a system has to be adapted to the context in which the artifact built using the architecture will reside. In physical architecture, this context includes the historical surroundings of the work, the climate in which it will exist, the ability of the local artisans and the available building materials, and the intended use of the building. For a software architecture, the context includes not only the applications that will use the architecture, but also the programmers who will build within that architecture and the constraints on the systems that will result.

In building the Darkstar architecture, the first thing we* realized is that any architecture for scaling would need to involve multiple machines. It is not clear that even the largest of mainframes could scale to meet the demands of some of today's online games (*World of Warcraft*, for example, is reported to have five million current subscribers, with hundreds of thousands of them active at any one time). Even if there were a single machine that could handle this load, it would be economically impossible to assume that a game would be so successful that it would require such a hardware investment at the beginning. This kind of application needs to be able to start small and then increase capacity as the user base increases, and then decrease capacity as interest in the game wanes. This maps well to a distributed system, where (reasonably small) machines can be added as demand increases and taken away when demand decreases. Thus we knew at the beginning that the overall architecture would need to be a distributed system.

We also knew that the system would need to exploit the current trends in chip architectures. MMOs and (to a lesser extent) virtual worlds have historically exploited Moore's law for scaling. As a processor doubles in speed, the world that can be created doubles in complexity, richness, and interactivity. No other area of computing has exploited the benefits of increased processor speed in quite the way the game world has. Personal computers designed for games are always pushing the limits of CPU speed, memory, and graphics capabilities. Game consoles push these limits even more aggressively, containing graphics systems far beyond those found in high-end workstations and building the entire machine around the specialized needs of the game player.

The recent change in chip evolution, from the constant increase in clock speeds to the construction of multicore processors, has changed the dynamic of what can be done in games. Rather than doing one thing faster, new chips are being designed to do multiple things at the same time. The introduction of concurrent execution at the chip level will give better total performance if the tasks being run by the chip can in fact be executed at the same time. Without

---

* In talking about the development of the Project Darkstar architecture, I will generally refer to what "we" did rather than speak about what "I" did. This is more than the use of the editorial "we." The design of the architecture was very much a collaborative project, started by Jeffrey Kesselman, Seth Proctor, and James Megquier, and put into its current form by Seth, James, Tim Blackman, Ann Wollrath, Jane Loizeaux, and me.

a change in clock speed, a chip with four cores ought to be able to do four times as much as a chip with a single core. In fact, the speed-up will not be quite linear, as there are other parts of the system that are not made concurrent in the same way. But increases in the overall performance of the system can be obtained by the use of concurrency, and building chips for such concurrent use is far simpler than building chips in which the clock speed is increased.

On the face of it, MMOs and virtual worlds ought to be reasonable candidates for multicore chips and distributed systems. Most of what goes on in an MMO or virtual world, like most of what goes on in the real world, is independent of the other things that are happening in that world. Players go on their own quests or decorate their own rooms. They battle monsters or design clothes. Even when they are engaged with another player or occupant of the world, they are interacting with only a very small percentage of the occupants of the world. This is the characterization of an embarrassingly parallel computational task, and that is just the sort of thing that multiple cores and multiple machines ought to be good at doing.

Although the tasks in these systems may be embarrassingly parallel, the programmers who work on such systems are not trained or experienced in the techniques of either distributed computing or concurrent programming. These are exceptionally subtle fields, difficult even for those who have been trained in them and who have considerable experience in using these techniques. To ask most game programmers to develop a highly concurrent, distributed game server would be asking them to go well outside of their area of expertise or experience.

▶ 伸缩性（scaling）设计需要从系统的多个层次去考虑。以一个典型基于Web的大数据分析系统为例，就需要从应用服务、数据分析、数据存储三个层次去考虑支持系统的水平伸缩。这是一个整体架构概念。

The First Goal

This context gave us our first goal for the architecture. The requirements for scaling dictated that the system be distributed and concurrent, but we needed to present a much simpler programming model to the game developer. The simple statement of the goal is that the game developer should see the system as a single machine running a single thread, and all of the mechanisms that would allow deployment on multiple threads and multiple machines should be taken care of by the Project Darkstar infrastructure.

In the general case, hiding either distribution or concurrency from the application is not possible. But MMOs and virtual worlds are not the general case. The kind of hiding that we are trying to accomplish comes at the price of requiring a very specific and restricted programming model. Fortunately, it is just the sort of model that lends itself to the kind of programming already used in the server-side components of games and virtual worlds.

The general programming model that Project Darkstar requires is a reactive one, in which the server side of the game is written as a listener for events generated by the clients (that is, the machines being used by the game players, generally either a PC or a game console). When an event is detected, the game server should generate a task, which is a short-lived sequence of computations that includes manipulation of information in the virtual world and communication with the client that generated the original event and possibly with other clients. Tasks can also be generated by the game server itself, either as a response to some internal change or on a periodic, timed basis. In this way, the game server can generate characters in the game or world that aren't controlled by some outside player.

This sort of programming model fits well with games and virtual worlds, but is also used in a number of enterprise-level architectures, such as J2EE and web services. The need to build an architecture different from those enterprise mechanisms was dictated by the very different environment in which MMOs and virtual worlds exist. This environment is nearly a mirror image of the classic enterprise environments, which means that if you have been trained in the enterprise environment, almost everything you know is going to be wrong in this new world.

The classic enterprise environment is envisioned as a thin client connected to a thick server (which is itself often connected to an even thicker database server). The server will hold most of the information needed by the clients, and will act as a filter to the backend database. Very little state is held at the client; in the best case, the client has very little memory, no disk of its own, and is a highly competent display device for the server, which is where most of the real work occurs.

The Game World

The MMO and virtual world environment starts with a very thick client—typically a top-of-the-line PC with the most powerful CPU available, lots of memory, and a graphics card that is itself computationally excellent, or a game console that is specially designed for graphics-intensive, highly interactive tasks. As much as possible, data is pushed out to these clients, especially data that is unchanging, such as geographic information, texture maps, and rule sets. The server is kept as simple as possible, generally holding a very abstract representation of the world and the entities within that world. Further, the server is designed to do as little computation as possible. Most of the computation goes on at the client. The real job of the server is to hold the shared truth of the state of the world, ensuring that any variation in the view of the world held at the various clients can be corrected as needed. The truth needs to be held by the server, since those who control the clients have a vested interest in maximizing their own performance, and thus might be tempted to change the shared truth (if they could) in their favor. Put more directly, players will cheat if they can, so the server must be the ultimate source of shared truth.

The data access patterns of MMOs and virtual worlds are also quite different from those that are seen in enterprise situations. The usual rule of thumb within the enterprise is that 90% of data accesses will be read-only, and most tasks read a large amount of data before altering a small amount. In the MMO and virtual world environment, most tasks access only a very small amount of the state on the server, but of the data that they access, about half of it will be altered.

Latency Is the Enemy

But the biggest difference in the two environments traces back to the differences in what the users are doing. In an enterprise environment, the goal is to conduct business, and some lags in processing are acceptable if the overall throughput is improved. In the MMO and virtual world environment, the goal is to have fun, and latency is the enemy of fun. So the infrastructure for an MMO or virtual world needs to be designed around the requirement of bounding latency whenever possible, even at the cost of throughput.

Online games and virtual worlds have clearly found ways to scale to large numbers of users. The current mechanisms fall into two groups. The first of these is geographic in nature. The game is designed as a group of different areas, with each area designed to be run on a single server. It might be an island or room in a virtual world or a town or valley in an online game. The design of the game tries to make each geographic area independent, and scale the geographic area in such a way that the server will not be overwhelmed by too many users occupying the area. In practice, such areas are often self-limiting, as when the server is being overwhelmed, the play becomes less responsive and less interesting. As a result, players leave for more interesting areas, which makes the formerly overwhelmed area less occupied and improves response time.

The problem with scaling by assigning geographic areas to different servers is that the decision of what areas scale to a server must be made when the game is being written. Although new areas might be able to be added to a game or world fairly easily, changing the area that is assigned to a server is something that requires changing the code. The decision of what areas are the unit of scale has to be made as part of development.

A second way of dealing with areas that are overcrowded in a game or world is known as *sharding*. A shard is a copy of the area, run on its own server and independent of other shards, that presents the same portion of the game as the original area. Thus, a shard might present a different copy of a particular room or village, allowing twice as many players to occupy that part of the world. The drawback of shards is that they do not allow players in different shards to interact with each other. As games and worlds become more social experiences than simple game play, this disadvantage can be major. The goal of players is not only to be in the virtual world, but to occupy it with their (real or virtual) friends. Sharding interferes with that goal.

Thus, another major goal of the Darkstar architecture is to allow on-the-fly scaling in a way that does not require the game logic to become involved in the scaling. The infrastructure should allow the game to dynamically react to the load rather than make such reaction part of the design of the game.

The Architecture

Darkstar is built as a set of separate services available in the address space of the server side of a game or virtual world. Each service is defined by a small programming interface. Although not the original intention, the basic services provided by Project Darkstar are much like those of a classic operating system, allowing the server side of the game or virtual world to access persistent storage, schedule and run tasks, and perform communication with the client side of the game or virtual world.

Structuring the system as an interconnected set of services is an obvious way to begin the process of divide and conquer that is basic to the design of any large computer system. Each service can be characterized by an interface that protects those using the service from changes in the underlying implementation, and allows those implementations to be undertaken independently. Changes in the implementation of one service ought not affect the implementation of another, even if that other service makes use of the implementation being changed (assuming the interface and the semantics of the interface don't change).

We had other reasons to adopt the service decomposition approach. From the very beginning, Project Darkstar was envisioned as an open source project, with the hope that we could leverage the work of the core team by allowing other members of the community to build additional services that could enrich the functionality of the core. Running an open source community is complicated under any circumstance, and we believed that having the greatest level of isolation between the services that make up the infrastructure would allow a higher level of isolation between different service implementation levels. Additionally, it was not clear that there was a single set of services that would be just right for all MMOs and virtual worlds. By structuring the infrastructure as a set of independent services, different sets of those services could be used in different circumstances dictated by the needs of the particular project using the infrastructure. The services included in any particular Darkstar stack can be set by a configuration file.

▶ 只要设计好服务的边界，定义好服务接口，并规划好服务的版本演化，无论是传统的 SOA 架构，还是当下流行的微服务架构，都能有效地保证系统的解耦、重用和扩展。

▶ 着重强调。

The Macro Structure

Figure 3-1 shows the basic structure of a game or virtual world based on the Project Darkstar infrastructure. There will be some number of servers that form the backend of the game or virtual world. Each of these servers runs a copy of the selected set of services (labeled the Darkstar stack) and a copy of the game logic. Clients will connect to one of these servers to interact with the abstract representation of the world held by the server.

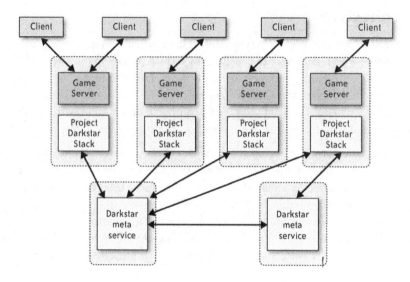

FIGURE 3-1. Project Darkstar high-level architecture

Unlike most replication schemes, the different copies of the game logic are not meant to process the same events. Instead, each copy can independently interact with the clients. Replication in this design is used primarily to allow scale rather than to ensure fault tolerance (although, as we will see later, fault tolerance is also achieved). Further, the game logic itself does not know or need to know that there are other copies of the server operating on other machines. The code written by the game programmer runs as if it were on a single machine, with coordination of the different copies done by the Project Darkstar infrastructure. Indeed, it is possible to run a Darkstar-based game on a single server if that is all the capacity the game needs.

Clients connect to the game logic using communication mechanisms that are part of the infrastructure. These mechanisms allow either direct client-to-server communication or a form of publish-subscribe channel, where any message sent on a channel is delivered to all of those subscribed to the channel.

The Darkstar stacks are coordinated by a set of meta-services—network-accessible services that are hidden from the game or virtual world programmer. These meta-services allow the various copies of the stack to coordinate the overall operation of the game. These meta-services will, for example, make sure that all of the separate copies continue to run and initiate failure recovery if some copy fails; keep track of the load on the copies and redistribute that load when needed; or allow new servers to be added at any time to increase the capacity of the whole. Since these services are completely hidden from the users of Project Darkstar, they can be changed or removed, or new ones can be added at any time without changing the code of the game or virtual world.

For the programmer building a game or virtual world in the Project Darkstar environment, the visible architecture is the set of services contained in the stack. The overall set of services is both changeable and configurable, but four basic services will always be present and form the core of the operating environment, as shown in Figure 3-2.

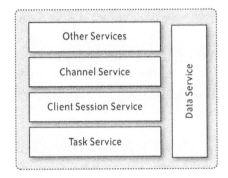

FIGURE 3-2. Darkstar stack

> ▶ DarkStar对服务的层次划分与关注点分离的设计原则，值得我们借鉴。我目前正在研发的产品运用了微服务模式，而在对服务进行设计时，也借鉴了这里的分层思想。当然，在设计服务时，还要有效地识别出跨服务的公共模块。此时，可以运用Kernel模式，将核心功能视为Kernel模块，而将公开在外的服务视为外部服务。

The Basic Services

The most basic of these stack-level services is the Data Service, which is used to store, retrieve, and manipulate all persistent data in the game or virtual world. The notion of persistence here is somewhat broader than might be found in other systems. In games or virtual worlds written in the Project Darkstar environment, any data that lasts longer than a single task is considered persistent and must be stored in the Data Service. Remember that we assume (and require) a programming model in which tasks are short-lived, so almost all of the data used to represent the server-side representation of the game or world will be persistent. The Data Service also knits together the separate copies of the game or world that are running on different servers, as all of these copies will share a single (conceptual) instance of the Data Service. All of the copies will have access to the same data, and all of the copies can read or change data stored in that service as needed.

Although the Data Store looks like a natural place for using a database, the requirements on the store are in fact very different from those that usually condition standard databases. There are very few static relations between the objects in the store, and there is no requirement within the game for any type of complex queries over the contents of the store. Instead, a simple naming scheme suffices, along with program-language-level references to the objects. The Data Store also has to be optimized for latency rather than throughput. The number of objects accessed by any particular task tends to be small (our preliminary measurements based on some prototype games and worlds suggest about a dozen objects per task), and about half of those objects that are accessed by any task are altered in the course of the task.

The second stack-level service is the Task Service, which is used to schedule and perform the tasks that are generated either in response to some event received from the clients or by the internal logic of the game or world server itself. Most tasks are one-time affairs, generated because of some action on the client, that read some data from the Data Service, manipulate that data, perhaps perform some communication, and then end. Tasks can also generate other tasks, or they can be generated as periodic tasks that will be run at particular times or intervals. All tasks must be short-lived; the maximal time for a task is a configured value, but the default is 100 milliseconds.

The game or world programmer sees a single task being generated either by an event or by the server logic itself, but under the covers the Darkstar infrastructure is scheduling as many simultaneous tasks as it can. In particular, tasks generated by the server logic will run in parallel with tasks generated in response to a client-initiated event, as will events generated in response to different clients.

Such concurrent execution leads to the possibility of data contention. To deal with such contention requires that the Task Service and the Data Service conspire. Under the covers and invisible to the server programmer, each task scheduled by the Task Service is wrapped in a transaction. This transaction ensures that either all of the operations in the task complete or none of them do. In addition, any attempts to alter values of objects held in the Data Service are mediated by that service. If more than one task attempts to alter the same data object, all but one of those tasks will be aborted and rescheduled to be performed later. The remaining task will run to completion. Once the running task has been completed, the other tasks can be run. Although it is possible for the server programmer to indicate that the data being accessed will be modified, this is not required. If a data object is simply read and then later modified, the modification will be detected by the Data Service before the task is committed. Indicating that modification is intended at the time of read is an optimization that allows early detection of conflicts, but the failure to indicate the intent to modify does not affect the correctness of a program.

Wrapping the tasks in a transaction means that the communication mechanisms must also be transactional, with messages sent only when the transaction wrapping the task that sends the messages commits. This is accomplished through the two remaining core services of the Darkstar stack.

Communication Services

The first of these is the Session Service, which mediates communication between a client and the game or world server. Upon login and authentication, a session is established between the client and the server. Servers listen for messages sent by the client on the session, parsing the contents of the message to determine what task to generate in response to the message. Clients listen on the channel to receive any responses from the server. These sessions mask the actual endpoints to both the client and the server, a factor that is important in the multimachine scaling strategy of Darkstar. The session is also responsible for ensuring that the order of messages is maintained. A message from a given client will not be delivered if the tasks that resulted from previous message deliveries have not completed. Having the session service order tasks in this way significantly simplifies the Task Service, which can assume that all of the tasks that it has at any time are essentially concurrent. The ordering of messages from a particular client is the only message-ordering guarantee made within the Darkstar framework; external observers might see an ordering of messages from multiple clients that is very different from that seen within the game or virtual world.

The second communication service that is always available in the Darkstar stack is the Channel Service. Channels are a form of one-to-many communication. Conceptually, channels can be joined by any number of clients, and any message that is sent on the channel will be delivered to all of the clients that have been associated with the channel. This might seem to be a perfect place to utilize peer-to-peer technologies, allowing clients to directly communicate with other clients without adding any load to the server. However, these sorts of communications need to be monitored by some code that is trusted to ensure that neither inappropriate messages nor cheating can take place by utilizing different client implementations. Since the client is assumed to be under the control of the user or player, the code that is on that client cannot be trusted, because it is easy to swap out the original client code for some other, "customized" version of the client. So, in fact, all channel messages have to go through the server, after being (possibly) vetted by the server logic.

One of the complexities of both Sessions and Channels is that they must obey the transactional semantics of tasks. Thus the actual transmission of a message on either a Session link or a Channel cannot happen when the call is made to the appropriate send() method; it can happen only when the task in which that method occurs commits.

Supplying these communication mechanisms gives us some of the pieces that are needed for the second part of our scaling mechanism. Since all communication must go through the Darkstar Session or Channel abstractions, and since those abstractions do not reveal the actual endpoints of the communication to the client or the server, there is a layer of abstraction between the entities communicating and the actual locations that are the start and end to that communication. This means that we can move the endpoint of the server communication from one machine in the Darkstar system to another without changing the way the client views the communication. From the client's point of view, all communication happens on a particular session or channel. From the point of view of the game or virtual world logic, communication is also through a single session or channel. But the underlying infrastructure can move the session or channel from one machine to another as needed to balance load as that load changes over time.

▶ 通过 Session Service保证进入会话的 task message 的有序性，使得 Task Service的实现可以不用考虑任务处理的并发性，而是利用Session Service的排队机制来约束，让Task Service的开发变得简单。
Session Service 的这种方式让我想起AKKA中的Actor模式。每个Actor都是自治的，并且都维护了能够保证消息顺序的MailBox，这就使得Actor之间的通信变得简单，开发者不用去考虑消息的并发执行，也避免了传统Shared Memory 方式因为引入锁带来的死锁问题。若要了解AKKA更多知识，可以访问其官方网站：http://akka.io。实际上，AKKA参考的是Erlang的OTP。

▶架构设计时，安全是我们需要引起重视的风险，尤其针对分布式系统，传递的信息应该避免被篡改，恶意攻击等攻击手段。

Task Portability

The core of the ability to balance load is that, given the programming model we require and the basic stack services that must be used, tasks that are performed in response to a client-generated or game-internal event are portable from any of the machines running a copy of the game or world logic on a Darkstar stack to any other machine running such a copy. The tasks themselves are written in Java,† which means that they can be run on any of the other machines as long as those (physical) machines have the same Java Virtual Machine as part of the runtime stack. All data read and manipulated by the task must be obtained from the Data Service, which is shared by all of the instances of the game or virtual world and the Darkstar stack on all of the machines. Communication is mediated by the Session Service or by Channels, which abstract the actual endpoints of the communication and allow any particular session or channel to be moved from one server to another. Thus, any task can be run on any of the instances of the game server without changing the semantics of the task.

This makes the basic scaling mechanism of Darkstar seemingly simple. If there is a machine that is being overloaded, simply move some of the tasks from that machine to one that is less loaded. If all of the machines are being overloaded, add a new machine to the group running a copy of the game or virtual world server logic on top of a Darkstar stack, and the underlying load-balancing software will start distributing load to that new machine.

The monitoring of the load on the individual machines and the redistribution of the load when needed is the job of the meta-services. These are network-level services that are not visible to the game or virtual world programmer, but are seen by and can themselves observe the services in the Darkstar stack. These meta-services observe, for example, which machines are currently running (and if any of those machines fail), what users are associated with the tasks on a particular machine, and the current load on the different machines. Since the meta-services are not visible to the game or virtual world programmer, they can be changed at any time without having an impact on the correctness of the game logic. This allows us to experiment with different strategies and approaches to dynamically load balance the system, and allows us to enrich the set of meta-services as required by the infrastructure.

The same mechanism that we have used for scaling over multiple machines is used to obtain a high degree of fault-tolerance in the system. Given the machine-independent nature of the data that is used by a task and the communication mechanisms, it may be clear that it is possible to move a task from one machine to another. But if a machine fails, how can we recover the tasks that were on that machine? The answer is that the tasks themselves are persistent objects, stored in the Data Service for the overall system. Thus, if a machine fails, any of the tasks that were being performed by that machine will be treated as aborted transactions, and will be

† More precisely, all of the tasks consist of sequences of bytecodes that can be executed on the Java Virtual Machine. We don't care what the source-level language is; all we care about is that the compiled form of that source language can be run on any of the environments that make up the distributed set of machines running the game or virtual world.

▶认识到meta-service这个概念是非常重要的。我认为这个概念实则是借用了meta-data的隐喻，在这里用以形容管理、协调、监听系统服务的服务。这种meta-service主要是为了保证各个运行中服务的健康度，并有利于运维人员更快发现运行问题。在我负责开发的一个基于RESTful理念的产品中，我的同事就为整个系统建立了一个Health Service。它同样是一个REST服务，并不为客户所知，也不产生业务价值，但运维人员可以非常轻松地通过Health Service了解当前产品中所有REST服务的运行状态。这个Health Service也可以视为Meta Service。

rescheduled on different machines. Although the latency of such rescheduling may be greater than the rescheduling of an aborted transaction that stays on the same machine, the correctness of the system will be the same. At most, the user of the system (the game player or virtual world inhabitant) will notice a momentary lag in response time. Such a lag may be irritating, but it is far less extreme than the current impact of a server crash in game or virtual world environments, where the crash at least results in logging out the player, with the possibility of losing a considerable amount of game play state.

Thoughts on the Architecture

Perhaps the first question anyone asks of an architecture and its implementation is how well it performs. Although optimizing an architecture prematurely is the source of a multitude of sins, it is also possible to design an architecture that cannot be implemented in a way that performs well. Due to one of the basic choices in the Darkstar architecture, this worry is quite real. And because of the nature of the game industry, determining the performance of a server infrastructure is difficult to do.

The difficulty in determining the performance of a game or world server infrastructure is an outgrowth of the simple fact that there are no benchmarks or commonly accepted examples for a large-scale MMO or virtual world. The lack of benchmarks is not surprising, given that the server components of most games or virtual worlds are built from the ground up for a particular instance of the game or virtual world. There are only a few general infrastructures that are offered as reusable building blocks, and these are generally extracted from a particular game or world after the fact and offered to others who are building similar games. Whether it is the relative youth of the game industry or an accident of the historical emergence of the technology from the entertainment industry, no commonly accepted benchmarks are available to test a new infrastructure or to allow the comparison of different infrastructures.

There is also little or no information available concerning the expected computation, data manipulation, and communication loads for a game or virtual world server that would allow for the construction of benchmarks or performance tests. This is partly an outgrowth of the custom nature of the servers that have been produced. Each of these is built for a particular game or virtual world and thus is specialized for the particular workload characteristics of that game or world. Even more, it is an outgrowth of the intensely secretive nature of the game industry, in which any information about a game in development is jealously guarded, and information about the way in which a released game was implemented is both tightly guarded and, to many in the industry, considered uninteresting. Much more thought and discussion is given to the artwork, the storyline, or the player interaction patterns that make a new game interesting or fun than is given to the way in which the server for the game was designed or to the mechanisms used to scale the game to its current population of players (a statistic that is also closely guarded). So just getting information about the kinds of loads that current games or virtual worlds place on a server is difficult.

In our experience, even when we can get developers to talk about the loads placed on the server by their game or virtual world, they are often incorrect in their reports. This is not because they are attempting to maintain some commercial advantage by misreporting what their server actually does, but because they genuinely don't know themselves. There is very little instrumentation placed in game servers that would allow them to gather information on how the server is actually performing or what it is doing. The analysis of such servers is generally experiential at best. Programmers work on the server until it allows game play to be fun, which is achieved in an iterative manner rather than by doing careful measurements of the code itself. There is far more craft than science in these systems.

This is not to say that the servers backing such games and virtual worlds are shoddily constructed pieces of code or that they are badly built. Indeed, many of them are marvels of efficiency that demonstrate clever programming techniques and the advantages of one-time, special-purpose servers for highly demanding applications. However, the custom of building a new server for each game or world means that little knowledge of what is needed for those servers has developed, and there is no commonly accepted mechanism for comparing one infrastructure to another.

Parallelism and Latency

This lack of information about what is needed for acceptable performance in the server is of particular concern to the Darkstar team, as some of the core decisions that we have made fly in the face of the lore that has developed around how to get good performance from a game or virtual world server. Perhaps the most radical difference between the Darkstar architecture and common practice is the refusal in the Darkstar architecture to keep any significant information in the main memory of the server machine. The requirement that all data that lasts longer than a particular task be stored persistently in the Data Store is central to the functionality of the Darkstar infrastructure. It allows the infrastructure to detect concurrency problems, which in turn allows the system to hide those problems from the programmer while still allowing the server to exploit multicore architectures. It is also a key component to the overall scaling story, as it allows tasks to be moved from one machine to another to balance the load over a set of machines.

Storing the game state persistently at all times is heresy in the world of game and virtual world servers, where the worry over latency is paramount. The received wisdom when writing such servers is that only by keeping all of the information in main memory will the latency be kept small enough to allow the required response times. Snapshots of that state may be taken on occasion, but the need for interactive speeds means that such long-term operations must be done rarely and in the background. So it appears on the face of it that we have based our architecture on a premise that will keep that architecture from ever performing well enough to serve the needs of its intended audience.

Although it is certainly true that requiring data to be persistent is a major difference in the architecture, and that accessing data through the Data Store will introduce considerable latencies into the architecture, we believe that the approach we have taken will be more than competitive for a number of reasons. First, we believe that we can make the difference between accessing data in main memory and accessing it through the data store much smaller than is generally believed. Although conceptually every object that lasts longer than a single task needs to be read from and written to persistent storage, the implementation of such a store can utilize the years of research in database caching and coherence to minimize the data access latencies incurred by the approach.

This is especially true if we can localize the access to particular sets of objects on a particular server. If the only tasks that are making use of a particular set of objects are run on a single server, then the cache on that server can be used to give near main-memory access and write times for the objects (subject to whatever durability constraints need to be met). Tasks can be identified with particular players or users in the virtual world. And here we can utilize the requirement that data access and communications go through services provided by the infrastructure to gather information about the data access patterns and the communication patterns taking place in the game or world at a particular time. Given this information, we believe that we can make very accurate estimations of which players should be co-located with other players. Since we can move players to any server that we wish, we can maximize the co-location of players in an active fashion, based upon the runtime behavior that we observe. This should allow us to make use of standard caching techniques that are well-known in the database world to minimize the latencies of accessing and storing the persistent information.

This sounds very much like the geographic decomposition that is currently used in large-scale games and virtual worlds to allow scaling. There, the server developers decompose the world into areas that are assigned to servers, and the various areas act as localization devices for the players. Players in the same area are more likely to interact than those in other areas, and so co-location on a server is enhanced. The difference is that current geographic decompositions occur as part of the development of the game and are reified in the source code to the server. Our co-location is based on runtime information, and can be dynamically tuned to the actual patterns of play or interaction that are occurring at the time of placement. This is analogous to the difference between compile-time optimization and just-in-time optimization. The former seeks to optimize for all possible runs of a program, whereas the latter attempts to optimize for the current run.

We don't believe that we can make the difference between main-memory access and persistent access disappear, but we also don't think that this is necessary in order to end up with performance that is better than that of infrastructures that make use of main memory. Remember that by making all of the data persistent, we are enabling the use of multiple threads (and therefore the multiple cores) within the server. Although we don't believe that the concurrency will be perfect (that is, that for each additional core we will get complete use of that core), we do believe (and preliminary results encourage this belief) that there is a significant amount of parallelism that can be exploited in games and virtual worlds. If the amount of concurrency that we can exploit is greater than the amount of latency that we might introduce, the overall performance of the game or virtual world will be better.

Betting on the Future

Our reliance on multithreading from multiple cores is essentially a bet on the way processors will evolve in the future. Currently servers are built with processors offering between 2 and 32 cores; we believe that the future of chip design will center around even more cores rather than on making any existing core run at a higher clock rate. When we began this project some years ago, this bet seemed far more speculative than it now appears. At that time, we often presented our designs as an exercise in "what if," saying that we were experimenting with an architecture that would be viable if the performance of chips became more a function of the number of threads supported than the clock speed of a single thread. This is one of the advantages of doing such a project in a research lab, where it is acceptable to take a much higher risk in your approach to a design as a way of exploring an area that might turn out to be commercially viable. Current trends in chip design make the decision to build an architecture centered on multithreading look far more prescient than it appeared at the time the decision was made.[‡]

Even if we can get only 50% of perfect concurrency, we could hit a performance break-even point if we can reduce the penalty of using persistent storage to between 2 and 16 times that of main memory. We believe we can do better in both the dimension of concurrency and in the dimension of reducing the difference between accessing the persistent state and keeping everything in memory. But much will depend on the usage patterns of those building upon the infrastructure (which, as we noted earlier, are difficult to discover).

Nor should we think of minimizing latency as the only goal of the infrastructure. By keeping all the server game or world objects in the Data Store, we minimize the amount of data that would be lost in the event of a server failure. Indeed, in most cases a server failure will be noticed only as a short increase in latency as the tasks (which are themselves persistent objects) are moved from the server that failed to an alternate server; no data should be lost. Some caching schemes might result in the loss of a few seconds of play, but even this case is far better than the current schemes used by online games and virtual worlds, where occasional snapshots are the main form of persistence. In such infrastructures, hours of game play might be lost if a server crashes at just the wrong time. As long as latencies are acceptable, the greater reliability of the persistence mechanism used by Darkstar can be an advantage for both the developers of the system built on the infrastructure and the users of that system.

[‡] Showing, once again, that very little is as important as luck in the early stages of a design.

Simplifying the Programmer's Job

Indeed, if minimizing latency while allowing scale were the only goal of the server developer, that developer would be best served by writing his own distributed and multithreaded infrastructure customized for the particular game. But this would require that the server developer deal with the complexities of distributed and concurrent programming. Before getting too obsessed with the need for speed, we should remember that a second, but equally important, goal of Darkstar is to allow the production of multithreaded, distributed games while providing the programmer a model of writing on a single machine in a single thread.

To a considerable extent, we have succeeded in this goal. By wrapping all tasks in transactions and detecting data conflicts within the Data Service, programmers get the benefits of multiple threads without needing to introduce locking protocols, synchronization, or semaphores into their code. Programmers do not have to worry about how to move a player from one server to another, since Darkstar handles the load balancing transparently for them. The programming model, although stylized and restrictive, has been found by early members of the community to be natural for the kinds of games and virtual worlds that they are building.

Unfortunately, we have found that we can't hide everything from the programmer. This became apparent when the very first game to be written on top of Darkstar showed very little parallelism (and exceptionally poor performance). On examination of the source code, it did not take us long to find the explanation. The data structures in the game had been written in such a way that any change of state in the game involved a single object, which was used as a coordinator for everything. The use of this single object effectively serialized all of the actions within the game, making it impossible for the infrastructure to find or exploit any concurrency.

Once we saw this, we had a long discussion with the game developers about the need to design their objects with concurrent access in mind. An audit of the data objects in the game showed a number of similar cases where concurrency was (unintentionally) precluded by choices made in the data design. Once these objects were redesigned, the performance of the overall system increased by multiple orders of magnitude.

This taught us that it is not possible for the developers using Darkstar to be completely ignorant of the underlying concurrent and distributed nature of the system. However, their knowledge of these properties of the system need not include the usual problems of concurrency control, locking, and dealing with communication between the distributed parts of the system. Instead, they are confined to the design activity of ensuring that their data objects are defined in such a way that concurrency can be maximized. Such design usually takes the more general form of ensuring that the objects defined are self-contained and do not depend on the state of other objects for their own operations, which is not a bad design principle in any system.

▶这里的设计实际上引入了一个 Mediator 对象，用以协调其他对象的职责，使得被协调对象之间的依赖变得更简单。但是，这种协调对象确实会存在文中所说的性能瓶颈。书中给出的解决方案是对游戏开发者进行约束，通过避免并发访问来规避性能瓶颈。这可以理解为框架的约束，又或者是框架设计者的态度。它同时给我们一个启示，有时候技术上的问题并不一定需要通过技术来解决。

（未完见下页）

> (接上页)
> 我想说明的是架构师需要更开阔的眼界，需要设计的创新，而不要让自己局限在思维的死胡同里。我知道有一个CMS系统因为架构设计可伸缩性方面的问题，导致企业内部在每月一度的报表提交过程中，由于并发数过大，报表数据过多，使得系统无法支撑。最后，这一问题竟然是通过改变企业内部的报表提交流程（调整各个部门的报表提交日期，以减小并发量）得以轻松解决。我们当然希望架构能够更好地支持高并发，并在设计上做出更多考量，但采用改变业务流程的方式也未尝不是一种可选的解决方案。

There is still much about the Darkstar architecture that we have not tested or that we don't fully understand. Although we have produced a system that allows multiple machines to run a game or virtual world utilizing multiple threads in a way that is (mostly) transparent to the server programmer, we have not yet tested the ability of the architecture to add other services beyond the core. Given the transactional nature of Darkstar tasks, this may turn out to be more complex than we first imagined, and our hope is that the additional services will not need to be participants in the core service transactions. We have also just begun to experiment with various ways of gathering information about the load on the system and balancing that load. Fortunately, since the mechanisms that do this balancing are completely hidden from the programmers using the system, we can pull out old approaches and introduce new ones without affecting those using Darkstar.

As an architecture, Darkstar presents a number of novel approaches that make it interesting. It is one of the few attempts to build a game or virtual world infrastructure with the same reliability and dependability properties as enterprise software while also meeting the latency, communication, and scaling requirements of the game industry. By trying to gain efficiency by using more machines and more threads, we hope to offset the increases in latency we introduce by the use of a persistent storage mechanism. Finally, the very different world of games and virtual environments, in which the clients are thick and the servers are thin, presents a contrast to the usual environment in which highly concurrent, distributed systems are generally built. It is too early to tell whether the architecture is going to be successful, but we believe that it is already interesting.

Principles and properties		Structures	
	Versatility	✓	Module
✓	Conceptual integrity	✓	Dependency
✓	Independently changeable		Process
✓	Automatic propagation		Data access
✓	Buildability		
	Growth accommodation		
	Entropy resistance		

CHAPTER FOUR

Making Memories

Michael Nygard

SINCE THE EARLIEST TINTYPES AND DAGUERREOTYPES, we have always seen photographs as special, sometimes even magical. A photograph captures a fleeting moment in time, in a way that our fallible memories cannot. But the best portraits do more than just preserve a moment; they illuminate it. They catch a certain glance or expression, a characteristic pose that lets the subject's personality shine through.

If you've had children in a U.S. school, you probably already know the name Lifetouch. Lifetouch photographs most elementary school, middle school, and high school students in the United States every single year. What you may not know is that Lifetouch also runs high-quality portrait studios. Lifetouch Portrait Studios (LPS) operates in major retail stores across the country, along with the "Flash!" chain of studios in shopping malls. In these studios, LPS's photographers take portraits that last a lifetime.

Digital photography has transformed the entire photography industry, and LPS is no exception. Giant rolls of film and frame-mounted cameras are disappearing, replaced with professional-grade DSLRs and flash memory cards. Unfettered photographers can move around, try different angles, and get closer than ever to their subjects. In short, they have more freedom to take those great portraits. The photographer works with the camera to turn photons into electrons, but somehow, somewhere, some system has to turn those electrons into atoms of ink and paper.

In 2005, my colleagues and I from Advanced Technologies Integration (ATI) in Minneapolis worked together with developers from LPS to roll out a new system to do exactly that.

Capabilities and Constraints

▶ Constraint（约束）对架构往往会产生一种设计的驱动力，而非负面的影响。因为对约束的识别，一方面满足了客户的需求，另一方面也明确了解决方案。例如，产品约束要求能够支持系统的水平伸缩。若系统牵涉到Web应用、数据存储，那么在应用服务层，我们就要求服务是无状态的，而在数据存储层，则需要支持分布式存储，例如选择NoSQL、PostgreSQL或者HDFS。Roy Fielding在其论文《架构风格与基于网络的软件架构设计》中也提到了约束的重要性："我工作的动机是希望理解和评估基于网络应用的架构设计，通过有原则地使用架构约束，从而从架构中获得所希望的功能、性能和社

（未完见下页）

Two dynamics drive a system's architecture: What must it do? What boundaries must it work within? These define the problem space.

We create, and simultaneously explore, the solution space by resolving these forces, navigating the positive pole of required behavior and the negative one of limitations. Sometimes we can create elegance, and even beauty, when the answers to individual constraints mesh together into a coherent whole. I'm happy to say that the Creation Center project did just that.

On this project, we faced several incontrovertible facts. Some are just the nature of the business; others could change, but not within our scope. Either way, we regarded these as immutable. These facts make up the left column in Figure 4-1.

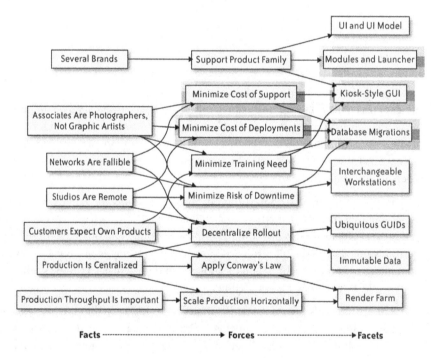

FIGURE 4-1. Facts, forces, and facets of Creation Center's architecture

Several brands

LPS supports multiple brands today and could add more in the future. At a minimum, Creation Center would have two visually distinct skins, and adding skins should not require extensive effort.

Associates are photographers, not graphic artists
>Photographers are trained to use the camera, not Photoshop. When an inexperienced user sits down at Photoshop, the most likely result is a lousy image. It's a power tool for power users, and there should be no need for a photographer in a portrait studio to get up the Photoshop learning curve. Photoshop and its cousins would also slow down studio workflow. Instead, studio associates need to create beautiful images rapidly.

Studios are remote
>Studios are geographically dispersed, with little to no local technical support. Hardware deliveries or replacements require shipping components back and forth.

Networks are fallible
>Some studios have no network connections. Even for the ones that do, it's not acceptable to halt the studio if the connection goes down.

Customers expect their own products
>Customers should receive their photos with their designs and text.

Production is centralized
>High-quality photographic printers are becoming more common, but making products that can last for decades requires much more expensive equipment.

Production throughput is important
>The same printers are also the constraint in the production process. Therefore, every other step in the process must be subordinated to the constraint.

These facts lead to several forces that we must balance. It's common to perceive the forces as fundamental, but they aren't. Instead, they emerge from the context in which the system exists. If the context changes, then the forces might be nullified or even negated.

We chose a handful of constructs to resolve these forces. The rightmost column of Figure 4-1 shows these facets of the architecture. Of course, these aren't the only Creation Center features worth discussing, but these facets of the architecture are of general interest. I also think they simultaneously illustrate a nice separation of concerns and mutually supporting structures.

Before digging into the specific features, we need to fill in one more piece of context: the system's workflow.

(接上页)
会学几方面的属性。当给定一个名称时，一组协作的架构约束就成为了一种架构风格。"在其论文的第5章，他利用约束从无到有地推导出了REST架构风格。

Workflow

The typical studio has two to four camera rooms, stocked with professional lighting, backdrops, and props. The photographers take pictures—each picture is called a "pose"—in the camera room. Outside of the camera room, photographers also handle customer service, scheduling, and customer pickups.

When the photographer finishes taking the pictures for a session, she sits down at any of several workstations to load the photographs from the camera's memory card.

> 正如本书第1章所述，架构师需要关注系统的质量。设计时，可以考虑列出这些至为关键的关注点，并排定优先级，寻找合理的解决方案。

我们可以考虑列出架构因素表，项目包括：因素、度量和质量场景、可变性、该因素对架构的影响、优先级、困难或风险。

例如因素可以是"当请求远程服务失败时，如何恢复"，而度量和质量场景则为"在生产环境下，当访问远程服务失败时，要能通过本地缓存临时提供服务，并及时通知服务失败，并在服务恢复后，能在1分钟内重建连接"。诸如此类的架构因素列表，可以帮助我们识别风险，细化方案。一旦问题出现，才能有的放矢，快速解决。

After loading a session, the photographer deletes any obviously bad photographs: ones with closed eyes, sour expressions, babies looking away, and so on. After deleting the bad ones, the rest become "base images." She then creates a number of enhancements from those base images. Enhancements range from simple tonal applications, such as black and white or sepia, to elaborate compositions of multiple photos. For example, a photographer might take a group portrait of three children and embed it in a design with three "slots" for individual portraits of the children.

After creating these enhancements, the photographer helps the customer order various sizes and combinations of prints. These include everything from 8" × 10" portraits to "sheets" of smaller sizes: 5" × 7", 3" × 5", or wallet sizes. Then there are the large formats. Customers can order portraits in sizes up to 24" × 30", made for framing and hanging on the wall.

After completing the customer's order, the photographer moves on to the next session.

At the end of each day, the studio manager creates a DVD of the day's orders, which she sends to the printing facility.

In the printing facility, hundreds of DVDs arrive each day. (I'll talk about the contents of the DVDs later.) The DVDs contain orders and photographs that need to be printed and shipped back to the studio, so the customer can pick them up. Before they can be printed, however, the final print-resolution photographs must be rendered as images. These print-ready images are immense. A 24" × 30" portrait rendered for high-quality printing, has over 100 million pixels, each in 32-bit color. Every single pixel is composited according to the design the photographer created in the studio. Depending on the composition, the rendering pipeline can be anywhere from 6 to 10 steps long. A simple rendering takes two to five minutes, but complex compositions for large formats churn for ten minutes or more.

At the same time, the printers spit out several finished prints per minute. Keeping the printers busy is the duty of the Production Control System (PCS), a complex system that handles job scheduling and orchestrates the render farm, manages image storage, and feeds the print queues.

When the finished order reaches the studio, the manager lets the customer know that she can come in to pick it up.

This workflow partly came from LPS's business context and partly from our choices about how to partition the system. Now let's look at the different facets from Figure 4-1.

Architecture Facets

Reducing the structure of a multidimensional, dynamic system into a linear narrative form is always a challenge, whether we are communicating our vision of a system that doesn't exist or trying to explain the interacting parts of one that we've already built. Hypertext might make it easier to approach the elephant from several perspectives, but paper doesn't yet support hyperlinks very well.

As we look at each of these facets, keep in mind that they are different ways of looking at the overall system. For instance, we used a modular architecture to support different deployment scenarios. At the same time, each module is built in a layered architecture. These are orthogonal but intersecting concerns. Each set of modules follows the same layering, and each layer is found across all the modules.

Indeed, we all felt deeply gratified that we were able to keep these concerns separated while still making them mutually supportive.

Modules and Launcher

All along, we were thinking "product family" rather than "application" because we had to support several different deployment scenarios with the same underlying code. In particular, we knew from the beginning that we would have the following configurations:

Studio Client
    A studio has between two and four of these workstations. The photographers use them for the entire workflow, from loading images through to creating the orders.

Studio Server
    The central server inside each studio runs MySQL for structured data such as customers and orders. The server also has much more robust storage than the workstations, using RAID for resiliency. The studio server also burns the day's orders to DVD.

Render Engine
    Once in production, we decided to build our own render engine. By using the same code for rendering to the screen in the studio and to the print-ready images in production, we could be absolutely certain that the customer would get what they expected.

At first, we thought these different deployment configurations would just be different collections of *.jar* files. We created a handful of top-level directories to hold the code for each deployment, plus one "Common" folder. Each top-level folder has its own *source*, *test*, and *bin* directories.

It didn't take long for us to become frustrated with this structure. For one thing, we had one giant */lib* directory that started to accumulate a mixture of build-time and runtime libraries. We also struggled with where to put noncode assets, such as images, color profiles, Hibernate configurations, test images, and so on. Several of us also felt a nagging itch over the fact that we had to manage *.jar* file dependencies by hand. In those early days, it was common to find entire packages in the wrong directory. At runtime, though, some class would fail to load because it depended on classes packaged into a different *.jar* file.

▶ Agile Architecture 非常强调架构的演进，始终保持最小的架构，避免对未来做出过多的假设。这基于一个前提，即"未来是不可预测的"。因此Neal Ford 提出Emergency Design，将重要的架构和设计决定推迟到last responsible moment。我赞同这个观点，但对于运用这一理念持谨慎态度。首先，这种推迟决定的设计决策应该针对未来不可预知的部分，若已明确知道未来的需求变化，还是应该事先设计，因为架构的重构可谓牵一发而动全身，非常困难，且成本高，对架构师的能力也提出了更高的要求。其次，在架构演进时，还需要一些管理手段和技术手段来支持，例如自动化测试、技术债管理等。然而无论如何，架构设计必须是迭代的，这完全符合敏捷架构的精神。

The breaking point came when we introduced Spring* about three iterations into the project. We were following an "agile architecture" approach: keep it minimal and commit to new architecture features only when the cost of avoiding them exceeds the cost of implementing them. That's what Lean Software Development calls "the last responsible moment." Early on, we had only a casual knowledge of Spring, so we chose not to depend on it, though we all expected to need it later.

When we added Spring, the *.jar* file dependency problems were multiplied by configuration file problems. Each deployment configuration needs its own *beans.xml* file, but well over half of the beans would be duplicated between files—a clear violation of the "don't repeat yourself" principle†—and a sure-fire source of defects. Nobody should have to manually synchronize bean definitions in thousand-line XML files. And, besides, isn't a multi-thousand-line XML file a code smell in its own right?

We needed a solution that would let us modularize Spring beans files, manage *.jar* file dependencies, keep libraries close to the code that uses them, and manage the classpath at build time and at runtime.

ApplicationContext

Learning Spring is like exploring a vast, unfamiliar territory. It's the NetHack of frameworks; they thought of *everything*. Wandering through the javadoc often yields great rewards, and in this case we hit pay dirt when I stumbled across the "application context" class.

The heart of any Spring application is a "bean factory." A bean factory allows objects to be looked up by name, creates them as needed, and injects configurations and references to other beans. In short, it manages Java objects and their configurations. The most commonly used bean factory implementation reads XML files.

An application context extends the bean factory with the crucial ability to make a chain of nested contexts, as in the "Chain of Responsibility" pattern from *Design Patterns* (Gamma et al. 1994).

The `ApplicationContext` object gave us exactly what we needed: a way to break up our beans into multiple files, loading each file into its own application context.

Then we needed a way to set up a chain of application contexts, preferably without using some giant shell script.

---

* http://www.springframework.org/

† See *The Pragmatic Programmer* by Andrew Hunt and David Thomas (Addison-Wesley Professional).

Module dependencies

Thinking of each top-level directory as a module, I thought it would be natural to have each module contain its own metadata. That way the module could just declare the classpath and configuration files it contributes, along with a declaration of which other modules it needs.

I gave each module its own manifest file. For example, here is the manifest file for the StudioClient module:

```
Required-Components: Common StudioCommon
Class-Path: bin/classes/ lib/StudioClient.jar
Spring-Config: config/beans.xml config/screens.xml config/forms.xml
        config/navigation.xml
Purpose: Selling station. Workflow. User Interface. Load images. Burn DVDs.
```

This format clearly derives from *.jar* file manifests. I found it useful to align the mental function "manifest file" with a familiar format.

Notice that this module uses four separate bean files. Separating the bean definitions by function was an added bonus. It reduced churn and contention on the main configuration files, and it provided a nice separation of concerns.

Our team strongly favored automatic documentation, so we built several reporting steps into the build process. With all the module dependencies explicitly written in the manifest files, it was trivial to add a reporting step to our automated build. Just a bit of text parsing and a quick feed to Graphviz generated the dependency diagram in Figure 4-2.

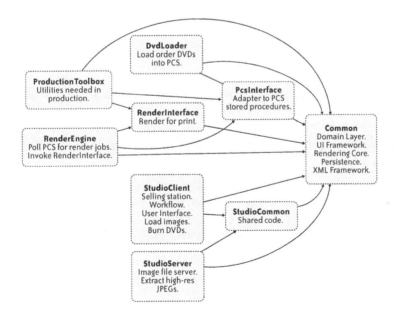

FIGURE 4-2. Modules and dependencies

With these manifest files, we just needed a way to parse them and do something useful. I wrote a launcher program, imaginatively called "Launcher," to do just that.

Launcher

I've seen many desktop Java applications that come with huge shell or batch scripts to locate the JRE, set up environment variables, build the classpath, and so on. Ugh.

Given a module name, Launcher parses the manifest files, building the transitive closure of that module's dependencies. Launcher is careful not to add a module twice, and it resolves the set of partial orderings into a complete ordering. Figure 4-3 shows the fully resolved dependencies for `StudioClient`. `StudioClient` declares both `StudioCommon` and `Common` as dependencies, but Launcher gives it only one copy of each.

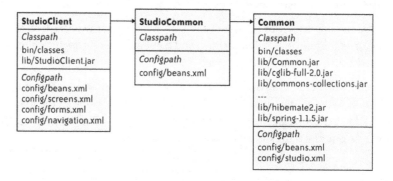

FIGURE 4-3. Resolved dependencies for StudioClient

To avoid classpath "pollution" from the host environment—ANT on a build box, or the JRE classpath on a workstation—Launcher builds its own class loader from the combined classpaths. All application classes get loaded inside that class loader, so Launcher uses that class loader to instantiate an initializer. Launcher passes the configuration path into the initializer, which creates all the application context objects. Once the application contexts are constructed, we're up and running.

Throughout the project, we refactored the module structure several times. The manifest files and Launcher held up with only minor changes throughout. We eventually arrived at six very different deployment configurations, all supported by the same structure.

The modules all share a similar structure, but they don't have to be identical. That was one of the side benefits of this approach. Each module can squirrel away stuff that other modules don't care about.

WHAT ABOUT OSGI?

When we started this project in late 2004, the OSGi framework was just beginning to gain broader visibility—thanks largely to Eclipse's adoption of it. We looked at it briefly, but were put off by the lack of widely available knowledge, expertise, and guidance.

OSGi's purpose, though, is a perfect fit for the problems we faced. Supporting multiple deployment configurations with a common codebase, managing the dependencies among modules, activating them in the correct sequence...clearly solving the same problem.

I suppose the fact that we didn't use OSGi was partly a quirk of timing and partly our own reluctance to take on what we perceived as more technical risk. I usually come down on the side of "acquire and integrate" rather than "roll your own," but there seems to be a tipping point: lightly supported open source projects with weak communities are more of a risk than well-understood, widely adopted ones. Likewise, I tend to avoid quasi-open frameworks that are actually vendor consortia. The community they serve is usually the community of vendors, not the community of users.

It wasn't clear to us which camp OSGi would fall into. If we were doing the project today, I think we probably would use OSGi instead of rolling our own.

▶ Java SE 9 的 Jigsaw 项目支持模块化系统，在语言层面上直接支持了模块化。OSGI确实推动了业界对模块管理的认识，但遗憾的是，真正在生产环境中使用OSGI的还是少数。我认为，要提高软件的质量，关键不在于对框架或平台的使用，而是如何对模块进行划分和设计，设计出优良的模块。

Kiosk-Style GUI

Studio associates are hired for their ability to work well with the camera and the families, especially children, not for their computer skills. At home, they might be Photoshop gurus, but in the studio, nobody expects them to become power users. In fact, during the busy season, a studio might bring on a number of seasonal associates. Consequently, fast ramp-up is critical.

One of the architects also served as our UI designer. He always had a clear vision of the interface, even if we didn't always agree on how much was feasible to implement. He wanted the user interface to be friendly and visible. There would be no menus. Users would interact with images through direct manipulation. Large, candy-coated buttons made all options visible. In short, the workstation should look like a kiosk.

That left the decision about what technology to use for the display itself.

One of our team made a survey of the Java rich UI technologies available, mainstream and fringe. We hoped to find a good declarative UI framework, something to help us avoid an endless slog through Swing tweaks. The results shocked us all.

In 2005, even after a decade of Java, two basic choices dominated the mainstream: XML hell or GUI builder spaghetti. The XML variants map more or less directly from Swing components to XML entities and attributes. This made no sense to us. GUI changes require a code release, whether the changes are implemented in straight Java code or in XML files. Why keep two languages in your head—Java plus the XML schema—instead of just Java? Besides, XML makes a clumsy programming language.

GUI builders had burned all of us before. Nobody wanted to end up with business logic woven into action listeners embedded in JPanels.

Reluctantly, we settled on a pure Swing GUI, but with some ground rules. Over a series of lunches at our local Applebee's, we hashed out a novel way of using Swing without getting mired in it.

UI and UI Model

The typical layered architecture goes "Presentation," "Domain," and "Persistence." In practice, the balance of code ends up in the presentation layer, the domain layer turns into anemic data containers, and the persistence layer devolves to calls into a framework.

At the same time, though, some important information gets duplicated up and down the layers. For instance, the maximum length of a last name will show up as a column width in the database, possibly a validation rule in the domain, and as a property setting on a JTextField in the UI.

At the same time, the presentation embeds logic such as "if *this* checkbox is selected, then enable *these* four other text fields." It sounds like a statement about the UI, but it really captures a bit of business logic: when the customer is a member of the Portrait Club, the application needs to capture their club number and expiration date.

So within the typical three-layer architecture, one type of information is spread out across layers, whereas another type of important information is stuck inside GUI control logic.

Ultimately, the answer is to invert the GUI's normal relationship to the domain layer. We put the domain in charge by separating the visual appearance of a screen from the logical manipulation of its values and properties.

Forms

In this model, a form object presents one or more domain objects' attributes as typed properties. The form manages the domain objects' lifecycles as well as calling down to the facades for transactions and persistence. Each form represents a complete screen full of interacting objects, though there are some limited cases where we use subforms.

The trick, though, is that a form is *completely* nonvisual. It doesn't deal with UI widgetry, only with objects, properties, and interactions among those properties. The UI can bind a Boolean property to any kind of UI representation and control gesture: checkbox, toggle button, text entry, or toggle switch. The form doesn't care. All it knows is that it has a property that can take a true/false value.

Forms never directly call screens. In fact, most of them don't even know the concrete class of their screens. All communication between forms and screens happens via properties and bindings.

Properties

Unlike typical form-based applications, the properties that a `Form` exposes are not just Java primitives or basic types like `java.lang.Integer`. Instead, a `Property` contains a value together with metadata about the value. A `Property` can answer whether it is single-valued or multivalued, whether it allows null values, and whether it is enabled. It also allows listeners to register for changes.

The combination of `Forms` and their `Property` objects gave us a clean model of the user interface without yet dealing with the actual GUI widgetry. We called this layer the "UI Model" layer, as shown in Figure 4-4.

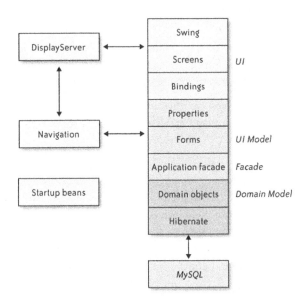

FIGURE 4-4. Layered architecture

Each subclass of `Property` works for a different type of value. Concrete subclasses have their own methods for accessing the value. For instance, `StringProperty` has `getStringValue()` and `setStringValue(String)`. Property values are always object types, not Java primitives, because primitives do not allow null values.

It might seem that property classes could proliferate endlessly. They certainly would if we created a property class for each domain object class. Most of the time, instead of exposing the domain object directly, the `Form` would expose multiple properties representing different

第 4 章 记忆留存 73

aspects of the domain object. For example, the customer form exposes `StringProperty` objects for the customer's first name, last name, street address, city, and zip code. It exposes a `DateProperty` for the customer's club membership expiration date.

Some domain objects would be awkward to expose this way. Connecting a slider that controls dilation of the image or embedded image in a design to the underlying geometry would have required more than half a dozen properties. Having the `Form` juggle this many properties just to drag a slider seemed like a pretty clear code smell. On the other hand, adding another type of property seemed like the path to wild type proliferation.

Instead, we compromised and introduced an object property to hold arbitrary Java objects. The animated discussion before that class appeared included the phrases "slippery slope" and "dumping ground." Fortunately, we kept that impulse in check—one of the perils of a type-checked language, I suppose.

We handled actions by creating a "command property," which encapsulates command objects but also indicates enablement. Therefore, we can bind command property objects to GUI buttons, using changes in the property's enablement to enable or disable the button.

The UI Model allowed us to keep Swing contained within the UI layer itself. It also provided huge benefits in unit testing. Our unit tests could drive the UI Model through its properties and make assertions about the property changes resulting from those actions.

So, forms are not visual themselves, but they expose named, strongly typed properties. Somewhere, those properties must get connected to visible controls. That's the job of the bindings layer.

Bindings

Whereas properties are specific to the types of their values, bindings are specific to individual Swing components. Screens create their own components, and then register bindings to connect those components to the properties of the underlying `Form` objects. An individual screen does not know the concrete type of form it works with, any more than a form knows the concrete type of the screen that attaches to it.

Most of our bindings would update their properties on every GUI change. Text fields would update on each keystroke, for instance. We used that for on-the-fly validation to provide constant, subtle feedback, rather than letting the user enter a bunch of bad data and then yelling at them with a dialog box.

Bindings also handle conversion from the property's object type to a sensible visual representation for their widgets. So, the text field binding knows how to convert integers, Booleans, and dates into text (and back again). Not every binding can handle every value type, though. There's no sensible conversion from an image property to a text field, for example. We made sure that any mismatch would be caught at application startup time.

An interesting wrinkle developed after we had built the first iteration of this property-binding framework. The first screen we tried it out on was the customer registration form. Customer registration is fairly straightforward, just a bunch of text fields, one checkbox, and a few buttons. The second screen, the album screen, is much more visual and interactive. It uses numerous GUI widgets: two proof sheets, a large image editor, a slider, and several command buttons. Even here, the form makes all the real decisions about selections, visibility, and enablement entirely through its properties. So the album form knows that the proof sheets' selections affect the central image editor, but the screen is oblivious. Keeping the screens "dumb" helped us eliminate GUI synchronization bugs and enabled much stronger unit testing.

IS ONE ENOUGH?

On some screens, proof sheets allow multiple selections; on others, only single selection. Worse yet, some actions are allowed only when exactly one thumbnail is selected. What component would decide which selection model to apply or when to enable other commands based on the selection? That's clearly logic about the UI, so it belongs in the UI Model layer. That is, it belongs in a form. The UI Model should never import a Swing class, so how can forms express their intentions about selection models without getting tangled up in Swing code?

We decided that there was no reason to restrict a GUI component to just one binding. In other words, we could make bindings that were specific to an aspect of the component, and those bindings could attach to different form properties.

For instance, we often had separate bindings to represent the content of a widget versus its selection state. The selection bindings would configure the widget for single- or multiselect, depending on the cardinality of its bound property.

Although it takes a long time to explain the property-binding architecture, I still regard it as one of the most elegant parts of Creation Center. By its nature, Creation Center is a highly visual application with rich user interaction. It's all about creating and manipulating photographs, so this is no gray, forms-based business application! Yet, from a small set of straightforward objects, each defined by a single behavior, we composed a very dynamic interface.

The client application eventually supported drag-and-drop, subselections inside an image, on-the-fly resizing, master-detail lists, tables, and double-click activation. And we never had to break out of the property-binding architecture.

Application facade

There's a classic pitfall in building a strong domain model. The presentation layer—or in this case, the UI Model—often gets too intimate with the domain model. If the presentation traverses relationships in the domain, then it becomes difficult to change the domain model. Like any agile team, we needed to stay flexible, and there was no way we would make design choices that would lead to less flexibility over time.

Martin Fowler's "Application Facade" pattern fit the bill (see the "References" section at the end of this chapter). An application facade presents only a portion of the domain model to the presentation layer. Instead of walking through graphs of domain objects, the presentation asks the application facade to assist with traversal, life cycle, activation, and so on.

Each form defined a corresponding facade interface. In fact, following the dictum that consumers—rather than their providers—should define interfaces we put the facade interface in the form's package. The form asks the facade to look up domain objects, relate them, and persist them. In fact, the facades managed all database transactions, so the forms were never aware of transaction boundaries.

The interfaces at this boundary, between forms and facades, also became an ideal place to isolate objects for unit testing. To test a particular form, the unit test creates a mock object that implements the facade's interface. The test trains the mock object to feed the form with some set of expected results, including error conditions that would be very difficult to reproduce with the real facade. I think we all regarded mock objects as a two-sided compromise: although they made unit tests possible, something still felt wrong about tying the tests so closely to the forms' implementations. For example, mock objects have to be trained with the exact sequence of method calls to expect, and the exact parameters. (Newer mock object frameworks are more flexible.) As a result, changes in the internal structure of the forms would cause tests to fail, even though no externally visible behavior changed. To a certain extent, this is just the price you pay for using mock objects.

All the Creation Center applications, both in the studio and in the printing facility, used the same stack of layers. Removing the GUI from the driver's seat kept the team from spending endless cycles in Swing tweaking. This inversion of control also provided a uniform structure that every application, and every pair, could follow. Even though we created more than the usual "three-layer cake," our stack was quite effective at separating concerns: Swing was limited to the UI, domain interaction in the forms, and persistence in the facades.

Interchangeable Workstations

When a photographer finishes a session, she grabs any open workstation. Depending on how busy the studio is, she'll usually finish with the customer at that time. It's common, though, for customers to come back later, maybe even on a different day. It would be ridiculous to permanently attach a customer to a single workstation—not just unworkable for scheduling, but also risky. Workstations break!

So any workstation in the studio must be interchangeable, but "interchangeable" presents some problems. The images for a single session can consume close to a gigabyte.

We briefly contemplated building the workstations as a peer-to-peer network with distributed replication. Ultimately, we opted for a more traditional client-server model, as shown in Figure 4-5.

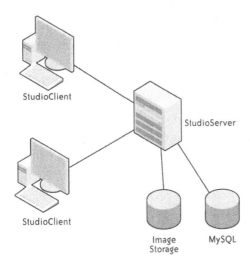

FIGURE 4-5. Studio deployment

The server is equipped with larger disks than the clients, and they are RAIDed for resilience. The server runs a MySQL database to hold structured data about customers, sessions, and orders. Most of the space, however, is devoted to storing the customers' photographs.

Because the studios are remote and the associates are not technically adept, we knew it would be important to make the "plumbing" invisible. Associates should never have to look at filesystems, investigate failures, or restart jobs. They should certainly never log into the database server! At worst, if a network cable should be bumped loose, once it is plugged back in, everything should work as normal and also should automatically recover from that temporary problem.

With that end in mind, we approached the system and application architecture.

Image repositories

To make the workstations interchangeable, the most essential feature would be automatic transfer of images, both from the workstation where the photographer loaded them to the server and from the server to another workstation.

The studio client and studio server both use a central component called an image repository. It deals with all aspects of storing, loading, and recording images, including their metadata. On the client side, we built a local, caching, write-behind proxy. When a caller asks for an image, this client image repository either returns it directly from local cache or downloads the file into local cache, and then returns it. Either way, callers remain blissfully ignorant.

Likewise, when adding images on the client, the client image repository uploads it to the server. We use a pool of threads to run background transfers so the user doesn't have to wait on uploads.

Both the client and server repositories are heavily multithreaded. We created a system of locking called "reservations." Reservations are a soft form of collaborative locking. When a client wants to add an image to the repository, it must first request and hold a "write reservation." This way, we can be sure that no other thread is reading the image file when we issue the reservation. Readers have to acquire a "read reservation," naturally.

Although we did not implement distributed transactions or two-phase commit, in practice there is only a small window between when the client image repository grants a write reservation and when the server side grants a corresponding write reservation. When that second reservation is granted, we can be confident that we will avoid file corruption.

In practice, even lock contention is rare. It requires two photographers at two different workstations to access exactly the same customer's session. Still, there are several workstations in every studio, and each workstation has many threads, so it pays to be careful.

NIO image transfer

Obviously, that leaves the problem of getting the images from the client to the server. One option we considered and rejected early was CIFS—Windows shared drives. Our main concern here was fault-tolerance, but transfer speed also worried us. These machines needed to move a lot of data back and forth, while photographers and customers were sitting around waiting.

In our matrix of off-the-shelf options, nothing had the right mix of speed, parallelism, fault-tolerance, and information hiding. Reluctantly, we decided to build our own file transfer protocol, which led us into one of the most complex areas of Creation Center. Image transfer became a severe trial, but we emerged, at last, with one of the most robust features of the whole system.

I had some prior experience with Java NIO, so I knew we could use it to build a blazing-fast image transfer mechanism. Building the NIO data transfer itself wasn't particularly difficult. We used the common leader-follower pattern to provide concurrency while still keeping NIO selector operations on a single thread.

Although the protocol wasn't difficult to implement, there were a number of nuances to deal with:

- Either end can close a socket, particularly if the client crashes. Sample code never deals with this properly.

▶Leader-Follower Pattern是常见的并发架构模式中的一种，它通常采用线程池来实现。在线程池中每次只有一个leader等待请求的到来。一旦请求到达，就将线程池中的另一个线程提升为leader，而自己则作为后台线程并发地处理刚才发来的请求，使得请求不至于因为长时处理而发生阻塞。在 Pattern-Oriented Software Architecture 系列的第2卷与第4卷对该模式有详细的介绍。

- While handling an IO event, the SelectionKey will still signal that it's ready. This can result in multiple threads calling into the same handler if you don't clear that operation from the key's interest set.
- The leader must perform all changes to a SelectionKey's interest set or else you get race conditions with the Selector, so we had to build a queue of pending SelectionKey changes that the leader thread would execute before calling select.

Handling these tricky details led to quite a bit more coupling between the various objects than I initially expected. If we had been building a framework, this whole area would have needed much more attention to loose coupling. For an application, however, we felt it was acceptable to regard the collection of collaborating objects in the server as a cohesive unit.

One particularly interesting effect showed up only when we ran a packet sniffer to see if we were really getting the maximum possible throughput. We weren't. At first, when the reactor read from a socket that had data available, it would read one buffer full and then return. We figured that it wouldn't take very long to get back around the loop if more than 8,192 bytes were available. It turns out that the studio network is fast enough to fill the server's TCP window before the next thread could get back into the handler, so virtually every transfer would stall for about half of the total transfer time. We added a loop inside the reactor, so it would keep reading until the buffer was drained. That cut the transfer time by nearly half, and reduced the amount of overhead in threading and dispatching. I found this particularly interesting because it works only for fast networks with low latency and only if the total number of clients is small. With higher network latency or more clients, looping that way would risk starving some clients. Again, it was a trade-off that made sense in our context.

UNIT TESTING AND CODE REVIEW

This NIO file server was the one time that I found it helpful to do a large group review, even on an agile project with complete pairing.

My pair and I worked on the threading, locking, and NIO mechanisms over most of an iteration. We unit tested what we could, but between the threading and low-level socket IO, we found it difficult to gain confidence in the code. So we did the next best thing: we got more eyes on it. I'd call that a special case, though. We were compensating for our inability to write sufficient unit tests.

In general, having two sets of eyes on the code all the time provides all the benefits of a code review. Combine that with automatic formatting and style checking, and there's just not enough remaining advantage of a code review to offset its cost. And if you can get the benefits without the cost, then why bother with the code review?

We kept a projector in our lab, connected to two machines through an A/B switch. Whenever we had a technique to illustrate or a design pattern to share, we'd take a few minutes after lunch to fire up the projector and walk through some code. This was particularly handy during the early stages, when the architecture and design were more fluid, and we were learning how to deal with Spring and Hibernate. It helped homogenize Eclipse practices and tricks, too.

The projector was also handy for iteration demos. We could have all the stakeholders in the room, without crowding around a single screen.

(Not to mention how helpful it was for projecting funny YouTube clips up on the wall.)

I knew it wouldn't be hard at all to build something fast but fragile. The real challenge would be making it robust, especially when the whole network would exist in a studio hundreds of miles away. One with no ability to log in remotely to debug problems or clean up after failures. One with small children, distracted parents, and servers sitting at toddlers' eye level. Talk about a hostile environment! Moving bits across the wire would not be enough; we needed atomic file transfer with guaranteed delivery.

The first layer of defense was the protocol itself. For a "put" operation—uploading a file from client to server—the first packet of the request includes the file's MD5 checksum. Once the client sends the last packet, it waits for a response from the server. The server responds with one of several codes: OK, TIMEOUT, FAILED_CHECKSUM, or UNKNOWN_ERROR. On anything but an OK, the client resends the entire file in what we call a "fast retry." The client gets three fast retries before the transfer fails.

▶ 只要牵涉到分布式的通信，就有必要为其提供retry的机会，这是站在robust的角度来考虑的；若是站在性能与可用性的角度来考虑，则还需要提供timeout机制，以避免失效的请求一直占用连接，无法释放资源。当然，这里额外提供的slow retry机制为robust又加了一道砝码，不过这种方式确实需要针对项目情况做出权衡。不考虑那种极端要求健壮性的系统，更多的方式是利用dead letter，或者通过event source提供可能的弥补手段。

Problems with file transfer will come in two varieties. One type is the "fast transient," a quick problem that will clear itself up, such as network errors. The other type requires human intervention. That means problems will either be cleared up in a few milliseconds, or they will take minutes to hours to correct. There's no point in retrying a fast file transfer over and over again. If it didn't work after the first few attempts, it's not likely to work for quite a while.

> Therefore, if the client exhausts all the fast retries, it puts the file transfer job in a queue. A background job wakes up every 20 minutes looking for pending file transfer jobs. It tries each job again, and if it fails again, it goes right back into the queue. Using Spring's scheduling support made this "slow retry" almost trivial to implement.

This mix of fast and slow retries lets us decouple maintenance and support on the server from the clients. There's no need to "cold boot" an entire studio for upgrades or replacements.

Fast and robust

The local and remote image repository and their associated file transfer mechanics became a seriously tough slog. Once it was done, though, the whole thing could upload images to the server faster than they could be read from the memory card. Downloading them on another machine was fast enough that users never perceived any activity at all. The client would download all the thumbnails for an album during the transition from one screen to the next. Downloading the screen-sized images for full-size display could be done during a mouse click. This speed let us avoid the user frustration of "loading" dialogs.

Database Migrations

Imagine operating 600 remote database servers across four time zones. They might as well be on a desert island, and digitally speaking, they are. If a database administrator needed to apply changes by hand, he would have to travel to hundreds of locations.

In such circumstances, one option would be to get the database design exactly right before the first release, and then never change it again. There may still be a few people who think that's possible, but certainly none of them were on my team. We expected and even counted on change at every level, including the database.

Another option would be to send release notes out to the field. The studio managers always called the service desk for a verbal walkthrough when they executed the installs. Perhaps we could include SQL scripts in documents on the release CDs for them to type in or copy-and-paste. The prospect of dictating any command that starts with, "Now type `mysqladmin -u root -p`..." gives me cold sweats.

Instead, we decided to automate database updates. Ruby on Rails calls these "database migrations," but in 2005 it wasn't a common technique.

Updates as objects

The studio server defines a bean called a database updater. It keeps a list of database update objects, each representing an atomic change to the database. Each database update knows its own version and how to apply itself to the database.

At startup time, the database updater checks a table for the current version of the database. If it doesn't find the table, it assumes that no updates exist or have been applied. Accordingly, the very first update bootstraps the version table and populates it with one row. That single row contains a version number and a lock field. To avoid concurrent updates, the database updater first updates this row to set the lock field. If it cannot, then it assumes some other machine on the network is already applying updates.

We used this migration ability to apply some simple changes and some sophisticated ones. One of the simple ones just added indexes to a couple of columns that were affecting performance. One of the updates that made us really nervous changed all the table types from MyISAM to InnoDB. (MyISAM, the default MySQL table type, does not support transactions or referential integrity. InnoDB does. If we had known that before our first release, we could have just used InnoDB in the first place.) Given that we had deployed databases with production data, we had to use a sequence of "alter table" statements. It worked beautifully.

After a few releases had gone out to the field, we had about 10 updates. None of them failed.

Regular exercise

Every time we run a build, we reset the local development database to version zero and roll forward. That means we exercise the update mechanism dozens of times every day.

▶ Rails提供的数据库迁移机制确认让我们眼睛一亮。对于版本维护来说，数据库的版本维护甚至比代码的版本维护更难，也更要命。尤其是产品已经发布，在生产环境下产生了生产数据之后进行数据库迁移，挑战更大。若是从一开始就合理地管理数据库脚本的版本，并实现脚本包括迁移脚本和回退脚本的自动化，无疑会降低数据库迁移的难度。我们的项目使用了flyway来实现数据库迁移的自动化。

We also unit test every database update. Each test case makes some assertions about the state of the database prior to the update. It applies the update and then makes some assertions about the resulting state.

Still, these tests all work with "well-behaved" data. Weird things happen out in the field, though, and real data is always messier than any test data set. Our updates create tables, add indices, populate rows, and create new columns. Some of these changes can break badly if the data isn't what we expect. We worried about the risky time during the updates and looked for ways to make the process more resilient.

Safety features

Suppose something goes wrong with one of the updates. A studio could be shut down until Operations found a way to restore the database, and if the update really goes wrong, it might leave the database corrupted or in some intermediate state. Then the studio wouldn't even be able to roll back to the previous version of the application. To avoid that disaster scenario, the database updater makes a backup copy of the database before it starts applying the updates. If it can't make the backup copy, then it halts the update process.

If errors occur during the updates, the updater automatically attempts to reload from that backup copy. If even that step fails, well, at least there's a copy onsite so a support technician can talk the studio manager through a manual restore.

In fact, in the absolute worst case, the printing facility always has a copy of the database that's no more than one day old. We used some of the extra space on the daily DVD to send a complete copy of the database every day. There's something to be said for a small database and a lot of storage space.

Field results

The time we invested in automated database updates paid off in several ways. First, we improved performance and reliability through some early updates. Feedback from the user community was immediate and positive after that release. Second, the operations group greatly appreciated the easy deployment of new releases. Previous systems had required the studios to ship removable hard drives back and forth, with all the attendant logistics problems. Finally, having the update mechanism allowed us to focus on "just sufficient" database design. We did not peer into the crystal ball or overengineer the database schema. Instead, we just designed enough of the schema to support the current iteration.

Immutable Data and Ubiquitous GUIDs

In working with customers, the studio associate creates some compositions that use multiple photographs, inset into a design. These designs come from a design group at company headquarters. Some designs are perennial, others are seasonal. Christmas cards in a wide variety of designs are a big seller, at least in the weeks before Christmas. Not surprisingly, demand drops precipitously after that.

A particular design includes some imagery for the background and a description of how many openings there are for base images, and the geometry of those openings. The associate can be very creative in filling those openings with photographs and with other compositions.

We found some interesting challenges dealing with these designs and the base images that go in them. For instance, what happens when a customer places an order, but then a new version of the design gets rolled out to the studio? At a smaller scale, what do you do if the associate nested one design within another—such as a sepia-tinted photograph inside a border—and then changes or deletes the original design?

At first, this looked like a nightmare of reference counting and hidden linkages. Every scheme we considered created a web of object references that could lead to gaps, missing images, or surprising changes. As a team, we all believed in "The Rule of Least Surprise," so hidden linkages causing changes to ripple from one product to another just wasn't going to work.

When our lead visionary came up with a simple, clear answer, it didn't take more than 30 seconds to sell the rest of us on it. The solution incorporated two rules:

1. Don't change anything after creating it. Designs and compositions would be immutable.
2. Copy, don't reference, the original.

Taken together, this means that selecting a design actually copies that design into the working space. If the associate adds the resulting composition to the album, it's actually a complete and self-contained copy of the design that gets added. Likewise, nesting one enhanced image into another makes a copy of the original and grafts it into the new composition. From the moment that graft happens, the original composition and the new one are completely independent of each other.

These copies are not just a trick of object references in memory. The actual XML description of the composition contains a complete copy of the design or the embedded compositions. This description lives in the studio's database, and it's the same description that gets sent on the DVD. When the studio manager burns the day's orders to DVD, the StudioServer packs in everything needed to create the final render: source images, backgrounds, alpha masks, and the instructions about how to combine them into the final image.

Having the complete description of the whole composition—including the design itself—on DVD became a huge advantage for production.

Previous systems kept the designs in a library, and orders just referenced them by ID. That meant the designers had to coordinate design IDs between the studios and the centralized printing facility. Therefore, designs had to be "registered" in production before they could be rolled out to the field. Should the IDs get out of sync, as sometimes happened, the wrong design would be produced and customers would not get the products they expected. Likewise, whenever the designers updated a design, there would be a few days' worth of DVDs in the pipeline made with the old version of the design. Sometimes it would come out OK, and sometimes it wouldn't.

Under the new system, designs never have to be registered. Whatever comes through in the XML is what gets produced, which frees the designers to make much more frequent changes and roll them out however they want. New revisions of designs don't affect orders in the pipeline, because each order is self-contained. Once the new revision gets out to the studios, then it starts showing up in the order stream.

The only parts that weren't copied were the image files themselves. They're too large to copy, and so instead we assign every image—whether part of a design or taken in the studio—its own GUID. As a rule, once something gets a GUID, it is officially immutable. When it's getting ready to burn orders to DVD, the StudioServer walks through the orders collecting GUIDs (using the controversial Visitor pattern). It adds every image it finds to the DVD, including both the customers' photographs and the design backgrounds.

Render Farm

The StudioClient helps associates create enhanced portraits from the basic images. Those enhanced portraits can be as simple as a sepia or black and white effect to make the portrait look more dramatic, or they can be as complex as a multilayered structure with alpha-composited backgrounds, text, and soft focus. Whatever the effect, the workstations in the studio do not produce the final rendered image. The printing facility has a variety of printers, supporting different sizes and resolutions. They're free to change printers or move jobs between printers at any time. The studios just don't know enough to produce the print-ready images.

> ▶ Conway's Lay（康威定律）可以帮助我们确定团队工作的边界。在微服务构架中，Conway's Law可以结合 DDD 中的 Boundecl Context，帮助我们识别上下文，进而设计服务的边界。

When those daily DVDs arrive, they get loaded into the production control system (PCS). PCS makes all the decisions about when to render the images for an order, when to print them, and what printers to send them to. A separate team, in a separate location and in a separate time zone, develops PCS. Previous projects had run into tremendous friction when trying to integrate too closely with PCS. All parties worked with good intentions, but the communication difficulty slowed both teams down. We needed to avoid that friction, and so we decided to apply Conway's Law (defined in the next section) proactively, by explicitly creating an interface in the software where we knew the team boundary would be.

Conway's Law, applied

Conway's Law is often invoked after the fact, to explain what might otherwise appear to be arbitrary divisions within a product. It speaks to a fundamental truth about development teams: anywhere there is a team boundary, you will find a software boundary. This emerges from the need to communicate about interfaces.

We felt it was important enough to keep the DVD format and layout under complete control of Creation Center that we added a program to our own scope: the DvdLoader. DvdLoader runs in the production facility, reading DVDs and calling various stored procedures within PCS to add orders, compositions, and images. PCS treats the composition instructions as an opaque string, and we were careful to avoid any decisions that would have PCS "opening up" the XML in that string. That sometimes means we duplicate information, such as dependencies on the base images themselves, but that is an acceptable trade-off for maintaining a clear boundary.

Similarly, we defined an interface that let the RenderEngine pull render jobs from PCS while keeping the XML description of the rendering itself under Creation Center's control.

We worked out written specifications of those interfaces, and then used FIT running on our development server to "nail down" the precise meaning. In effect, we used FIT as an executable specification of the interfaces. That turned out to be vital because even the people who negotiated the interface still found discrepancies between what they thought they agreed to and what they actually built. FIT let us eliminate those discrepancies during development rather than during integration testing, or worse, in production.

INCREMENTAL ARCHITECTURE

One of the recurring questions in the agile community is, "How much architecture should you create up front?" Some of the leading agile thinkers will tell you, "None. Refactor mercilessly and the architecture will emerge." I've never been in that camp.

Refactoring improves the design of code without changing its functionality. But, to refactor your way to better design, you must first be able to recognize good and bad design. We have a good catalog of "code smells" to guide us there, but I don't know of any equivalent for "architecture smells." Second, it must be possible to change things continuously even across interface boundaries. This has always led me to believe that a system's fundamental architecture must be in place at the start of development.

Now, after the Creation Center project, I'm much less confident in that answer. We added major pieces of the architecture relatively late in the project. Here are some examples:

- Hibernate: Added after two or three iterations. We didn't need the database before this.
- Spring: Added nearly one-third of the way to release 1.0. It quickly became central to our architecture. I don't remember how we got along without it, but we did.
- FIT: Added halfway to release 1.0.
- DVD-burning software: Purchased and added near the end of initial development.
- Support for windowed UIs: Added in the final two iterations before launch.

▶ 架构师在进行技术选型时，应尽可能考虑得更长远，因为很多技术选型一旦做出决策，实施之后就很难改变，或者说改变的成本太高。对架构的重构往往是"不得已而为之"，应尽量避免出现这种情况。倘若确定无法预测，就需要在封装和抽象上多下工夫，尽量使未来改变的成本变低。可以参考引入DDD中"防腐层"概念，又或者Facade或Adapter模式，在重用的同时隔离变化。

In each case, we took the approach of exploring options thoroughly before making decisions. We would make a decision at the "last responsible moment," that point where the cost of *not* deciding outweighed the cost of implementing the feature. Although there were a few things that we might have done differently if Spring had been there from the start, we were not harmed by adding it later. In those early iterations, we focused on uncovering what the application wanted to be rather than how Spring wants us to build applications.

DVD loading

The `DvdLoader` program, which runs in the printing facility, is really a batch processor that reads orders from DVDs and loads them into PCS. As with everything else, we focused on robustness. `DvdLoader` reads an entire order, verifying that the DVD includes all the constituent elements, before it adds the order to PCS. That way it doesn't leave partial or corrupted orders in the database.

Because images can appear on many DVDs, the loader checks to see whether there's already an image loaded with that GUID. If not, the loader adds it. Orders can therefore be resent from the studio whenever necessary, even if PCS has already purged the order and its underlying images. This also means that the background images used in a design get loaded the first time an order for that design arrives.

The DVDs are therefore self-contained and idempotent.

Render pipeline

For the render engine itself, we drew on the classic pipes and filters architecture. "Pipeline" is a natural metaphor for rendering images, and separating the complex sequence of actions into discrete steps also made unit testing simple.

On pulling a job from PCS, the render engine creates a `RenderRequest`. It passes the `RenderRequest` into the rendering pipeline, where each stage operates on the request itself. One of the final stages in the pipeline saves the rendered image to the path specified by PCS. By the time the request exits the pipeline, it holds only a result object with a success indicator and an optional collection of problems.

Each step in the pipeline has its own opportunity to report problems by adding an error message to the result. If any step reports errors, the pipeline aborts and the engine reports the problem back to PCS.

Fail fast

Every system has failure modes; the only question is whether you design them in or just let them happen. We took care to design in "safe" failures, particularly in the production process. There was no way we wanted our software to be responsible for stopping the production line.

There's another aspect, too. When the customer picks up his order, it should be the right one! That is, the product we deliver really needs to match the product the customer ordered. It seems like a trivial statement, but it is very important to render the production scale images in the same way that the on-screen image was rendered. We worked hard to ensure that exactly the same rendering code would be used in production as in the studio. We also made sure that the rendering engine would use the same fonts and backgrounds in production.

In our render engine, we adopted a philosophy of "Fail Fast, Fail Loudly." As soon as the render engine pulls a job from PCS, it checks through all the instructions, validating that all the resources the job requires are actually available. If the job includes text, the render engine loads the font right away. If the job includes some background images or an alpha mask, the render engine loads the underlying images right away. If anything is missing, it immediately notifies PCS of the error and aborts that job. Out of the 16 steps in the rendering pipeline, the first 5 all deal with validation.

After several months in production, we finally found one error that the render engine didn't detect early: it didn't reserve disk space for the rendered image up front. One day when PCS filled its storage volumes, render jobs started to fail late instead of failing early. In all the preceding time, there were no remakes due to bad renders.

Scale out

Each render engine operates independently. PCS doesn't keep a roster of the render engines that exist; each engine just pulls jobs from PCS. In fact, engines can be added or removed as needed. Because each engine looks for a new job as soon as it finishes the previous one, we automatically get load balancing, scaled to the horsepower of the individual engines. Faster render engines just consume jobs at a higher rate. Heterogeneous render engines are no problem.

The only bottleneck would be PCS itself. Because the render engines call stored procedures to pull jobs and update status, each render engine generates two transactions every three to five minutes. PCS runs on a decent-sized cluster of Microsoft SQL Server hosts, so it is in no danger of limiting throughput anytime soon.

User Response

Our first release was installed at two local studios, both within easy "drive-and-debug" distance. The associates' feedback was immediate and very positive. One studio manager estimated that the new system was so much faster and easier to use that she would be able to handle 50% more customers during the holiday season. One customer was reported to ask where she could buy a copy of the software. We commonly heard reports of customers taking the mouse directly and making their own enhancements. You can imagine that customers are much more likely to order products they've created themselves.

We had a few kinks in the production process, but those were corrected very quickly. Thanks to the resilience we built into the loader and render farm, the printing facility has been able to scale up to handle the volume from many more studios than originally expected, while also enjoying higher production quality.

Conclusion

I could spend much more time and space with fond descriptions of every class, interaction, or design decision, with the devotion of a new parent describing his infant's every burp and wobble. Instead, this chapter condenses a year's worth of effort, exploration, blood, and sweat. It illustrates how the structure and dynamics of the Creation Center architecture emerged from fundamental forces about the business and its context. By keeping concerns well separated and guiding the incremental design and development, Creation Center balanced those forces in a pleasing way.

References

Buschmann, Frank, Kevlin Henney, and Douglas C. Schmidt. 2007. *Pattern-Oriented Software Architecture: A Pattern for Distributed Computing*, vol. 4. Hoboken, NJ: Wiley.

Fowler, Martin. 1996. *Analysis Patterns: Reusable Object Models.* Boston, MA: Addison-Wesley.

Fowler, Martin. "Application facades." *http://martinfowler.com/apsupp/appfacades.pdf.*

Gamma, Erich, et al. 1994. *Design Patterns: Elements of Reusable Object-Oriented Software.* Boston, MA: Addison-Wesley.

Hunt, Andrew, and David Thomas. 1999. *The Pragmatic Programmer.* Boston, MA: Addison-Wesley.

Lea, Doug. 2000. *Concurrent Programming in Java,* Second Edition. Boston, MA: Addison-Wesley.

Martin, Robert C. 2002. *Agile Software Development, Principles, Patterns, and Practices.* Upper Saddle River, NJ: Prentice-Hall.

Principles and properties		Structures	
	Versatility		Module
✓	Conceptual integrity		Dependency
	Independently changeable		Process
	Automatic propagation	✓	Data access
	Buildability		
✓	Growth accommodation		
✓	Entropy resistance		

CHAPTER FIVE

Resource-Oriented Architectures: Being "In the Web"

Brian Sletten

Architecture is inhabited sculpture.

—*Constantin Brâncusi*

IN THIS CHAPTER, WE WILL OBSERVE THAT AN INFORMATION-FOCUSED ARCHITECTURE in the Enterprise demonstrates some of the same positive properties as the Web: scalability, flexibility, architectural migration strategies, information-driven access control, and so on. In the process, it empowers the business side of the house to make capital investment and software development decisions based on business needs, not simply because fragile technology choices require them to pay for flux.

Introduction

It is with great shame that we as an IT industry must acknowledge this embarrassing fact: it is easier for most organizations to find information on the Web than it is to find information in their own systems. Think about that for a moment. It is easier for them to locate data, through third parties, on a global information system than to do so within environments in which they have complete control and visibility. There are many reasons for this travesty, but the biggest problem is that we tend to use the wrong abstractions internally, overemphasizing our software and services and underemphasizing our data. This wrong-headed approach is a big part of why our business units are so perturbed with our IT departments. We forget that companies do not care about software except for the features and functionality it enables. What the business really wants are easier ways to manage the data they have collected, build upon it, and reuse it to support their customers and core functions.

How is it that organizational information management is so radically different from the Web? Unfortunately, the answer has as much to do with corporate politics as it does technology choices. We have legacy systems that complicate modern interaction idioms. We attempt to leverage solutions from vendors whose interests are not always aligned with our own. We want silver bullets that will solve all of our problems (even though Dr. Brooks disabused us of that notion years ago*). Even if you somehow happen to land in an organization with a perfectly matched technology infrastructure, data stewards and data consumers are often in territorial land grab battles that discourage information sharing. This is one of the reasons companies do not function as cleanly as the Web: there does not seem to be suitable incentive to share, even though there is clearly a need to do so. The take-home message is that not all problems are technical. To some extent, Web techniques will help us route around political problems, too, because you do not always need special permission to expose links to information that is available to you in other forms.

The good news is that we can look to the Web for guidance on what makes it such a splendid environment for finding information. Applying these concepts within an organization can help solve this problem and allow similar benefits, such as low-cost data management, strategies for architectural migration, information-driven access control, and support for regulatory compliance. The Web's success is largely due to the fact that it has raised the possibilities for information sharing while also lowering the bar. We have created tools and protocols that simultaneously support knowledge transfer between the leading scientific minds of the world as well as allowing our grandmothers to connect to their families and find content and communities that interest them. This is no small feat, and we would do well to consider the confluence of ideas that led to these realities. We have to live within the architectures we build, so we should build architectures that simultaneously satisfy and inspire us.

▶ Leonard Richardson 定义了Web的成熟度模型。模型将大多数基于Web服务的服务使用单个URI来标识一个endpoint，并且使用HTTP Post来转移基于SAOP的负载的方式划归为第零级服务。

Conventional Web Services

Before we begin looking at a new architecture for our information-driven environments, we should take a brief look at how we have been building similar systems recently and see what might be done better. We have been pitched a dominant vision for Enterprise Architecture for the last (nearly) 10 years that is built around the notion of reusable business services. We need to remind ourselves that Web Services were intended to be a business strategy, a way to enable functionality to be defined in a handful of places, accessed anywhere, from any language,

* http://en.wikipedia.org/wiki/No_Silver_Bullet

asynchronously. We wanted to be able to upgrade a service without affecting the clients that use it. Unfortunately, the unending and ever-changing technology stack associated with this goal has confused people and not solved the problems we face in real architectures for real organizations. Our goal in this new vision is not simply to be different, but to add value and improve the status of the Service-Oriented Aggravation we have seen.

We have a collection of technologies that comprises our basic understanding of Web Services: SOAP for service invocation, WSDL for contract description, and UDDI for service metadata publishing and discovery. SOAP grew out of different traditions including the remote procedure call (RPC) model and the asynchronous XML messaging model (doc/lit). The first approach is brittle, does not scale, and did not really work out all that well under its previous names of DCOM, RMI, and CORBA. The problems are neither caused nor solved by angle brackets; we simply tend to build systems in this manner at the wrong level of granularity and prematurely bind ourselves to a contract that clearly will not remain static. The second advances the art and is a fine implementation strategy, but does not quite live up to the interoperability hype that has hounded it from the beginning. It complicates even simple interactions because its processes are influenced by the goal of solving larger interaction problems.

The doc/lit style allows us to define a request in a structured package that can be forwarded on, amended, processed, and reprocessed by the loosely coupled participants in a workflow. Like an itinerant pearl, this message accretes elements and attributes as it is handled by intermediaries and endpoints in a potentially asynchronous style. We achieve horizontal scalability by throwing ever more message handlers at a tier. We can standardize interaction styles across partner and industry boundaries and business processes that cannot be contained by a single context. It represents a decontextualized request capable of solving very difficult interaction patterns.

When strict service decomposition and description alone (i.e., SOAP and WSDL) proved insufficient to solving our interaction needs, we moved up the stack and introduced new business processing and orchestration layers. A proliferation of standards and tools has thus complicated an already untenable situation. When we cross domain and organizational boundaries we run into conflicting terms, business rules, access policies, and a very real Tower of WS-Babel. Even if we commit to this vision, we have no real migration strategies and have fundamentally been lied to about the potential for interoperability. Clay Shirky has famously categorized Web Services interoperability as "turtles all the way up."[†]

The problem is, when most people want to invoke reusable functionality in a language- and platform-independent way, these technologies are overkill, they are too complicated, and they leak implementation details. In order to invoke this functionality, you have to speak SOAP. That is a fine implementation choice, but in this world of loosely coupled systems, we do not always like to advertise or require our clients to know these details for simple interaction styles.

† *http://en.wikipedia.org/wiki/Turtles_all_the_way_down*

The big picture idea for SOAP involves decontextualized requests that maintain transactional integrity in an asynchronous environment. In the business realities of real systems, however, the context has to be put back into the request. First, we must associate identity with the request, and then credentials, and then sign the message and encrypt sensitive information, and on and on. The "simple" burden of issuing SOAP requests gets encumbered by the interaction style and our business needs. If someone in an organization wants to retrieve some information, why can't they just ask for it? And, once these questions have been answered once, why do 10 (or 100 or 1,000) people asking the same question have to put the same burden on the backend systems every time they issue the same query?

These questions highlight some of the abstraction problems that exist with the conventional Web Services technology stack and offer at least a partial explanation for the WS-Dissatisfaction that pervades the halls of IT departments around the world. These technologies are implementation techniques for decomposing our invocation of behavior into orchestrated workflows of services, but we cannot express the full vocabulary of an organization's needs with only the notion of services. We lose the ability to identify and structure information out of the context in which it is used in a particular invocation. We lose the ability to simply ask for information without having to understand the technologies that are used to retrieve it. When we tie ourselves to contract-bound requests running on a particular port of a particular machine, we lose loose-coupling and asynchronous interaction patterns as well as the ability to embrace changing views of our data. Without the ability to uniquely identify the data that passes through our services, we lose the ability to apply access control at an information level. This complicates our already untenable problem of protecting access to sensitive, valuable, and private information in an increasingly networked world.

SOAP and WSDL are not the problems here, but neither are they complete solutions. We will very likely use SOAP in the doc/lit style in the resource-oriented architectures I am about to describe; we just do not have to accept them as the only solution. Nor will we always need to advertise that we are using them behind the scenes if there is no need to do so. In order to take this next step, we need to look at the Web and why it has been so successful as a scalable, flexible, evolvable information-sharing platform. Implementation details are often not relevant to our information consumers.

The Web

The prevailing mental model for the Web is document-centric. In particular, when we think about the Web, we think about consuming documents in web browsers because that is how we experience it. The real magic, however, is the explicit linkage between publicly available information, what that linkage represents, and the ease with which we can create windows into this underlying content. There is no starting point, and there is no end in sight. As long as we know what to ask for, we can usually get to it. Several technologies have emerged to help us know what to ask for, either through search engines or some manner of recommendation system.

We like giving names to things because we are fundamentally name-oriented beings; we use names to disambiguate "that thing" from "that other thing." One of our earliest communication acts as children is to name and point to the subjects that interest us and to ask for them. In many ways, the Web is the application of this childlike wonder to our collective wisdom and folly. As creatures with insatiable knowledge appetites, we simply decide what we are interested in and begin to ask for it. There is no central coordination, and we are free to document our wandering by republishing our stories, thoughts, and journeys as we go. We think of the Web as a series of one-way links between documents (see Figure 5-1).

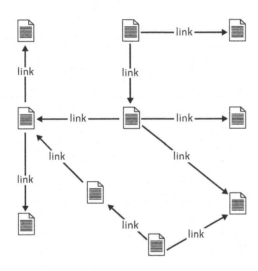

FIGURE 5-1. Conventional notion of the Web

Linked documents are only part of the picture, however. The vision for the Web always included the idea of linked data as well. This content can be consumed through a rendered view or directly referenced and manipulated in preferred forms in different contexts. You can imagine a middle-tier layer asking for information as an XML document while the presentation tier prefers a JSON object via an AJAX call. The same name refers to the same data in different forms. By allowing the data to be addressed like this, it is easy to build layered applications that have consistent views, even if they are asking for different levels of detail or wish to have the data styled in a particular way. Applications and environments that produce and consume data in this loosely linked style are no longer simply "on the Web," they are "in the Web." We are moving toward a Web of Data that connects people, documents, data, services, and concepts, as in Figure 5-2.

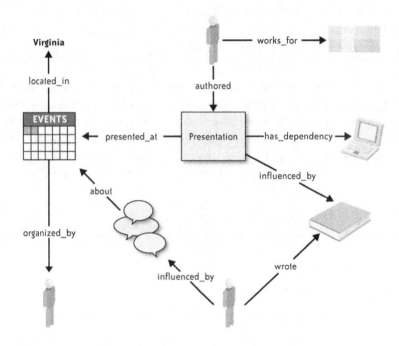

FIGURE 5-2. Web of Data

The basic interaction in this environment is a logical client-server request. We have an address for information of interest. The name, a Uniform Resource Locator (URL), is a type of identifier that not only disambiguates a reference in a global address space, but also tells us how to resolve the request. At no point during the process are we required to understand the technologies in place to satisfy the request. This keeps the process simple and resilient in the face of backend changes. As our favorite sites migrate from static to dynamic data production or change application server vendors, these facts are hidden from us. Although many sites do not effectively handle content negotiation in the process, we at least have the potential to receive different representations for the same named entity. We may wish to get something back in a different format, depending on whether we are making the request on a computer or a phone. Later in this discussion we will also see how to take advantage of this property to control the level of detail for access control and regulatory compliance.

The naming schemes used on the Web allow us to identify our documents, our data, our services, and now, even our concepts. We have historically had a difficult time differentiating between a reference to, say, Abraham Lincoln and a document about him. For example, the site *http://someserver/abrahamlincoln* could be either. The W3C Technical Architecture Group (TAG) has produced a recommendation[‡] that non-network addressable resources (i.e., things

---

[‡] *http://lists.w3.org/Archives/Public/www-tag/2005Jun/0039*

that do not actually live on the Web but are still interesting to us) can be indicated by the use of a 303 response code rather than the usual 200. This is a hint to a client that, "Yes, the thing you asked for is legitimate and of interest, but does not actually live on the Web. You can find more information here...."

Web addresses begin with a reference to the HTTP protocol followed by the name of the server that will respond to the request. After that, there is a hierarchical scheme that should reflect a path through an information space. This is a logical name describing something about the structure of the data. Multiple paths might resolve to the same resource but will have value in different scenarios. *http://server/order/open* might return a list of open orders at a particular point in time, and *http://server/order/customer/112345* might reflect all open orders for a particular customer. Clearly, there would be overlap between the results returned from either of these logical requests. When we do not know what specifically to ask for, we might go the more general route. When we want to inquire as to the status of a specific customer, we would go the more direct route. We retrieve these logical URL references either from some other part of the system or generate them based on input from the client entering data through a user interface.

The separation of concerns here is among the key abstractions of the interaction style. We isolate the things we are interested in discussing, the actions by which we manipulate those things, and the forms we choose to send and receive them in. This is demonstrated in Figure 5-3 drawn from the discussion at RESTWiki.§ In the REpresentational State Transfer (REST)‖ architectural style, we refer to the resources (nouns), the verbs, and the representation of the response. The resources can be anything we can address (including concepts!). The verbs are GET (retrieve), POST/PUT (create/update), and DELETE (remove). GETs are constrained to have no consequences. This is called an idempotent request. The semantics of this interaction will contribute to the potential for caching. POSTs are generally used when there is no central authority to respond to the request (e.g., submitting news articles to a Usenet community) or we do not yet have the means of addressing our resource. We cannot identify an order before we create it, because the server application is responsible for creating order IDs. Therefore, we tend to POST these requests to a bit of functionality (e.g., a servlet) that accepts the request on our behalf and generates an ID in the process. PUTs are used to update and overwrite the existing state of a named resource. DELETE has no great use on the public Web (thankfully!), but in the context of an internally controlled, resource-oriented environment, indicating that we no longer need or care about particular resources is an important part of managing their life cycles. The REST style fundamentally works by separating the concerns of logically naming the resources we care about, the means by which we manipulate them, and the formats in which we choose to represent them, as shown in Figure 5-3.

▶ REST很好地遵循了关注点分离原则。通过Heep Verb来抽象与统一对Resource的操作接口，使得REST API更加稳定，也更能够拥抱变化。

---

§ *http://rest.blueoxen.net/cgi-bin/wiki.pl?RestTriangle*
‖ *http://en.wikipedia.org/wiki/REST*

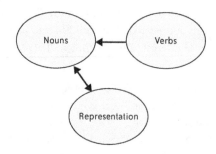

FIGURE 5-3. REST separation of concerns

This separation contrasts sharply with the contractual nature of a SOAP service invocation, where the structure of the request, the behavior being invoked, and the form of the return type are often bound to a contract through the Web Services Definition Language (WSDL). Contracts are not bad things; they are useful until we want to get out of them. One of the primary goals of the Web Services technology stack was to reduce coupling and introduce an asynchronous processing model where the handler of a message could be updated to reflect new business logic without affecting the client. The WSDL-binding approach took note of this goal and did precisely the opposite. We usually cannot change the backend binding on the same port without affecting the client (which is what we were explicitly trying to avoid!).

▶ 我在参与某系统技术栈迁移时，就采用REST来包装遗留系统。这其实是DDD 9种Context Map中的一种，即Big Ball of Mud（大泥球）。它并不是提出这种反模式（auti-pattern），而是说将整个大泥球一般的遗留系统视为一种特殊的Boundecl Contiext。

The resource-oriented approach allows us to enforce contracts if and when we want to, but it does not require us to do so. By separating out the name of the thing from the structure of the form we accept, we can reuse the same logical name to support multiple types of interaction. We can upgrade the back-end without necessarily breaking existing clients. If we move from a model where all existing clients POST messages to a URL with the version 1 of the message schema, we can add support on the backend for version 2 of the schema while allowing business to proceed as usual if it makes sense to do so. If we ever want to reject an older schema, we can, but again we can choose when to do so. This flexibility is one of the reasons resource-oriented architectures help put the business back in control: backend system changes do not necessarily force frontend updates. If we wrap a legacy system with a RESTful interface, we can continue to use it until there is a compelling business reason to change it. Certainly other technologies allow us to wrap legacy systems in this way. It is the general approach to using the logical names that gives us a greater opportunity for avoiding middleware flux that makes the difference here.

In an effort to promote horizontal scalability, the RESTful style requires that requests be stateless. This means that any information that is needed to respond to a request comes in as part of the request. This allows us to use load balancers to bounce the request handling among any number of backend servers. In the face of increased load, more hardware can be thrown at the problem, and any of the servers can pick up and handle the request. Although scalability was the goal of this architectural constraint, another important consequence emerges from the application of stateless requests to the semantics of the GET request: we can start to imagine the potential to cache the results from arbitrary requests. The address of the responder (the main portion of the URL) plus the full state of the request (URL hierarchy plus query parameters) becomes a compound hash key into a result set (e.g., database query, applying a transformation to another piece of data, etc.). You will not get these caching benefits for free, but environments that leverage the potential suddenly become easily imaginable. One of the many compelling features of the NetKernel resource-oriented environment[#] is that it deeply and fully takes advantage of this potential to enable a form of architectural memoization,[*] with almost no effort on your part. We will see more about this later in the "Applied Resource-Oriented Architecture" section.

With a common naming scheme for all the items that interest us and a logical request process that allows the form of the thing to change over time or in a different context, we almost have the infrastructure necessary to turn our organizational information management on its head. The final tool we need is the ability to express metadata about the things we are addressing. This is where the Resource Description Framework (RDF) comes in. This W3C Recommendation uses a graph model to allow open-ended expressions of information about our named entities. Who created it? When was it created? What is it about? What is it related to? The ability for us to name and address existing data stored in relational databases allows us to describe any data we like without having to convert it into a new form. This is a common expectation and complaint about RDF that does not hold up in practice. We will usually leave the data where it is and integrate at a level that makes sense.

In the following listing, we see an N3 expression of some RDF to describe the creator, title, copyright date, and license associated with a particular resource. The example shows the use of three terms from the Dublin Core Metadata Initiative[†] and one term from the Creative Commons[‡] community. We are free to reuse terms from whatever vocabularies exist, or to create new ones where we need to describe new terms:

```
@prefix dc: <http://purl.org/dc/elements/1.1/> .
@prefix cc: <http://creativecommons.org/ns/> .
<http://bosatsu.net/team/brian/index.html> dc:creator
        <http://purl.org/people/briansletten> .
<http://bosatsu.net/team/brian/index.html> dc:title
        "Brian Sletten's Homepage" .
<http://bosatsu.net/team/brian/index.html> dc:dateCopyrighted
        "2008-04-26T14:22Z" .
<http://bosatsu.net/team/brian/index.html> cc:license
        <http://creativecommons.org/licenses/by-nc/3.0/> .
```

[#] *http://1060.org*

[*] *http://en.wikipedia.org/wiki/Memoization*

[†] *http://dublincore.org*

[‡] *http://creativecommons.org/ns*

Not only do we now have the ability to use whatever terms we would like to, we can add new terms and relationships at any point in the future without affecting the existing relationships. This schemaless approach is tremendously appealing to anyone who has ever modified an XML or RDBMS schema. It also represents a data model that not only survives in the face of inevitable social, procedural, and technological changes, but also embraces them.

This RDF would be stored in a triplestore or other database, where it could be queried through SPARQL or a similar language. Most semantically enabled containers support storing and querying RDF in this way now. Examples include the Mulgara Semantic Store,[§] the Sesame Engine,[||] the Talis Platform,[#] and even Oracle 10g and beyond. Nodes in the graph can be selected based on pattern-matching criteria, so we could ask questions of our resources such as "Who created this URL?", "Show me everything that Brian has created," or "Identify any Creative Commons–licensed material produced in the last six months." The terms that mean "created by," "has license," etc. are expressed in the relevant vocabularies, but are easily translated into our stated goals. The flexibility of the data model coupled with the expressiveness of the query language makes describing, finding, and invoking RESTful services reasonably straightforward. It is certainly more pleasant than trying to find and invoke services through lobotomized and high-impedance technologies such as UDDI.

With the ability to address and resolve arbitrary resources, the ability to retrieve them in different forms and the ability to describe them in Open World and mixed-vocabulary ways, we are now ready to apply these ideas in the Enterprise. We will describe an information-driven architecture that supports "surfing" webs of data like you might "surf" the Web of documents.

Resource-Oriented Architectures

The resource-oriented style is marked by a process of issuing logical requests for named resources. These requests are interpreted by some kind of engine and turned into a physical representation of the resource (e.g., HTML page, XML form, JSON object, etc.). See Figure 5-4.

The basic interaction style in a resource-oriented architecture (ROA) is demonstrated in this figure. A logical request is named, resolved, and transferred back to the requestor in some form by a resource-oriented engine. The named resource is likely to resolve to a database query or some bit of functionality that manages information (e.g., a RESTful service). What responds to the request, which is largely irrelevant to the person interested in the information, is potentially a servlet, a Restlet,[*] a NetKernel module, or some other bit of addressable

---

[§] http://mulgara.org

[||] http://openrdf.org

[#] http://talis.com

[*] http://restlet.org

functionality that will interpret the request. This logical step hides a whole world of possibilities and technology choices, and simply does not leak unnecessary details to the client. It does not support all interaction styles, but you may be surprised at how much can fit comfortably behind a URL.

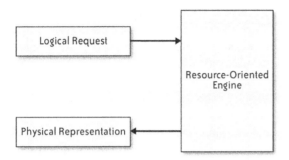

FIGURE 5-4. Resource-oriented architectures

Consider the address *http://server/getemployees&type=salaried*. Many people who think they are doing REST create URLs that look like this. Unfortunately, it is not a good REST service name (most Restafarians would argue it isn't REST at all!), because it conflates the nouns and verbs. That is what I like to call "addressing behavior through URLs" or "RPC through URLs." There is nothing magic about the REST way of separating nouns from verbs; it just allows us to identify the thing we care about. The aforementioned URL cannot be reused to update the employee list, because POSTing an employee record to "/getemployees" does not make any sense. If the URL were, instead, *http://server/employee/salaried*, then issuing a GET request to it will result in the same information, but this becomes a longer-lived address for the business concept of "salaried employees," just as *http://server/employee/hourly* can refer to employees who are paid by the hour. We may not choose to update these information resources, as they represent queries into whatever our backing store is. However, it is consistent within the */employee* information space, which we may choose to navigate in other ways. *http://server/employee/12345678* represents an employee with a particular ID, and *http://server/employee* might represent all employees. POSTing a record to this latter URL could represent hiring someone. PUTing a record to the specific employee ID URL could represent updating an employee record after a move, a raise, a promotion, etc. DELETEing the same address could indicate that the named resource is no longer of interest in the organization (i.e., either they quit or were fired).

▶ REST除了作为一种架构风格，还可以视为是架构设计的驱动力，它驱动人们站在"Resource（资源）"的角度去思考业务模型。我们可以从DDD的Domain Model与前端MVC模式结合着Resource-Oriented Architeeture设计思想来考虑领域建模。

This highlights one of the main distinctions between REST and SOAP that causes confusion when people conflate the intent of the two styles. SOAP is a fine technology for invoking behavior, but it falls down as a means of managing information. REST is about managing information, not necessarily invoking arbitrary behavior through URLs. When people start scratching their heads and wondering if four verbs are enough to do what they want to do, they are probably not thinking about information; they are thinking about invoking behavior. If you are doing RPC through URLs, you might as well use SOAP. If you are treating important business concepts as addressable information resources that can be manipulated and represented in different forms in different contexts, you are taking advantage of REST and are likely to see some of the same benefits we see on the Web. Even if your backend systems use SOAP to satisfy a request, you can imagine benefiting from a RESTful interface. Not only does providing that kind of an address allow users to "surf for data," you also potentially introduce the ability to cache results and eliminate some of the pain of a changed WSDL contract. The clients would go through a logical coupling that gets translated into a SOAP message and response being generated. The content of the response could be stripped out of what we get back. We simply do not need to advertise that fact, and can gain architectural migration strategies in the process.

As we see in Figure 5-5, the same named resource might be returned in different physical forms in different contexts while retaining the same identity. We can imagine some type of company report organized into an information space that can be traversed through a time facet (e.g., year and then month). As long as there is only one type of report, *http://server/report/2008/02* is a reasonably good, long-lived name. At no point in the future will we change the fact that we had a report for February 2008. We may wish to access the data as XML in one scenario, as an Excel spreadsheet in another, or as a rendered JPEG image for inclusion in a summary report. We do not want different names for each of the scenarios, so we leverage content negotiation to specify our preference. The resource-oriented engine needs to know how to respond to a request type, but that is easy enough to enable. Some future data format might emerge that no current clients support. The clients will not need to be modified simply because we add support on the server and some other client takes advantage of it. This resilience in the face of change was designed into the Web and is something that we will want to take advantage of in the Enterprise as well. The client and server can negotiate a particular form for a named resource during the resolution process. This allows the same named resource to be structured differently in different contexts (e.g., XML during the middle tier, JSON in the browser, etc.). The structured forms can be cached by the server if it chooses to in each form.

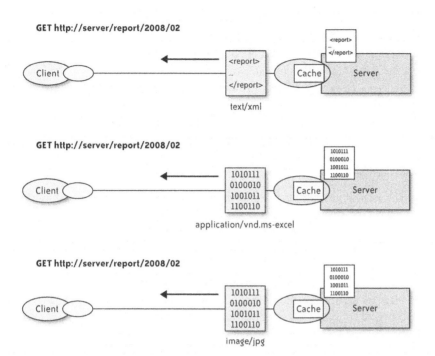

FIGURE 5-5. Negotiating content in a resource-oriented environment

In addition to picking the physical representation within the context of resolving a request, we might also enable the server to decide how much of the referenced data set to return based on the identity of the user, the application being used, etc. We can imagine a scenario where a call center agent using a relevant application needs to access sensitive information to resolve an issue. This could include Social Security numbers, credit card numbers (or hopefully only the last four digits), home addresses, etc. There is a specific business need to justify the agent accessing this information, so we could have a declarative policy in place that lets it happen. The same employee using a different application in a different context (perhaps a marketing analysis package) is unlikely to have a business need to access that sensitive information, although we may still want to resolve a reference to the same customer to access her demographics and purchase history. In this case, the context would not support access to the sensitive data, and we could enforce an automatic filtering process to remove or encrypt the sensitive information. The decision of which approach to take would depend upon where the data needed to go next. Encrypted data requires access to keys, which becomes another management burden. It is just as easy to remove the sensitive data but include it in a different resolution context when it is needed.

Managing single-point access control might not be a big problem for conventional Enterprise architectures. However, given the increased presence of workflows, explicitly modeled business processes, and the like, we have plenty of opportunities to consider a user of one application needing to invoke a capability or service in more than one context. If we are passing actual data between systems, the application developers become responsible for knowing about the access control issues when crossing application boundaries. If we instead pass a reference to the data, the initial application is no longer responsible, and we can keep the information-driven centralized access control working for us. Many existing SOA systems restrict access to services based on identity or role, but they rarely support restrictions to the specific data that passes through these services. This limitation is part of what makes conventional web services simultaneously confusing and insufficiently secure. Access policies should be applied to behavior and data within a context, but without being able to name the data and contain the context, that becomes very difficult.

When people first begin exploring resource-oriented architectures, they get concerned about exposing sensitive information through links. Somehow, returning blobs of data behind opaque queries seems more secure. They have difficulty separating the act of identifying and resolving something from the context in which it is done. That context contains sufficient information to decide whether or not to produce the information for a particular user. It is orthogonal to the request itself and will be met by the extant authentication and authorization systems in place for an organization. Anything from HTTP Basic Auth, to IBM's Tivoli Access Manager, to OpenID or other federated identity systems can be leveraged to protect the data. We can audit who had access to what and encrypt the transport with one- or two-way SSL to prevent eavesdropping. Addressability does not equal vulnerability. In fact, passing references around instead of data is a safer, more scalable strategy. The resource-oriented style is not less secure because it has fewer complicating security features (e.g., XML Encryption, XML Signature, XKMS, XACML, WS-Security, WS-Trust, XrML, etc.), but is arguably more secure because people can actually understand the threat model and how the protection strategies are applied.

These ideas become phenomenally important when we are faced with the daunting and very serious realities of demonstrating regulatory compliance. Credit card companies, health care watchdog organizations, corporate governance auditors, and the like can bring real teeth to a corporate audit demanding proof that only employees whose job function requires access to sensitive information can get to it. Even if your organization is in compliance, if it is difficult to demonstrate this ("First, look in this log on this system and then trace the message flowing through these intermediaries, where it is picked up and processed into a query, as you can see in this other log…"), it can be an expensive process. Employing declarative access control policies against the resolution of logical references makes it explicit (and simple to follow) who knew what, and when.

Data-Driven Applications

Once an organization has gone to the trouble of making its data addressable, there are additional benefits beyond enabling the backend systems to cache results and migrate to new technologies in unobtrusive ways. Specifically, we can introduce entirely new classes of data-driven applications and integration strategies. When we can name our data and ask for it in application-friendly ways, we facilitate a level of exploration, business intelligence, and knowledge management that will make most analysts drool when they see it. The Simile Project,[†] a joint effort between the W3C and the MIT CSAIL group, has produced a tremendous body of work demonstrating these ideas and how much drool can actually be produced.

Consider the scenario of tracking the efficacy of various marketing strategies on website traffic and sales. We might need to pull information in from a spreadsheet, a database, and several log files or reports from web analytics software. Although tying these things together now is not exactly rocket science, it does require a nontrivial level of effort to find, request, convert, and republish the results. If we simply produce a spreadsheet summary and email it around, we effectively lose the ability to retrieve the results at some future point without searching our already clogged inboxes. Adopting a CMS or other document management system that we can link to will increase the amount of time necessary to produce the result. Whatever the frequency is for generating these reports, we will have to repeat the process every time.

In a resource-oriented architecture, we could simply address the source of each of the data elements and ask for them as JSON files so they could be easily consumed in a browser-based environment. The Exhibit project[‡] from Simile with a Timeline view[§] almost gives us this ability. Throw in a little bit of work to convert Excel spreadsheets to JSON objects, and we have a reusable environment that, when in place, would allow us to assemble and republish these marketing reports in a matter of seconds. Now consider that the same infrastructure could enable the ability to bring other forms of data together as easily for different types of analysis and reporting, and you begin to realize the value of a web of addressable data. These kinds of environments are emerging in the Enterprise; if your organization cannot tie its data together this easily, it should be able to do so.

Applied Resource-Oriented Architecture

Recently, I built a resource-oriented system on the rearchitecture work my company did for the Persistent URL (PURL) system. The original PURL[‖] implementation was done close to 15 years ago. It was a forked version of Apache 1.0, written in C and reflecting the state of the art at the time.[#] It has been a steady piece of Internet infrastructure since then, but it was showing its age and needed modernization, particularly to support the W3C TAG's 303 recommendation and higher volumes of use. Most of the data was accessible through web pages or ad hoc CGI-

▶ 运用REST架构风格，设计会变得更简单。对于服务只需要确定如下几个要素：
*资源；
*消息的结构；
*符合REST设计规范的URI；
*定义状态迁移。

其余的工作基本都由HTTP基础设计帮助我们完成了，这样做的好处是接口变得更加统一和一致，服务变得更稳定。

[†] *http://simile.mit.edu*

[‡] *http://simile.mit.edu/exhibit*

[§] *http://simile.mit.edu/timeline*

[‖] *http://purl.org*

[#] This codebase formed the basis of the very successful TinyURL (*http://tinyurl.com*) service.

bin scripts because at the time, the browser seemed like the only real client to serve. As we started to realize the applicability of persistent, unambiguous identifiers for use in the Semantic Web, life sciences, publication, and similar communities, we knew that it was time to rethink the architecture to be more useful for both people and software.

The PURL system was designed to mediate the tension between good names and resolvable names. Anyone who has been publishing content on the Web over time knows that links break when content gets moved around. The notion of a Persistent URL is one that has a good, logical name that maps to a resolvable location. For example, a PURL could be defined that points from *http://purl.org/people/briansletten* to *http://bosatsu.net/foaf/brian.rdf* and returns a 303 to indicate a "see also" response. I am not a network-addressable resource, but my Friend-of-a-Friend (FOAF) file[*] is a place to find more information about me. I could pass that PURL around to anyone who wants to link to my FOAF file. If I ever move to some other company, I could update the PURL to point to a new location for my FOAF file. All existing links will remain valid; they will just 303 to the new location. This process is described in Figure 5-6. The PURL Server implements the W3C Technical Architecture Group (TAG) guidance that 303 response codes can be used to provide more information about non-network addressable resources.

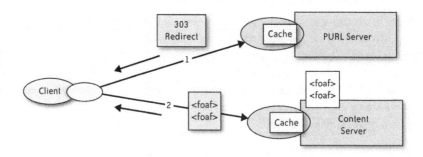

FIGURE 5-6. PURL "See Also" redirect

In addition to supporting the PURL redirection, we wanted to treat each major piece of data in the PURL system as an addressable information resource. Not only does this simplify the interaction with the user interface, it allows for unintended potential reuse of the data beyond what we originally planned. Manipulation of the resource requires ownership credentials, but

---

[*] *http://foaf-project.org*

anyone is allowed to fetch the definition of a PURL. There is the direct resolution process of hitting a PURL such as *http://purl.org/employee/briansletten* (which will result in the 303 redirect), as well as the indirect RESTful address of the PURL resource *http://purl.org/admin/purl/employee/briansletten*, which will return a definition of the PURL that currently looks something like the following:

```
<purl status="1">
    <id>/employee/briansletten</id>
    <type>303</type>
    <maintainers>
        <uid>brian</uid>
    </maintainers>
    <seealso>
        <url>http://bosatsu.net/foaf/brian.rdf</url>
    </seealso>
</purl>
```

Clients of the PURL server can "surf" to the data definition as a means of finding information about a PURL resource without actually resolving it. No code needs to be written to retrieve this information. We can view it in a browser or capture it on the command line with curl. As such, we can imagine writing shell scripts that use data from our information resources to check whether a PURL points to something valid and is returning reasonable results. If not, we could find the owner of the PURL and fire off a message to the email address associated with the account. Addressable, accessible data finds its way into all manner of unintended orchestrations, scripts, applications, and desktop widgets because it is so easy and useful to do so.

In the interest of full disclosure, we failed to support JSON as a request format in the initial release, which complicated the AJAX user interface. JavaScript XML handling leaves a lot to be desired. Even though we use the XML form internally, we should have gone to the trouble of exposing the JSON form for parsing in the browser. You can be sure we are fixing this oversight soon, but I thought it was important to highlight the benefits we could have taken advantage of if we had gotten it right in the first place. You do not need to support all data formats up front, but these days supporting both XML and JSON is a good start.

As an interesting side note, we could have chosen several containers and tools to expose this architecture as expressed so far. Anything that responds to HTTP requests could have acted as our PURL server. This represents a shallow but useful notion of RESTful interfaces and resource-oriented architecture, as demonstrated in Figure 5-7. Any web server or application server can act as a shallow resource-oriented engine. The logical HTTP requests are interpreted as requests into servlets, Restlets, and similar addressable functionality.

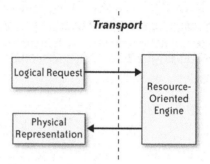

FIGURE 5-7. Shallow resource-oriented architectures

We chose to use NetKernel as the foundation for this architecture because it is the embodiment of resource-oriented architectures and has a dual license, allowing its use with both open source and commercial projects. The idea of a logical coupling between layers with different representations is baked into the software architecture and offers similar benefits of flexibility, scalability, and simplicity. The linkage between the layers is through asynchronously resolved, logical names. This deeper notion of resource-oriented architectures looks something like Figure 5-8. NetKernel is an interesting software infrastructure because it takes the idea of logically connected resources inside so that HTTP logical requests can be turned into other logical requests. This architecture reflects the properties of the Web in a runtime software environment.

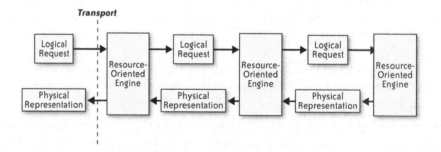

FIGURE 5-8. Deep resource-oriented architectures

The external URL *http://purl.org/employee/briansletten* gets mapped through a rewrite to a piece of functionality called an accessor.† Accessors live in modules that export public URI definitions representing an address space they will respond to. The convenience here is that it is possible to radically change the implementation technologies in a newer version of a module and simply update the rewrite rules to point to the new implementation. The client needs to

---

† *http://docs.1060.org/docs/3.3.0/book/gettingstarted/doc_intro_code_accessor.html*

be none the wiser as long as we return compatible responses. We can approximate this flexibility in modern object-oriented languages through the use of interfaces, but that still constrains us to a "physical" coupling to the interface definition. With the logical-only binding, we still need to support expectations from existing clients, but beyond that we are not coupled to any particular implementation detail. This is the same value we see communicating through URIs on the Web, but in locally running software!

Internally, we use the Command Pattern[‡] associated with the method type of the request to implement the accessor. An HTTP GET method is mapped to a `GetResourceCommand` that maintains no state. When the request comes in, we pull the command out of a map and issue the request to it. The REST stateless style ensures that all information needed to answer the request is contained in the request, so we do not need to maintain state in the command instance. We can access that request state through the context instance in the following code. This code looks relatively straightforward to Java developers. We are calling methods on Java objects, catching exceptions, the works. An important thing to note is the use of the IURAspect interface. We are essentially saying that we do not care what form the resource is in. It could be a DOM instance, a JDOM instance, a string, or a byte array; for our purposes it does not matter. The infrastructure will convert it into a bytestream tagged with metadata before responding to the request. If we had wanted it in a particular form supported by the infrastructure, we could have simply asked for it in that form. This declarative, resource-oriented approach helps radically reduce the amount of code that is necessary to manipulate data and allows us to use the right tool for the right job:

```
if(resStorage.resourceExists(context, uriResolver)) {
    IURAspect asp = resStorage.getResource(context, uriResolver);

    // Filter the response if we have a filter
    if (filter!=null) {
        asp = filter.filter(context, asp);
    }

    // Default response code of 200 is fine
    IURRepresentation rep = NKHelper.setResponseCode(context, asp, 200);
    rep = NKHelper.attachGoldenThread(context, "gt:" + path , rep);
    retValue = context.createResponseFrom(rep);
    retValue.setCacheable();
    retValue.setMimeType(NKHelper.MIME_XML);
} else {
    IURRepresentation rep = NKHelper.setResponseCode(context,
            new StringAspect("No such resource: "
                    + uriResolver.getDisplayName(path)), 404);
    retValue = context.createResponseFrom(rep);
    retValue.setMimeType(NKHelper.MIME_TEXT);
}
```

[‡] *http://en.wikipedia.org/wiki/Command_pattern*

Most of the information resources will return a 200 when a GET request is issued. Obviously, PURLs override that behavior to return 302, 303, 307, 404, etc. The interesting resource-oriented tidbit is revealed when we inspect the PURL-oriented implementation of the resStorage.getResource() method:

```
INKFRequest req = context.createSubRequest("active:purl-storage-query-purl");
req.addArgument("uri", uri);
IURRepresentation res = context.issueSubRequest(req);
return context.transrept(res, IAspectXDA.class);
```

In essence, we are issuing a logical request through the active:purl-storage-query-purl URI with an argument of ffcpl:/purl/employee/briansletten. Ignore the unusual URI scheme; it is simply used to represent an internal request in NetKernel. We do not know what code is actually going to be invoked to retrieve the PURL in the requested form, nor do we actually care. In a resource-oriented environment, we simply are saying, "The thing that responds to this URI will generate a response for me." We are now free to get things going quickly by serving static files to clients of the module while we design and build something like a Hibernate-based mapping to our relational database. We can make this transition by rewriting what responds to the active:purl-storage-query-purl URI. The client code never needs to know the difference. If we change the PURL resolution away from a local persistence layer to a remote fetch, the client code can still not care. These are the benefits we have discussed in the larger notion of resource-oriented Enterprise computing made concrete in a powerful software environment.

Not only are our layers loosely coupled like this, but we get the benefit of an idempotent, stateless request in this environment as well. The earlier code snippet that fetches the PURL definition gets flattened internally to an asynchronously scheduled request to the URI active:purl-storage-query-purl+uri@ffcpl:/purl/employee/briansletten. As we discussed earlier, this becomes a compound hash key representing the result of querying our persistence layer for the generated result. Even though we know nothing about the code that gets invoked, NetKernel is able to cache the result nonetheless. This is the architectural memoization that I mentioned before. The actual process is slightly more nuanced, but in spirit, this is what is going on. If someone else tries to resolve the same PURL either internally or through the HTTP RESTful interface, we could pull the result from the cache. Though this may not impress anyone who has built caching into their web pages, it is actually a far more compelling result when you dig deeper. Any potential URI request is cacheable in this way, whether we are reading files in from disk, fetching them via HTTP, transforming an XML document through an XSLT file, or calculating pi to 10,000 digits. Each of these invocations is done through a logical, stateless, asynchronous result, and each has the potential to be cached. This resource-oriented architectural style gives us software that scales, is efficient, is cacheable, and works through uniform, logical interfaces. This results in substantially less brittle, more flexible architectures that scale, just like the Web and for the same reasons.

Conclusion

The resource-oriented architecture approach walks a series of fine lines. On the one hand, to initiates of convention, the approaches might seem a little strange and untried. People concerned about their resumes want to stick with tried and true approaches. On the other hand, to those who have studied the Web and its fundamental building blocks, it makes perfect sense and represents the largest, most successful network software architecture ever imagined and implemented. In one light, it requires a radically different way of thinking. In another light, it allows a powerful mechanism for wrapping and reusing existing code, services, and infrastructure with logically named interfaces that do not leak implementation details for many forms of interaction. We have the freedom to be resilient in what we accept on the server without breaking existing clients. We can support new structural forms for the same data over time. We are able to migrate backend implementations without necessarily affecting our clients. Additionally, important properties such as scalability, caching, information-driven access control, and low-ceremony regulatory compliance fall out of these design choices.

Software developers do not usually care about data; they care about algorithms, objects, services, and other constructs such as this. We have some fairly specific recommended blueprints and technologies for our J2EE, .NET, and SOAP-based architectures. Unfortunately, most of these blueprints ignore information as a first-class citizen. They tie us into specific bindings that make it hard to make changes without breaking existing clients. This is the flux treadmill we have been on for years, and the business units are tired of paying for it. Web Services were supposed to be an exit strategy, but inappropriate levels of abstraction and overly complicated edge-case use cases have made the whole process an entirely WS-Unsatisfying experience. It is time to take a step away from software-centric architectures and start to focus on information and how it flows. We will still write our software using the tools we know and purport to love; it will just not be the focus of our architectural bindings.

The resource-oriented approach offers compelling bridges between business units and the technology departments that support them. There are real efficiencies and business value propositions offered by an information-centric view on how our systems are connected. Rather than starting from scratch with each new Big Idea from our vendors, we can learn valuable lessons from the Web on how its architectural style elicits important properties. Architecture is inhabited sculpture; we are forced to endure the choices that we make for quite some time. We should take the opportunity to imbue our architecture with functionality, beauty, and a resilience to change to make our time in it more useful and pleasant.

Principles and properties		Structures	
✓	Versatility	✓	Module
	Conceptual integrity	✓	Dependency
✓	Independently changeable		Process
✓	Automatic propagation	✓	Data access
	Buildability		
✓	Growth accommodation		
✓	Entropy resistance		

CHAPTER SIX

Data Grows Up: The Architecture of the Facebook Platform

Dave Fetterman

> Show me your flowcharts and conceal your tables, and I shall continue to be mystified. Show me your tables, and I won't usually need your flowcharts; they'll be obvious.
>
> —*Fred Brooks*, The Mythical Man-Month

Introduction

MOST CURRENT STUDENTS OF COMPUTER SCIENCE INTERPRET that Fred Brooks quote to mean "show me your *code* and conceal your *data structures*...." Information architects have a solid understanding that *data* rather than *algorithms* sit at the center of most systems. And with the rise of the Web, data that the user produces and consumes motivates the use of information technologies more than ever. Glibly, web users don't navigate to QuickSort. They visit a storehouse of data.

This data may be universal, like a phone directory; proprietary, like an online store; personal, like a blog; open, like local weather conditions; or tightly guarded, like online bank records. In any case, the user-facing functionality of almost any web presence boils down to delivering an interface to a set of site-specific core data. This information forms the core value of most any website, whether generated by a top-notch research team on staff or contributed by users around the world. Data motivates the product that users enjoy, so architects build the rest of the traditional "n-tier" software stack (the *logic* and *display*) around it.

This is the story of Facebook's data and how it has evolved with the creation of the Facebook Platform.

Facebook (*http://facebook.com*) stands as an example of an architecture built around data of great utility, including user-submitted personal relationship mappings, biographical information, and text or other media content. Facebook's engineers built the rest of the site's architecture with an eye to displaying and manipulating this social data. Most of the site's business logic closely depends on this social data, such as the flow and access patterns of various pages, implementation of search, surfacing of News Feed content, and application of visibility rules to content. To the user, the value of the site springs directly from the value of the data he and his social connections have contributed to the system.

"Facebook the social website" is conceptually a standard n-tier stack, where a user's request fetches data from Facebook's internal libraries, which is then transformed through Facebook's logic, to be output through Facebook's display. Facebook's engineers then recognized the usefulness of this data beyond the confines of its container. The creation of the Facebook Platform markedly changed the gestalt of Facebook's data access system, accommodating a vision much broader than the isolated functionality of the n-tier stack, to integrate outside systems in the form of *applications*. With the user's social data at the center of the architecture, the Platform has developed into an array of web services (Facebook Platform Application Programming Interface, or Facebook API), a query language (Facebook Query Language, or FQL), and a data-driven markup language (Facebook Markup Language, or FBML) to wed application developers' systems with Facebook's.

As given sets of data become more widely available and users demand unified use of their data across multiple web and desktop products, the architect reading this chapter will likely find herself either a consumer of such a platform or the producer of a similar platform surrounding her own site's data. This chapter takes the reader on the journey of opening up Facebook's data to outside stacks in a controlled way, the architectural choices that follow from each step of the data evolution, and the process of reconciling this with the unique privacy requirements that permeate social systems. It includes:

- Motivating scenarios for these types of integrations
- Moving data functions from an internal stack call to an externally visible web service (the Facebook API)
- Authorizing access to this web service with an eye to maintaining the privacy of the social system
- Creating a data query language to ease the burden of new clients of this web service (Facebook FQL)

- Creating a data-driven markup language to both integrate application display back into Facebook and enable use of otherwise inaccessible data (Facebook FBML)

And once we've evolved the architecture of an application significantly from a separate stack:

- Building technologies that bridge the gap between the Facebook experience and the external application's experience

For the data platform consumer, this chapter illustrates the design decisions made and the rationale behind them. Notions such as user sessions and authentication, web services, and the various ways of handling the application's logic will constantly appear as themes in these types of platforms all over the Web. Understanding the thought behind them provides a great exercise in data architecture, and proves relevant when thinking about the kinds of features and forms these platform producers will likely create in the future.

The data platform producer is encouraged to keep in mind his own data set, and learn from the ways Facebook has opened its data model. Some of the design choices and reconciliations remain specific to Facebook, or at least to handling social data guarded by privacy, and may not be wholly applicable to a given data set. Nonetheless, at each step we present a product problem, a data-driven solution, and the solution's high-level implementation. With each new solution, we are essentially creating a new product or platform, so at all points we must reconcile this new product with the expectations of users. In turn, we create new technologies to accompany each step of the evolution, and sometimes change the web architecture surrounding the application itself.

An open source version of the Facebook Platform is available at *http://developers.facebook .com/*. Like much of that release, the code in this chapter is written in PHP. Feel free to follow along, noting that the samples here are abbreviated for clarity.

We start with the motivation for these types of integrations with an example of "external" application logic and data (a book store), Facebook's social data (user information and "friend" relationships), and the case for integrating the two.

Some Application Core Data

Web applications, even those that do not produce or consume a data platform of any sort, are still motivated largely by their internal data. As an example, take *http://fettermansbooks .com*, a hypothetical website that provides information on books (and likely, the ability to purchase these titles if the mood struck). The site's features may include a searchable index of inventory, basic information about each of the products, and even user-contributed reviews about each title. Access to this specific information forms the core of the application and motivates the rest of the architecture. The site may employ Flash and AJAX, be accessible through mobile devices, and provide an award-winning user interface. However, *http:// fettermansbooks.com* exists essentially to provide visitors some kind access to core mappings like those in Example 6-1.

EXAMPLE 6-1. Example book data mappings

```
book_get_info : isbn -> {title, author, publisher, price, cover picture}
book_get_reviews: isbn -> set(review_ids)
bookuser_get_reviews: books_user_id -> set(review_ids)
review_get_info: review_id -> {isbn, books_user_id, rating, commentary}
```

All of these are ultimately implemented as something very similar to simple sets and fetches from an indexed data table. Any such book site worth its salt would likely implement other functions that are not so simple, such as the simple "search" in Example 6-2.

EXAMPLE 6-2. A simple search mapping

```
search_title_string: title_string -> set({isbn, relevance score})
```

Each key in the domain of these functions generally would justify at least one web page on *http://fettermansbooks.com*—a unique set of logic surrounding the range data, rendered through a unique display path. For instance, to see a selection of reviewer X's submissions, a *http://fettermansbooks.com* user would likely be directed to visit a page like *fettermansbooks.com/reviews.php?books_user_id=X*, or to see all info about a particular book with ISBN *Y* (including hops to individual review pages), he would visit *http://fettermansbooks.com/book.php?isbn=Y*.

A notable property of sites such as *http://fettermansbooks.com* is that nearly every piece of data is available to every user. It generates all the content in, say, the `book_get_info` mapping to aid users in discovering as much information about a book as possible. This may be optimal in the case of a site trying to sell books, but visibility restrictions govern much of the architectural considerations of the data access layer in the following example using social data.

Some Facebook Core Data

With the rise in popularity of the network of technologies called Web 2.0, the centrality of data within systems has only grown more obvious. The central themes of Web 2.0 presences are that they are data-driven, and that users themselves provide the majority of that data. Facebook, like *http://fettermansbooks.com*, primarily comprises a set of core data mappings that motivate the feel and functionality of its website. An extremely stripped-down set of these Facebook mappings could look like the set in Example 6-3.

EXAMPLE 6-3. Example social data mappings

```
user_get_friends: uid -> set(uids)
user_get_info: uid -> {name, pic, books, current_location,...}
can_see: {uid_viewer, uid_viewee, table_name, field_name} -> 0 or 1
```

uid here refers to a (numeric) Facebook user identifier, and the info returned from `user_get_info` refers to a user's profile content (see *users.getInfo* in Facebook's developer documentation), including perhaps titles of the user's favorite books as they are entered on *http://facebook.com*. This system differs little from *http://fettermansbooks.com* at its core, except that the epicenter of the data, and hence site functionality, revolves around users' connection to other users ("friends"), users' content ("profile information"), and visibility rules for that content ("can_see").

This can_see data set is special. Facebook has a very central notion of privacy around user-generated data—business rules for user X's view of user Y's information. Never directly viewable by itself, this data motivates very important considerations that will emerge again and again when we see examples of integrating external logic and data with Facebook's. By itself, Facebook's pervasive use of this data set differentiates it from a website like *http://fettermansbooks.com*.

The Facebook Platform and other social platforms are a recognition that these types of social mappings are useful, both within a site like *http://facebook.com*, and when *integrated* with the functionality of a site such as *http://fettermansbooks.com*.

▶ 对于数据驱动的应用而言，数据权限控制至关重要。权限事实上牵涉到数据的访问权限。例如这里提及的 can_see 和 can_edit 等访问操作的控制。访问控制与 user 结合起来，则是权限控制规则。

Facebook's Application Platform

For a user of both *http://fettermansbooks.com* and *http://facebook.com*, the picture of Internet applications at this point looks something like Figure 6-1.

In the usual n-tier architecture, an application maps input (for the Web, the union of GET, POST, and cookie information) to requests for raw data likely residing in a database. These are translated to in-memory data and passed to some *business logic* for intelligent processing. The output module translates these data objects for display, into HTML, JavaScript, CSS, and so on. Here, on the top of the figure, is an application's n-tier stack running on its infrastructure. Before the advent of applications in its Platform, Facebook operated wholly under the same architecture. Importantly, in both architectures, the business logic (including Facebook's privacy) is effectively executed according to rules established in some data component of the system.

More voluminous and relevant data means the business logic may deliver more personally tailored content, so the experience of browsing books to review, read, or purchase on *http://fettermansbooks.com* (or any such application) would be powerfully *augmented* by a user's social data from Facebook. Specifically, showing friends' book reviews, wish lists, and purchases could aid a user in making her own purchasing decisions, discovering new titles, or strengthening connections to other users. If Facebook's internal mapping user_get_friends were accessible to other external applications such as *http://fettermansbooks.com*, this would add powerful social context to these otherwise separate applications, and eliminate the application's need to create its own social network. Applications of all sorts could do well to integrate this data, since developers can apply these core Facebook mappings to countless other web presences where users produce *or* consume content.

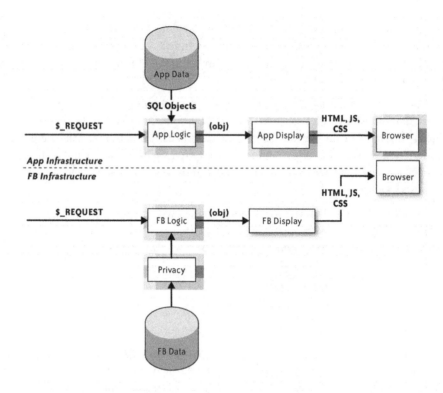

FIGURE 6-1. Separate Facebook and n-tier application stacks

The technologies of the Facebook Platform accomplish this through a number of evolutions in social web and data architecture:

- Applications can access useful social data through the Facebook Platform data services, adding social context to external web applications, desktop OS applications, and alternative device applications.

- Applications can publish their display using a data-driven markup language called FBML to integrate their application experience on the pages of *http://facebook.com*.

- With the change in architecture that FBML requires, developers can use Facebook Platform cookies and Facebook JavaScript (FBJS) to minimize the changes needed to add an application presence to *http://facebook.com*.

- And finally, applications can have these capabilities without sacrificing the *privacy* and expectations about *user experience* that Facebook has built around its user data and display.

▶ Facebook的应用平台重点是围绕数据为中心构建独立的自治的服务，并将数据进行隐私保护和适度封装，以服务为重用单位。

The last point is the most interesting. The architecture of the Facebook Platform is not always beautiful—it is largely considered a first-mover in the social platform universe. Most of the architectural decisions made to create universally available social context are shaped by this yin and yang: data availability and user privacy.

Creating a Social Web Service

Looking back on an example as simple as *http://fettermansbooks.com*, it becomes clear that most Internet applications would benefit from added social context for the data they present. However, we run into a product problem: the availability of that data.

PRODUCT PROBLEM: Applications could make use of a user's social data on Facebook, but this data is inaccessible.

DATA SOLUTION: Make Facebook data available through an externally accessible web service (Figure 6-2).

The addition of the Facebook API to Facebook's architecture begins the relationship between external applications and Facebook through the Facebook Platform, essentially adding Facebook's data to the external application's stack. For a Facebook user, this integration begins when he explicitly authorizes the outside application to obtain social data on his behalf.

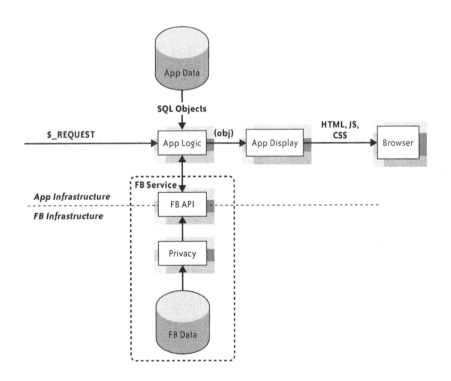

FIGURE 6-2. The application stack consumes Facebook data as web service

Example 6-4 shows what the code behind the landing page of *http://fettermansbooks.com* might look like without any Facebook integration.

EXAMPLE 6-4. Example book site logic

```
$books_user_id = establish_booksite_userid($_REQUEST);
$book_infos = user_get_likely_books($books_user_id);
display_books($book_infos);
```

This `user_get_likely_books` function operates entirely from the data that the book application controls, possibly using clever relevance techniques to guess at a user's interests.

However, imagine Facebook makes available two simple remote-procedure call (RPC) methods for users on sites outside its walls:

- `friends.get()`
- `users.getInfo($users, $fields)`

With these, and a mapping from *http://fettermansbooks.com*'s user identifiers to Facebook's, we can add social context to any content on *http://fettermansbooks.com*. Consider this new flow for Facebook users in Example 6-5.

EXAMPLE 6-5. Book site logic with social context

```
$books_user_id = establish_booksite_userid($_REQUEST);
$facebook_client = establish_facebook_session($_REQUEST,$books_user_id);

if ($facebook_client) {
  $facebook_friend_uids = $facebook_client->api_client->friends_get();
  foreach($facebook_friend_uids as $facebook_friend) {
    $book_site_friends[$facebook_friend]
        = books_user_id_from_facebook_id ($facebook_friend);
  }
  $book_site_friend_names = $facebook->api_client->
      users_getInfo($facebook_friend_uids, 'name');

  foreach($book_site_friends as $fb_id => $booksite_id) {
    $friend_books = user_get_reviewed_books($booksite_id);
    print "<hr>" . $book_site_friend_names[$fb_id] . "'s likely picks: <br>";
    display_books($friend_books);
  }
}
```

▶ establish_face book_session 相当于一个工厂，通过它可以获得Facebook的客户端。

The bolded parts of this example are where the book application harnesses the data of the Facebook Platform. If we could figure out the code behind the function `establish_facebook_session`, this architecture would make available much more *data* in order to turn this book-aware application into a fully user-aware application.

Let's examine how Facebook's API enables this. First, we'll check out a simple technical walkthrough of the web service wrapping Facebook data, created through use of appropriate metadata by a flexible code generator called Thrift. Developers can use these techniques outlined in the next section to effectively build web services of any kind, regardless of whether the data in the developer's storehouse is public or private.

But note that Facebook users do not consider their Facebook data to be fully public. So after our technical overview, we'll look at maintaining Facebook-level privacy through the main authentication means in the Platform API: user sessions.

Data: Creating an XML Web Service

In order to add basic social context to an example application, we've established the existence of two remote method calls, friends.get and users.getInfo. The internal functions accessing this data are likely sitting in a library somewhere in the Facebook code tree, serving similar requests on the Facebook site. Example 6-6 shows some examples.

EXAMPLE 6-6. Example social data mappings

```
function friends_get($session_user) { ... }
function users_getInfo($session_user, $input_users, $input_fields) { ... }
```

We now build a simple web service, transforming GET and POST input over HTTP to internal stack calls, and outputting the results as XML. In the Facebook Platform's case, the name of the destination method and its arguments are passed in the HTTP request, as well as some credentials specific to the calling application (an assigned "api key"), specific to a user-application pair (a "user session key"), and specific to the request instance itself (a request "signature"). We'll address the session key later in "A Simple Web Service Authentication Handshake." The high-level sequence for servicing a request to *http://api.facebook.com* is then:

1. Examine passed credentials ("A Simple Web Service Authentication Handshake") to verify the invoking application's identity, user's current authorization of the application, and authenticity of the request.

2. Interpret the incoming GET/POST as a method call with reasonable arguments.

3. Dispatch a single call to internal method and collect the result as in-memory data structures.

4. Transform these structures into a known output form (e.g., XML or JSON) and return.

The main points of difficulty in constructing interfaces consumed externally usually arise in steps 2 and 4. Consistently maintaining, synchronizing, and documenting the data interfaces for an external consumer is important, and constructing the skeleton code to ensure this consistency by hand is a thankless and time-consuming job. Additionally, we may need to make this data available to internal services written in many languages, or communicate results to an external developer in different web protocols such as XML, JSON, or SOAP.

Here, then, the beautiful solution is the use of metadata to encapsulate the types and signatures describing the APIs. Engineers at Facebook have created an open source cross-language

▶ 避免重复编写代码的常用方案就是利用模板结合元数据进行代码生成。现在的微服务架构亦多采用这种方式。

inter-process communication (IPC) system called Thrift (*http://developers.facebook.com/thrift*) that accomplishes this cleanly.

Diving right in, Example 6-7 shows an example "dot thrift" file for our sample API version 1.0, which the Thrift package turns into much of the machinery of the API.

EXAMPLE 6-7. *Web service definition through Thrift*

```
xsd_namespace http://api.facebook.com/1.0/
/***
 * Definition of types available in api.facebook.com version 1.0
 */
typedef i32 uid
typedef string uid_list
typedef string field_list

struct location {
 1: string street xsd_optional,
 2: string city,
 3: string state,
 4: string country,
 5: string zip xsd_optional
}

struct user {
 1: uid uid,
 2: string name,
 3: string books,
 4: string pics,
 5: location current_location
}

service FacebookApi10 {

 list<uid> friends_get()
  throws (1:FacebookApiException error_response),

 list<user> users_getInfo(1:uid_list uids, 2:field_list fields)
  throws (1:FacebookApiException error_response),
}
```

Each type in this example is a primitive (`string`), a structure (`location`, `user`), or a generic-style collection (`list<uid>`). Because each method declaration has a well-typed signature, code defining the reused types can be directly generated in any language. Example 6-8 shows part of the generated output for PHP.

EXAMPLE 6-8. *Thrift-generated service code*

```
class api10_user {

public $uid = null;
public $name = null;
public $books = null;
```

```php
public $pic = null;
public $current_location = null;


public function __construct($vals=null) {
  if (is_array($vals)) {
    if (isset($vals['uid])) {
      $this->uid = $vals['uid'];
    }
    if (isset($vals['name'])) {
      $this->name = $vals['name'];
    }
    if (isset($vals['books'])) {
      $this->books = $vals['books'];
    }
    if (isset($vals['pic'])) {
      $this->pic = $vals['pic'];
    }
    if (isset($vals['current_location'])) {
      $this->current_location = $vals['current_location'];
    }
   // ...
  }
 // ...
}
```

All internal methods returning the type user (such as the internal implementation of users_getInfo) create all needed fields and end with something like the line in Example 6-9.

EXAMPLE 6-9. Consistent use of generated type

```
return new api_10_user($field_vals);
```

For example, if the current_location is present in this user object, then `$field_vals['current_location']` is set to new api_10_location(...) somewhere before Example 6-9 is executed.

The names of the fields and types themselves actually generate the schema for the XML output, as well as the accompanying XML Schema Document (XSD). Example 6-10 shows an example of the actual XML output of the whole RPC flow.

EXAMPLE 6-10. XML output from web service call

```xml
<users_getInfo_response list="true">
 <users type="list">
  <user>
   <name>Dave Fetterman</name>
   <books>Zen and the Art, The Brothers K, Roald Dahl</books>
   <pic></pic>
   <current_location>
    <city>San Francisco</city>
    <state>CA</state>
    <zip>94110</zip>
```

```
        </current_location>
      </user>
    </users>
</users_getInfo_response>
```

> ▶ 由于微服务架构风格要求设计出更加细粒度的服务，可能导致大量重复的服务模板代码。此时可以借鉴 Thrift 工具的实现思想，通过利用代码生成工具来快速生成服务，降低开发成本。

Thrift generates similar code for declaring RPC function calls, serializing into known output structures, and turning internal exceptions into external error codes. Other toolsets such as XML-RPC and SOAP provide some of these benefits as well, perhaps with a greater CPU and bandwidth cost.

Employing a beautiful tool like Thrift provides recurring benefits:

Automatic type synchronization
   Adding `favorite_records` to the user type or turning uid into an i64 needs to happen across all methods consuming or generating these types.

Automatic binding generation
   All the messy work of reading and writing types is gone, and translating function calls into XML-generating RPC methods requires the function declaration, type checking, and error handling that Thrift does automatically.

Automatic documentation
   Thrift generates a public XML Schema Document, which serves as unambiguous documentation to the outside world, usually much better than what one finds in "the manual." This document can also be used directly by some external tools to generate bindings on the client side.

Cross-language synchronization
   This service can be consumed externally by both XML and JSON clients, and internally over a socket by daemons in all types of languages (PHP, Java, C++, Python, Ruby, C#, etc.). This requires metadata-based code generation so the service designer isn't spending her time updating each of these with every small change.

We now have the data component of a social web service. Next we'll figure out how to establish these session keys to enforce the privacy model users expect on any extension of Facebook.

A Simple Web Service Authentication Handshake

A simple authentication scheme makes this data accessible within the Facebook user's notion of privacy. A user has a certain view into the data of the Facebook system, based on who the user is, his privacy settings, and the privacy settings of those connected to him. Users may authorize individual applications to inherit this view. What is externally visible to a user through an application is a significant portion of (but no more than) what the user could see on the Facebook site itself.

In the architecture for a separate application site (Figure 6-1), user authentication often takes the form of cookies sent from a browser, originally assigned to the user after a verifying action taken on the site. However, cookies normally part of Facebook usage are not available to Facebook anymore in Figure 6-2—the outside application requests information from the platform without the help of a user's browser. To fix this, we establish on Facebook a *session key* mapping, as shown in Example 6-11.

EXAMPLE 6-11. A session key mapping

get_session: {user_id, application_id} -> session_key

The client of the web service simply sends the session_key along with every request, to let the web service know on which viewer's behalf the request executes. If the user (or Facebook) disables this application or has never used this application, this fails the security checks, returning an error. Otherwise, the outside application site will carry this session key around with its own user records or in a cookie for that user.

But how does one obtain this key in the first place? The example function establish_facebook_session in the *http://fettermansbooks.com* application code (Example 6-5) is a placeholder for this flow. Every application has its own unique "application key" (also called an api_key) that begins the application authorization flow (Figure 6-3):

1. The user is redirected to the Facebook login with a known api_key.
2. The user enters her credentials on Facebook to authorize the application.
3. The user is redirected to a known application landing site with the session key and user ID.
4. The application is now authorized to make calls to the API endpoint on the user's behalf (until the session expiration or deletion).

To help the user initiate this flow, a link or button could be rendered:

    <a href="http://www.facebook.com/login.php?api_key=abc123">

with that application key (say, "abc123"). If a user agrees to authorize this application using the password form *on Facebook* (note that the password is the last piece of data Facebook would export), the user is directed back to this application site with a valid session_key and his Facebook user ID. This session key is quite private, so for further verification, each call made by the application is accompanied by a hash generated from a shared secret.

Assuming the developer has stashed his api_key and application secret, establish_facebook_session can be written quite simply from the flow in Figure 6-3. Though the details between these kinds of handshake systems can vary, it is important that no user can be authorized unless he enters his password in the crucial step on Facebook. Interestingly enough, some early applications simply used this authorization handshake as their own password system, not employing any Facebook data at all.

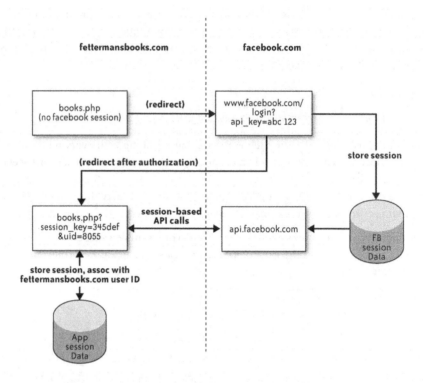

FIGURE 6-3. Authorizing access to the Facebook Platform API

However, some applications do not easily lend themselves to this second "redirect" step. "Desktop"-style applications or applications based on a device such as a mobile phone or built into a browser can be quite useful as well. In this case, we employ a slightly different scheme using a secondary authorization token. A token is requested by an application through the API, passed to Facebook on the first login, and then exchanged by the application for a session key and a per-session secret after on-site user authentication.

Creating a Social Data Query Service

We have expanded our internal libraries to the outside world by creating a web service with a user-controlled authentication handshake. With this simple change, Facebook's social data now drives any other application stack its users choose to authorize, creating new relations within that application's data through a universally interesting social context.

As seamless as this data exchange becomes in the mind of the user, the developer consuming these platform APIs knows the data sets are very distinct. The pattern the developer uses to access his own data is quite different than the one used to get Facebook's. For one, Facebook's data lives on the other side of an HTTP request, and making these method calls across many

HTTP connections adds latency and overhead to the developer's own pages. His own database also offers a higher granularity of access than do the few dozen methods in the Facebook Platform API. Using his own data and a familiar query language such as SQL allows him to select only certain fields of a table, sort or limit results, match on alternative indices, or nest queries. If the platform's API does not offer the developer the ability to do intelligent processing on the platform's server, the developer must often import a superset of the relevant data and then do these standard logical transforms on his own servers after receiving it. This can be a significant burden.

PRODUCT PROBLEM: Obtaining data from the Facebook Platform APIs incurs much more cost than obtaining internal data.

As more traffic or usage starts to flow through an application consuming an outside data platform, factors such as bandwidth consumption, CPU load, and request latency can start to add up quickly. Surely this problem must have at least a partial solution. After all, haven't we optimized this in the data layer of our own single application stack? Isn't there a technology that enables fetching multiple sets of data in one call? How about doing selection, limiting, and sorting in the data layer itself?

DATA SOLUTION: Implement external data access patterns using the same one employed for internal data: a query service.

Facebook's solution is called FQL, detailed later in the section "FQL." FQL bears a great deal of resemblance to SQL, but casts platform data as fields and tables rather than simply loosely defined objects in our XML schema. This gives developers the ability to use standard data query semantics on Facebook's data, which is probably the same way they get to their own data. At the same time, the benefit of pushing computation to the platform side mirrors the benefits of pushing operations to the data layer in SQL. In both cases, the developer consciously avoids paying the price in his application logic.

▶ 满足架构的一致性原则。

FQL represents yet another improved data architecture based on Facebook's internal data, and is the next step after standard black-box web services. But first, we mention an easy and obvious way for a platform developer to eliminate the round-trip load of many data requests, and show why this is ultimately insufficient.

Method Call Batching

The simplest solution to load problems is something akin to Facebook's `batch.run` API. This eliminates the round-trip latency of multiple calls to *http://api.facebook.com* over the HTTP stack by accepting input for multiple methods in one batch, and returning the outputted XML trees in one response. On the client side, this flow translates to something like the code in Example 6-12.

▶ 我们在设计跨进程调用的API，例如RESTful服务接口时，应借鉴这种方式。

▶ 使用类SQL语法的方式针对数据进行处理的一个好处是它基本满足SQL标准语法的定义，使得多数开发人员无须学习另一门语言就能快速上手，作为一个开放平台，降低了接入系统的开发难度。另一个好处是SQL语句可以是字符串，我们可以非常容易地通过一些元数据组装符合要求的SQL语句。目前，有许多支持数据分析的平台都支持类SQL语法，包括Hive、Impala和Spark SQL等，其目的也在于此。我正在开发的一个数据分析平台使用了Spark SQL，通过使用SQL，而非Data Frame的API，可以非常方便地完成分析语句的组装，再通过有效地利用UDF（User Defined Function）进行扩展，用以实现强大的数据分析功能。

EXAMPLE 6-12. Batching method calls

```
$facebook->api_client->begin_batch();
$friends = &$facebook->api_client->friends_get();
$notifications = &$facebook->api_client->notifications_get();
$facebook->api_client->end_batch();
```

In Facebook Platform's PHP5 client library, end_batch effectively initiates the request to the platform server, obtains all results, and updates the reference variables used for each result. Here we are *getting* data for the user in a batch from a single user session. Commonly, the batch query mechanism is used to group together many *setting* operations, such as mass Facebook profile updates or large user notification bursts.

The fact that these *set* operations are most effective here reveals the main problem with this batching style. Problematically, each call must be independent of the results of the other. Set operations for many different users usually enjoy this property, but a common case remains unaddressed: using the results of one call as inputs to the next. Example 6-13 features a common scenario that would *not* work with the batch system.

EXAMPLE 6-13. Improper use of batching

```
$fields = array('uid', 'name', 'books', 'pic', 'current_location');
$facebook->api_client->begin_batch();
$friends = &$facebook->api_client->friends_get();
$user_info = &$facebook->api_client->users_getInfo($friends, $fields); // NO!
$facebook->api_client->end_batch();
```

The content of $friends clearly does not exist at the time the client sends the users_getInfo request. The FQL model solves this and other problems elegantly.

FQL

FQL is a simple query language wrapper around Facebook's internal data. The output generally shares the same format as the Facebook Platform API, but the input graduates from a simple RPC library model to a query model reminiscent of SQL: *named tables and fields with a known relationship*. Like SQL, this technology adds the abilities to select on instances or ranges, select a subset of fields from a data row, and nest queries to push a greater amount of work to the data server, eliminating the need for multiple calls over the RPC stack.

An as example, if the desired output were the fields named 'uid', 'name', 'books', 'pic', and 'current_location' for all users who are my friends, in our pure-API model, we would run a procedure like that in Example 6-14.

EXAMPLE 6-14. Chaining method calls client-side

```
$fields = array('uid', 'name', 'books', 'pic', 'current_location');
$friend_uids = $facebook->api_client->friends_get();
$user_infos = users_getInfo($friend_uids, $fields);
```

This results in more calls to the data server (here, two calls), greater latency, and more points of possible failure. Instead, for viewing user number 8055 (yours truly), we render this in FQL syntax as the single call in Example 6-15.

EXAMPLE 6-15. Chaining method calls server-side with FQL

```
$fql = "SELECT uid, name, books, pic, current_location FROM profile
        WHERE uid IN (SELECT uid2 from friends where uid1 = 8055)";
$user_infos = $facebook->api_client->fql_query($fql);
```

We conceptually treat the data referred to by users_getInfo as a *table* with a number of selectable *fields*, based on an index (uid). If augmented appropriately, this new grammar enables a number of new data access capabilities:

- Range queries (for example, event times)
- Nested queries (SELECT fields_1 FROM table WHERE field IN (SELECT fields_2 FROM ....))
- Result limits and ordering

Architecture of FQL

Developers invoke FQL through the API call fql_query. The crux of the problem involves unifying the named "objects" and "properties" of the external API with named "tables" and "fields" in FQL. We still inherit the flow of the standard API: fetching the data through our internal methods, applying the rules normally associated with API calls on this method, and transforming the output according to the Thrift system from the earlier section "Data: Creating an XML Web Service." For every data-reading API method there exists a corresponding "table" in FQL that abstracts the data behind that query. For instance, the API method users_getInfo, which makes the name, pic, books, and current_location fields available for a given user ID is represented in FQL as the user table with those corresponding fields. The external output of fql_query actually conforms to the output of the standard API as well (if the XSD is modified to allow for omitted fields in an object), so a call to fql_query on the user table returns output identical to an appropriate call to users_getInfo. In fact, often calls such as user_getInfo are implemented at Facebook on the server side as FQL calls!

> **NOTE**
> At the time of this writing, FQL supports only SELECT rather than INSERT, UPDATE, REPLACE, DELETE, or others, so only *read* methods can be implemented using FQL. Most Facebook Platform API methods operating on this type of data are read-only at this point anyway.

Let's start with this user table as an example and build the FQL system to support queries on it. Underneath all the layers of data abstraction through the Platform (the internal calls, the users_getInfo external API call, and the new user table of FQL), imagine Facebook had a table named 'user' in its own database (Example 6-16).

EXAMPLE 6-16. Example Facebook data table

```
> describe user;
+---------------+---------------+-----+
| Field         | Type          | Key |
+---------------+---------------+-----+
| uid           | bigint(20)    | PRI |
| name          | varchar(255)  |     |
| pic           | varchar(255)  |     |
| books         | varchar(255)  |     |
| loc_city      | varchar(255)  |     |
| loc_state     | varchar(255)  |     |
| loc_country   | varchar(255)  |     |
| loc_zip       | int(5)        |     |
+---------------+---------------+-----+
```

Within the Facebook stack, suppose our method for accessing this table is:

```
function user_get_info($uid)
```

which returns an object in the language of our choice (PHP), usually used before applying privacy logic and rendering on *http://facebook.com*. Our web service implementation did much the same, transforming the GET/POST content of a web request to such a call, obtaining a similar stack object, applying privacy, and then using Thrift to render this as an XML response (Figure 6-2).

We can wrap user_get_info within FQL to programmatically apply this model, with tables, fields, internal functions, and privacy all fitting together in a logical, repeatable form.

Following are some key objects created in the FQL call in Example 6-15 and the methods that describe how they relate. Discussion of the entire string parsing, grammar implementation, alternative indexing, intersecting queries, and implementing the many different combining expressions (comparisons, "in" statements, conjunction, and disjunction) are beyond the scope of this chapter. Instead, we'll just focus on the data-facing pieces: the high-level specification of the data's corresponding field and table objects within FQL, and transforming the input statement to queries to each field's can_see and evaluate functions (Example 6-17).

EXAMPLE 6-17. Example FQL fields and tables

```
class FQLField {
  // e.g. table="user", name="current_location"
  public function __construct($user, $app_id, $table, $name) { ... }

  // mapping: "index" id -> {0,1} (visible or invisible)
  public function can_see($id) { ... }
```

```
    // mapping: "index" id -> Thrift-compatible data object
    public function evaluate($id) { ... }
}
class FQLTable {
    // a static list of contained fields:
    // mapping: () -> ('books' => 'FQLUserBooks', 'pic' ->'FQLUserPic', ...)
    public function get_fields() { ... }
}
```

The `FQLField` and `FQLTable` objects constitute this new method for accessing data. `FQLField` contains the data-specific logic transforming the index of the "row" (e.g., user ID) plus the viewer information (user and app_id) into our internal stack data calls. On top of that, we ensure privacy evaluation is built right in with the required `can_see` method. When processing a request, we create in memory one such `FQLTable` object for each named table (`'user'`) and one `FQLField` object for each named field (one for `'books'`, one for `'pic'`, etc.). Each `FQLField` object mapped to by one `FQLTable` tends to use the same data accessor underneath (in the following case, user_get_info), though it is not necessary—it's just a convenient interface.
Example 6-18 shows an example of the typical string field for the user table.

EXAMPLE 6-18. *Mapping a core data library to an FQL field definition*

```
// base object for any simple FQL field in the user table.
class FQLStringUserField extends FQLField {

    public function __construct($user, $app_id, $table, $name) { ... }

    public function evaluate($id) {
        // call into internal function
        $info = user_get_info($id);
        if ($info && isset($info[$this->name])) {
            return $info[$this->name];
        }
        return null;
    }

    public function can_see($id) {
        // call into internal function
        return can_see($id, $user, $table, $name);
    }
}

// simple string data field
class FQLUserBooks extends FQLStringUserField { }

// simple string data field
class FQLUserPic extends FQLStringUserField { }
```

▶ 总体而言，这里可以看作Interpreter模式的运用。在针对表达式树进行解析时，多采用这一模式。在这棵表达式树中，树的每个节点都定义了evaluate()函数，使其能够进行递归调用，如下页定义在FQLFieldExpression的evaluate()函数所示。

FQLUserPic and FQLUserBooks differ only in their internal property $this->name, set by their constructor during processing. Note that underneath, we call user_get_info for every evaluation we need in the expression; this performs well only if the system caches these results in process memory. Facebook's implementation does just that, and the whole query executes in time on the order of a standard platform API call.

Here is a more complex field representing current_location, which takes the same input and exhibits the same usage pattern, but outputs a struct-type object we've seen earlier (Example 6-19).

EXAMPLE 6-19. A more complex FQL field mapping

```
// complex object data field
class FQLUserCurrentLocation extends FQLStringUserField {
  public function evaluate($id) {
    $info = user_get_info($id);
    if ($info && isset($info['current_location'])) {
      $location = new api10_location($info['current_location']);
    } else {
      $location = new api10_location();
    }
    return $location;
  }
}
```

Objects such as api10_location are the generated types from "Data: Creating an XML Web Service," which Thrift and the Facebook data service know how to return as well-typed XML. Now we're seeing why even with a new input style, FQL's output does not need to be incompatible with that of the Facebook API.

The main evaluation loop of FQLStatement in the following example provides a high-level idea of FQL's implementation. Throughout this code we reference FQLExpressions, but in a simple query, we're mostly talking about FQLFieldExpressions, which wrap internal calls to the FQLField's own evaluate and can_see methods, as in Example 6-20.

EXAMPLE 6-20. A simple FQL expression class

```
class FQLFieldExpression {

  // instantiated with an FQLField in the "field" property
  public function evaluate($id) {
    if ($this->field->can_see($id))
      return $this->field->evaluate($id);
    else
      return new FQLCantSee(); // becomes an error message or omitted field
  }

  public function get_name() {
    return $this->field_name;
  }
}
```

To initiate the whole flow, the SQL-like string input is transformed via *lex* and *yacc* into the main FQLStatement's $select expression array and the $where expression. FQLStatement's evaluate() function returns the objects we've requested. The main statement evaluation loop in Example 6-21 goes through the following steps in this simple high-level sequence:

1. Get all constraints on indexes of the rows we wish to return. For example, when selecting on the user table, these would be the UIDs we want to query. If we were looking at an events table indexed by time, say, these would be the boundary times.

2. Translate these to the canonical IDs of the table. The user table is also queryable by field name; if an FQL expression used name, this function would use an internal user_name ->user_id lookup function.

3. For each candidate ID, see if it matches the RHS expression clause (Boolean logic, comparisons, "IN" operations, etc.). If not, toss it out.

4. Evaluate each expression (in our case, fields in the SELECT clause), and create an XML element of the form <COL_NAME>COL_VALUE</COL_NAME>, where COL_NAME is the name of the field in the FQLTable, and COL_VALUE is the result of the evaluation of the field through its corresponding FQLField's evaluate function.

EXAMPLE 6-21. The FQL's main evaluation flow

```
class FQLStatement {

  // contains the following members:
  // $select: array of FQLExpressions from the SELECT clause of the query
  //  corresponding to, say, "books", "pic", and "name"
  // $from: FQLTable object for the source table
  // $where: FQLExpression containing the constraints for the query.
  // $user, $app_id: calling user and app_id

  public function __construct($select, $from, $where, $user, $app_id) { ... }

  // A listing of all known tables in the FQL system.
  public static $tables = array(
    'user'      => 'FQLUserTable',
    'friend'    => 'FQLFriendTable',
  );

  // returns XML elements to be translated to service output
  public function evaluate() {

    // based on the WHERE clause, we first get a set of query expressions that
    // represent the constraints on values for the indexable columns contained
    // in the WHERE clause

    // Get all "right hand side" (RHS) constants matching ids (e.g. X, in 'uid = X')
    $queries = $this->where->get_queries();
```

第 6 章　数据增长：Facebook 平台的架构　131

```
    // Match to the row's index. If we were using 'name' as an alternative index
    // to the user table, we would transform it here to the uid.
    $index_ids = $this->from_table->get_ids_for_queries($queries);

    // filter the set of ids by the WHERE clause and LIMIT params
    $result_ids = array();

    foreach ($ids as $id) {
      $where_result = $this->where->evaluate($id);

        // see if this row passes the 'WHERE' constraints
        // is not restricted by privacy
        if ($where_result && !($where_result instanceof FQLCantSee))
          $result_ids []= $id;
    }

    $result = array();
    $row_name = $this->from_table->get_name(); // e.g. "user"

    // fill in the result array with the requested data
    foreach ($result_ids as $id) {
      foreach ($this->select as $str => $expression) { // e.g. "books" or "pic"
        $name = $expression->get_name();
        $col = $expression->evaluate($id); // returns the value
        if ($col instanceof FQLCantSee)
          $col = null;

        $row->value[] = new xml_element($name, $col);
      }

      $result[] = $row;
    }
    return $result;
}
```

FQL has some other subtleties, but this general flow illustrates the union of existing internal data access and privacy implementations with a whole new query model. This allows the developer to process his request more quickly and access data in a more granular way than the APIs, while still retaining the familiarity of SQL syntax.

As many of our APIs internally wrap corresponding FQL methods, our overall architecture has evolved to the state shown in Figure 6-4.

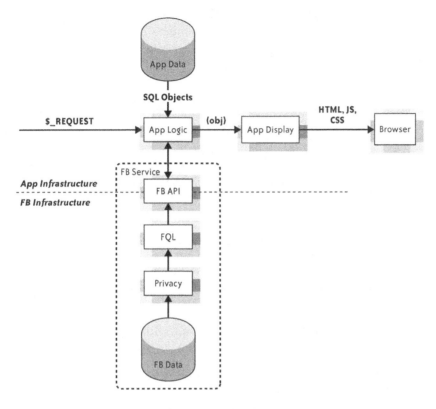

FIGURE 6-4. *The application stack consumes Facebook data as web and query service*

Creating a Social Web Portal: FBML

The services discussed earlier provide outside application stacks the ability to incorporate the social platform's data into their systems, which is a powerful step. These data architectures realize the promise of making the social platform's data more available: users in common between the external application (e.g., *http://fettermansbooks.com*) and the data platform (e.g., *http://facebook.com*) can share their social information between them, eliminating the need for a new social network with every new social application. However, even with these new capabilities, these applications don't yet enjoy the full power of a social utility like Facebook. The applications still need to be *discovered* by many users to become valuable. At the same time, not all of the internal data supporting the social platform can be made available to these external stacks. The platform creator needs to solve each of these problems, which we take in turn.

PRODUCT PROBLEM: For social applications to gain compelling critical mass, users on the supporting social graph must be made aware of other users' interactions with these applications. This suggests deeper integration of the application *into* the social site.

This problem has existed since the dawn of software: the difficulty of getting our data, product, or system out into general use. The lack of users becomes a particularly notable difficulty in the space of Web 2.0 because without users to consume and (especially) *generate* our content, how useful can our system ever become?

Facebook supports a large number of users who are interested in sharing information along social connections, and it can feature content from applications just as well as its own content. Giving external applications a presence on the Facebook site would make applications built by both large and small developers more discoverable, helping them gain the critical mass needed to support good social functionality.

Whatever solution we create, the applications need a distinct display presence on the Facebook site. The Facebook Platform makes this available to our application, reserving the URL path *http://apps.facebook.com/fettermansbooks/...* for that application's content rendered on Facebook. We'll see how the platform integrates the application's data, logic, and display shortly.

The second problem is another outgrowth of our data services built into the "Data: Creating an XML Web Service" and "FQL" sections, and is just as tricky.

PRODUCT PROBLEM: External applications cannot use certain core data elements that Facebook does not expose through its web services.

Facebook makes a great deal of data available to its users when producing the content of its website (*http://facebook.com*), but it chooses not to make every bit of this available through the external data services. Privacy information itself (the can_see mapping from "Some Facebook Core Data") is a good example—not explicitly visible to users on the Facebook site, the can_see mapping remains invisible to the data services as well. Yet enforcing the use of the privacy preferences users maintain on Facebook is the hallmark of a well-integrated application, and one that upholds the expectations of users on the social system. How are developers able to harness this data, which Facebook, to maintain the privacy of its users, has not released through its data services?

The most elegant solution to these problems will incorporate Facebook data with the external application's data, logic, and display, while still operating under a trusted environment for the user.

DATA SOLUTION: Developers create application content for execution and display on the social site itself through a data-driven markup language, interpreted by Facebook.

Applications using only the Facebook Platform elements of the "Data: Creating an XML Web Service" and "FQL" sections create a social experience *external* to Facebook, augmented by the use of Facebook's social data services. With the data and web architecture described in this section, applications themselves become a data service of a sort, supplying the content for Facebook to display under *http://apps.facebook.com*. A URL such as *http://apps.facebook.com/fettermansbooks/...* would no longer map to Facebook-generated data, logic, and display, but would query the service at *http://fettermansbooks.com* to generate the application's content.

We must simultaneously keep in mind our assets and our constraints. On one hand, we have a highly trafficked social system for users to discover external content, and a great deal of social data to augment such social applications. On the other hand, requests need to originate on the social site (Facebook), consume the application as a service, and render content such as HTML, JavaScript, and CSS, all without violating the privacy or expectations of users on Facebook.

▶ 这句话揭示了整个平台的核心设计思想。

First, we show some *incorrect* ways to attempt this.

Applications on Facebook: Directly Rendering HTML, CSS, and JS

Imagine an external application's configuration now include two new fields named `application_name` and `callback_url`. By entering in a name like "fettermansbooks" and a URL like *http://fettermansbooks.com/fbapp/* respectively, *http://fettermanbooks.com* declares that it will service user requests to URLs like *http://apps.facebook.com/fettermansbooks/PATH?QUERY_STRING* on its own servers, at *http://fettermansbooks.com/fbapp/PATH?QUERY_STRING*.

A request to *http://apps.facebook.com/fettermansbooks/...* then simply fetches the HTML, JS, and CSS contents on the application servers and displays this as the main content of the page on Facebook. This renders the external site as essentially an *HTML web service*.

This changes the n-tier model of an application significantly. Earlier, a stack consuming Facebook's content as a data service served direct requests to *http://fettermansbooks.com*. Now, the application maintains a tree under its web root that itself provides an HTML service. Facebook obtains content from an online request to this new application service (which may, in turn, consume Facebook's data services), wraps it in the usual Facebook site navigation elements, and displays it to the user.

However, if Facebook renders an application's HTML, JavaScript, or CSS directly in its pages, this allows the application to completely violate the user's expectation of the more controlled experience on *http://facebook.com*, and opens the site and its users up to all kinds of nasty security attacks. Allowing direct customization of markup and script from outside users is almost never a good idea. In fact, code or script injection is usually the *goal* of attackers, so it's not much of a feature.

Plus: no new data! Although this forms the basis of how an application's stack changes, this solution solves neither of our product problems fully.

Applications on Facebook: iframes

An obvious first stab for more *safely* displaying the content of one application in the visual context and flow of another site relies on a technology already incorporated into the browser itself: the *iframe*.

To reuse the mappings from the previous section, a request to *http://apps.facebook.com/ fettermansbooks/PATH?QUERY_STRING* would result in HTML output like this:

    <iframe src="http://fettermansbooks.com/fbapp/PATH?GET_STRING"></iframe>

The content of this URL would display in a frame inside the Facebook page, and in its own sandboxed environment could contain any type of web technology: HTML, JS, AJAX, Flash, and others.

This essentially results in the browser becoming the request broker rather than Facebook. An improvement on the model from the previous section, the browser also maintains the safety of the rest of the elements on the resulting page, so developers can create whatever experience they want inside this frame.

For applications whose developers want to invest minimal effort in moving their code from their site's logic to the Platform, the iframe still makes sense. In fact, Facebook continues to support the iframe model of full page generation. Although this solves the first product goal, incorporation into a social site, the second remains an open question. Even with the safety of the iframe-based request flow, these developers do not benefit from any new data beyond that exposed by the API service.

▶ FBML 与 HTML 的区别在于，FBML 在 Facebook 的范围内，可以被 Facebook 控制，它可以访问 Facebook 的隐私数据，经过处理后，返回满足调用者需求的业务数据，在保护了数据隐私的情况下，提供了不差于 HTML 的 Render 功能。

Applications on Facebook: FBML As Data-Driven Execution Markup

The straw-man solutions in the previous two sections each have their charm. The HTML solution takes the intuitive step of reframing applications themselves as web services, bringing contact back for display on the Facebook domain. The iframe model incorporates the benefit of running developer content in a separable (and safe) execution sandbox. The best solution would retain the application-as-service model and the safety and trust of the iframe, while enabling developers to use more social data.

The problem is that in order to provide the unique experience of their social application, developers must provide the data, logic, and display from their own stack. However, this output must be generated with user data that cannot leave the Facebook domain.

> The solution? Send back not HTML but specific markup that defines sufficient amounts of the application's logic and display, plus requests for protected data, and let Facebook render it entirely in its trusted server environment! This is the premise of FBML (Figure 6-5).

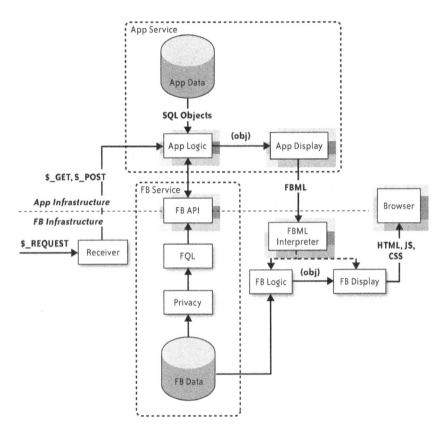

FIGURE 6-5. Applications as FBML services

In this flow, a request to *http://apps.facebook.com* is again transformed to an application request, and again, the application's stack consumes the Facebook data services. However, rather than returning HTML, the developer rewrites the application to return FBML, which incorporates many HTML elements but adds special Facebook-defined tags. When this request has returned its contents, Facebook's FBML interpreter transforms this markup into instances of its own data, execution, and display *while rendering the application page*. The user then receives a page composed of the usual web elements of Facebook pages, but infused with the data, logic, and feel of the application. No matter the FBML returned, FBML enables Facebook to enforce its notions of privacy and elements of good user experience technologically.

FBML is a specific instantiation of XML with many familiar tags from HTML, augmented with platform-specific tags for display on Facebook. FBML shares the high-level pattern of FQL: modifying a known standard (HTML, or in FQL's case, SQL) to defer execution and decisions to the Facebook Platform server. As shown in Figure 6-5, the FBML interpreter allows the developer himself to control the logic and display executed on the Facebook server through this FBML *data*. This is a great example of data at the center of execution: FBML is simply *declarative* execution rather than *imperative* flow (as is the case in C, PHP, etc.).

Now on to the specifics. FBML is an instantiation of XML, so it is composed of tags, attributes, and content. Tags fall into the following broad conceptual categories.

Direct HTML tags

If an FBML service returns the tag `<p/>`, Facebook would render this simply as `<p/>` on the output page. As the bedrock of web display, most HTML tags are supported, with the exception of a few that violate Facebook-level trust or design expectations.

So the FBML string `<h2>Hello, welcome to <i>Fetterman's books!</i></h2>` would be left essentially intact when rendered into HTML.

Data-display tags

Here's where some of the power of data comes in. Imagine that profile pictures were not transferred off-site. By specifying `<fb:profile-pic uid="8055">`, the developer can display more data to the Facebook user as part of their application, without requiring the user to fully trust this information to that developer.

For example:

```
<fb:profile-pic uid="8055" linked="true" />
```

translates to the FBML:

```
<a href="http://www.facebook.com/profile.php?id=8055"
 onclick="(new Image()).src = '/ajax/ct.php?app_id=...'">
  <img uid="8055" linked="true" src="..." alt="Dave Fetterman" title="Dave Fetterman" />
</a>
```

> **NOTE**
> The complicated `onclick` attribute is generated to restrict JavaScript while displayed in a Facebook page.

Note that even if information were protected, this content is never returned to the application stack, but only displayed to the user. Execution on the container side makes this data available to the viewer, without requiring that it pass through the hands of the application!

Data-execution tags

As an even better example of using hidden data, the user privacy restrictions accessible only through the internal `can_see` method (Example 6-3) are an important part of application experience, yet not accessible externally through data services. With the `<fb:-if-can-see>` tag and others like it, an application can specify a target user in the attributes such that the child elements are rendered only if the viewer can see that target user's specific content. Thus, the privacy values themselves are not exposed to the application, yet it allows an application to enforce privacy.

In this sense, FBML is a trusted declarative execution environment, in contrast to an imperative execution environment such as C or PHP. In a strict sense, FBML is not "Turing-complete" like these languages (for instance, no looping constructs are available). Much like HTML itself, no state can be saved during the execution except that implied by the tree traversal; for instance, <fb:tab-item> makes sense only within <fb:tabs>. However, FBML enables a great deal of the functionality that most developers want to provide to their users, through making data available to the user within the trusted system.

FBML effectively helps *define* the logic and display of the executing application, while still allowing the unique content of the application to begin on application servers.

Design-only tags

Facebook has been praised for its design ethic, so many developers choose to maintain the "look and feel" of Facebook through reusing Facebook design elements in some way. Often, they accomplish this by lifting JavaScript and CSS from *http://facebook.com*, but FBML brings with it something like a "design macro" library that meets the same need in a more controlled way.

For example, Facebook applies known CSS classes that render input such as <fb:tabs>...</fb:tabs> into a specific tab structure at the top of the developer's page. These design elements can incorporate execution semantics as well; for example, <fb:narrow>...</fb:narrow> will render its children's contents in FBML only if this execution shows up in the narrow column of a user's profile box.

Example 6-22 shows some FBML that uses design-only tags.

EXAMPLE 6-22. Example display-oriented FBML

```
<fb:tabs>
<fb:tab-item href="http://apps.facebook.com/fettermansbooks/mybooks.php"
 title='My Books' selected='true'/>
<fb:tab-item href="http://apps.facebook.com/fettermansbooks/recent.php"
 title='Recent Reviews' />
</fb:tabs>
```

This would be rendered as a set of visual tabs linking to the specified content, using Facebook's own HTML, CSS, and JavaScript packages.

Replacement HTML tags

HTML engenders little trust risk and no data exposure, so replacement tags in FBML are just for modifying or restricting a certain set of parameters, such as Flash autoplay. This is not strictly required by any display platform; they simply enforce that applications conform to the default display behavior of the container site. Still, these kinds of modifications become important as the ecosystem of applications grows to mirror the look and feel of the container site.

Consider this FBML example:

```
<fb:flv src=http://fettermansbooks.com/newtitles.flv height="400"
    width="400" title="New Releases">
```

This translates to quite a long string of JavaScript, rendering a video play module; this element is controlled by Facebook, intentionally disallowing such behaviors as autoplay.

"Functionality package" tags

Some Facebook FBML tags encompass entire suites of common Facebook application functionality. `<fb:friend-selector>` creates a type-ahead friend selector package common to many Facebook pages, incorporating Facebook data (friends, primary networks), CSS styling, and JavaScript for keypress actions. Tags such as this enable the container site to encourage certain design patterns and elements of commonality among applications, as well as enable developers to quickly implement the behavior they would like.

FBML: A small example

Recall the improvements we were able to make to our hypothetical external website with the introduction of the friends.get and users.getInfo APIs to the original *http://fettermansbooks .com* code. Now we'll show an example of how FBML can combine the social data, privacy business logic, and feel of a fully integrated application.

If we were able to obtain the reviews of a book using a database call book_get_all_reviews($isbn), we could combine friend data, privacy, and the "wall" feature to display reviews of the book using FBML on the container site through the code in Example 6-23.

EXAMPLE 6-23. Creating an application using FBML

```
// Wall-style social book reviews on Facebook
// FBML Tags used: <fb:profile-pic>, <fb:name>, <fb:if-can-see>, <fb:wall>

// from section 1.3
$facebook_friend_uids = $facebook_client->api_client->friends_get();
foreach($facebook_friend_uids as $facebook_friend) {
  if ($books_user_id = books_user_id_from_facebook_id($facebook_friend))
    $book_site_friends[] = $books_user_id;
}

// a hypothesized mapping, returning
// books_uid -> book_review object
$all_reviewers = get_all_book_reviews($isbn);

$friend_reviewers = array_intersect($book_site_friends, array_keys($all_reviewers));
```

```
echo 'Friends' reviews:<br/>';
echo '<fb:wall>';

// put friends up top.
foreach ($friend_reviewers as $book_uid => $review) {
  echo '<fb:wallpost uid="'.$book_uid.'">';
  echo '(' . $review['score'] . ')' . $review['commentary'];
  echo '</fb:wallpost>';
  unset($all_reviewers[$book_uid]); // don't include in nonfriends below.
}

echo 'Other reviews:<br/>';

// only nonfriends remain.
foreach ($all_reviewers as $book_uid => $review) {
  echo '<fb:if-can-see uid="'.$book_uid.'">'; // defaults to 'search' visibility
  echo '<fb:wallpost uid="'.$book_uid.'">';
  echo '(' . $review['score'] . ')' . $review['commentary'];
  echo '</fb:wallpost>';
  echo '</fb:if-can-see>';
}

echo '</fb:wall>';
```

Even though this takes the form of a service outputting FBML instead of a web call outputting HTML, the usual flow remains intact. Here, Facebook data enables the application to show more relevant book reviews (friends' reviews) before less relevant ones, and uses FBML to display the result using appropriate privacy logic and design elements on Facebook.

FBML Architecture

Transforming FBML provided by developers into the HTML shown on *http://facebook.com* requires a number of technologies and concepts working together: parsing the input string into a syntax tree, interpreting tags in this tree as internal method calls, applying the rules of FBML syntax, and maintaining the constraints of the container site. Like FQL, here we again focus primarily on the interaction of FBML with the platform's data, and detail only in broad strokes the other pieces of the technology puzzle. FBML handles a complex problem, and the full implementation details of FBML are quite voluminous—these include omitted topics such as FBML's error logging, the ability to pre-cache content for later rendering, signing the results of form submission for security, and so forth.

First, to the low-level issue of parsing FBML. In inheriting some of the roles of the browser, the Facebook platform also inherits some of its problems. For developer convenience, we do not require input to arrive as schema-verifiable or even well-formed XML—unclosed HTML tags, like <p> (as opposed to XHTML, like <p/>) break the assumption that the input could be parsed as true XML. Because of this, we need a way to first transform an input stream of FBML into a well-formed syntax tree with tags, attributes, and content.

▶ 我并不知道目前的Facebook应用是否还在使用FBML，然而对比同样是由Facebook推出的React中的FSX，虽然前者是更为特定的应用，而后者更为通用，但是在封装性、表现力以及简洁性方面，FSX显然更胜一筹。

For this we employ some open source from a browser codebase. This chapter takes this part of the process as a black box, so let's now assume that after receiving the FBML and sending it through this flow, we will have a tree-like structure called FBMLNode, which gives us the ability to query the tag, attribute key-value pairs, and raw content at any node in the generated syntax tree, and recursively query child elements.

Jumping to the highest level, note that FBML appears all over the Facebook site: application "canvas" pages, the content of News Feed stories, and the content of profile boxes, to name a few places. Each of these contexts or "flavors" of FBML defines constraints on the input; for instance, canvas pages allow iframes, whereas profile boxes do not. And naturally, because FBML maintains privacy of data in a way similar to the API, the execution context must include both the viewing user and the application ID of the application generating the content.

▶ Flavor是一个很好的隐喻。无论是架构层面，还是代码层面，引入恰当的隐喻可以起到意料之外的效果。

So before we actually engage with the payload of FBML, we start with the rules of our environment, encompassed in the FBMLFlavor class in Example 6-24.

EXAMPLE 6-24. The FBMLFlavor class

```
abstract class FBMLFlavor {

// constructor takes array containing user and application_id
  public function FBMLFlavor ($environment_array) { ... }
  public function check($category) {
    $method_name = 'allows_' . $category;
    if (method_exists($this,$method_name)) {
      $category_allowed = $this->$method_name();
    } else {
      $category_allowed = $this->_default();
    }
    if (!$category_allowed)
      throw new FBMLException('Forbidden tag category '.$category.' in this flavor.');
  }
  protected abstract function _default();
}
```

▶ check 是一个 template method，它与传统 template method模式的实现不同，它利用了PHP 的 method_exists() 方法，开放了对 category的检查。

The flow instantiates a child of this abstract flavor class that corresponds to the page or element rendering the FBML. Example 6-25 shows an example.

EXAMPLE 6-25. An instantiation of the FBMLFlavor class

```
class ProfileBoxFBMLFlavor extends FBMLFlavor {
  protected function _default() { return true; }
  public function allows_redirect() { return false; }
  public function allows_iframes() { return false; }
  public allows_visible_to() { return $this->_default(); }
 // ...
}
```

The flavor's design is simple: it contains the privacy context (user and application) and implements the check method, setting up the rules for the meaty logic contained in the FBMLImplementation class shown later. Much like the Platform API's implementation layer, the implementation class serves as the actual logic and data access portion of the service, with the rest of the code delivering access to these methods. Each Facebook-specific tag, such as `<fb:TAG-NAME>`, will have a corresponding implementation method fb_TAG_NAME (e.g., the class method fb_profile_pic will implement the logic for the `<fb:profile-pic>` tag). Each standard HTML tag has a corresponding handler as well, named tag_TAG_NAME. These HTML handlers often let the data go through untouched, but often, FBML needs to make checks and do transforms even on "normal" HTML elements.

Let's jump into the implementation of some of these tags, and then glue it all together. Each of these implementation methods accepts an FBMLNode returned from the FBML parser and returns some output HTML as a string. Here are example implementations for some direct HTML, data-display, and data-execution tags. Note that these listings use some functions not fully detailed here.

Implementing direct HTML tags in FBML

Example 6-26 contains the internal FBML implementation of the `<img>` tag. The image tag's implementation has some more logic, sometimes rewriting the image source URL to the URL of that image cached on Facebook's servers. This demonstrates the power of FBML: an application stack can return markup very similar to the HTML used to support its own site, yet Facebook can enforce the behavior required by the Platform through purely technical means.

▶ 标签的实现相当于一个细粒度的 Controller，控制对页面的 Render，生成符合要求的html。

EXAMPLE 6-26. Implementation of the fb:img tag

```
class FBMLImplementation {
  public function __construct($flavor) {... }


  // <img>: example of direct HTML tag (section 4.3.1)
  public function tag_img($node) {

    // images are not allowed in some FBML contexts -
    // for example, the titles of feed stories
    $this->_flavor->check('images');

    // strip of transform attribute key-value pairs according to
    // rules in FBML
    $safe_attrs = $this->_html_rewriter->node_get_safe_attrs($node);
    if (isset($safe_attrs['src'])) {
      // may here rewrite image source to one on a Facebook CDN
      $safe_attrs['src'] = $this->safe_image_url($safe_attrs['src']);
    }
    return $this->_html_rewriter->render_html_singleton_tag($node->
    get_tag_name(), $safe_attrs);
  }
}
```

Implementing data-display tags in FBML

Example 6-27 shows examples of using Facebook data through FBML. `<fb:profile-pic>` takes `uid`, `size`, and `title` attributes and combines these to produce output HTML based on internal data and according to Facebook's standard. In this case, the output is a profile picture with the specified user's name, linked to that user's profile page, shown only if that content is visible to the viewing user. This function lives within the `FBMLImplementation` class as well.

EXAMPLE 6-27. Implementation of the fb:profile-pic tag

```
// <fb:profile-pic>: example of data-display tag
public function fb_profile_pic($node) {
  // profile-pic is certainly disallowed if images are disallowed
  $this->check('images');

  $viewing_user = $this->get_env('user');
  $uid = $node->attr_int('uid', 0, true);
  if (!is_user_id($uid))
    throw new FBMLRenderException('Invalid uid for fb:profile_pic ('.$uid .')');

  $size = $node->attr('size', "thumb");
  $size = $this->validate_image_size($size);

  if (can_see($viewing_user, $uid, 'user', 'pic')) {
    // this wraps user_get_info, which consumes the user's 'pic' data field
    $img_src = get_profile_image_src($uid, $size);
  } else {
    return '';
  }
  $attrs['src'] = $img_src;
  if (!isset($attrs['title'])) {
    // we can include the user name information here too.
    // again, this function would wrap internal user_get_info
    $attrs['title'] = id_get_name($id);
  }

  return $this->_html_renderer->render_html_singleton_tag('img', $attrs);
}
```

Data-execution tags in FBML

The recursive nature of FBML parsing makes possible the `<fb:if-can-see>` tag, an example of FBML actually controlling *execution*, like an `if` statement in standard imperative control flow. Another method within the `FBMLImplementation` class, it is detailed in Example 6-28.

EXAMPLE 6-28. Implementation of the fb:if-can-see tag

```
// <fb:if-can-see>: example of data-execution tag
public function fb_if_can_see($node) {
  global $legal_what_values; // the legal attr values (profile, friends, wall, etc.)
  $uid = $node->attr_int('uid', 0, true);
  $what = $node->attr_raw('what', 'search'); // default is 'search' visibility
  if (!isset($legal_what_values[$what]))
```

```
    return ''; // unknown value? not visible

  $viewer = $this->get_env('user');
  $predicate = can_see($viewer, $uid, 'user', $what);
  return $this->render_if($node, $predicate); // handles the else case for us
}

// helper for the fb_if family of functions
protected function render_if($node, $predicate) {
  if ($predicate) {
    return $this->render_children($node);
  } else {
    return $this->render_else($node);
  }
}

protected function render_else($node) {
  $html = '';
  foreach ($node->get_children() as $child) {
    if ($child->get_tag_name() == 'fb:else') {
      $html .= $child->render_children($this);
    }
  }

  return $html;
}

public function fb_else($ignored_node) { return ''; }
```

If the can_see check passes for the specified viewer-object pair, the engine renders the children of the <fb:if-can-see> node recursively. Otherwise, the content below any optional <fb:else> children is rendered. Notice how fb_if_can_see directly accesses the <fb:else> children; if <fb:else> appears outside one of these "if-style" FBML tags, the tag and its children return no content at all. So FBML is not just a simple swap routine; it is aware of the structure of the document, and thus can incorporate elements of conditional flow.

Putting it all together

Each of the functions just discussed needs to be registered as a callback that is used while parsing the input FBML. At Facebook (and in the open source Platform implementation), this "black box" parser is written in C as an extension to PHP, and each of these callbacks lives in the PHP tree itself. To complete the high-level flow, we must declare these tags to the FBML parsing engine. As elsewhere, Example 6-29 is highly edited for simplicity.

EXAMPLE 6-29. The FBML main evaluation flow

```
// As input to this flow:
// $fbml_impl - the implementation instantiated above
// $fbml_from_callback - the raw FBML string created by the external application

// a list of "Direct HTML" tags
$html_special = $fbml_impl->get_special_html_tags();
```

```
  // a list of FBML-specific tags (<fb:FOO>)
$fbml_tags = $fbml_impl->get_all_fb_tag_names();

// attributes of all tags to rewrite specially
$rewrite_attrs = array('onfocus', 'onclick', /* ... */);

// this defines the tag groups passed to flavor's check() function
// (e.g. 'images', 'bold', 'flash', 'forms', etc.)
$fbml_schema = schema_get_schema();

// Send the constraints and callback method names along
// to the internal C FBML parser.
fbml_complex_expand_tag_list_11($fbml_tags, $fbml_attrs,
  $html_special,$rewrite_attrs, $fbml_schema);

$parse_tree = fbml_parse_opaque_11($fbml_from_callback);
$fbml_tree = new FBMLNode($parse_tree['root']);

$html = $fbml_tree->render_html($fbml_impl);
```

FBML augments browser parse technology with callbacks wrapping the data, execution, and display macros created and managed by Facebook. This simple idea allows full integration of applications, enabling use of data intentionally exposed through the API while maintaining the safety of the user experience. Almost a programming language in itself, FBML is data fully grown up: externally provided declarative execution safely controlling data, execution, and display on Facebook.

Supporting Functionality for the System

At this point, developer-created software is running on the Facebook services, incorporated as not just widgets but as full applications. Along the way, we've created a very different notion of a social web application. We started with the standard setup of isolated data, logic, and display of a typical web application, bereft of any social data except what users could be convinced to contribute. We've now fully progressed to an application consuming Facebook social data services while becoming *itself* an FBML service for full integration into the container site.

Facebook data has progressed a long way from the internal libraries discussed in the first section of this chapter. However, there are still a few important, common web scenarios and technologies that, up to this point, the Platform still does not support. In casting the application as a service returning FBML, instead of an HTML/CSS/JS endpoint consumed directly by a browser, we've stepped on the toes of some important assumptions about modern web applications. Let's see how the Facebook Platform has rectified some of these problems.

Platform Cookies

The new web architecture of applications cuts out some technologies built into the browser, upon which many web stacks rely. Perhaps most importantly, browser cookies used to store information about a user's interaction with the application stack are no longer available, since the consumer of the application's endpoint is not a browser but the Facebook Platform.

At first glance, sending cookies from the browser along with the request to the application stack might appear to be a good solution. However, the domain of these cookies is then "*http://facebook.com*", when, in fact, the cookie information pertains to the experience provided by the application domain.

The solution? Endow Facebook with the role of the browser, by duplicating this cookie functionality within Facebook's own stores. If an application's FBML service sends back headers attempting to set a browser cookie, Facebook simply stores this cookie information keyed on the (user, application_id) pair. Facebook then "recreates" these cookies as a browser would when sending subsequent requests to this application stack by this user.

This solution is simple and requires the developer to change very little of his assumptions when moving his HTML stack over to the FBML service role. Note that this information cannot be used when a user decides to navigate to an HTML stack that this application may provide. On the other hand, it can be useful to separate a user's application experience on Facebook from her experience on the application's HTML site.

FBJS

When the application stack is consumed as an FBML service rather than directly by the user's browser, Facebook has no opportunity to execute the browser-side script. Directly returning this developer content untouched (an insufficient solution, as presented at the beginning of the FBML section) could solve this, yet it violates the Facebook-imposed constraints on the display experience. For instance, Facebook does not want onload events shooting out pop-up windows when a user's profile page loads. However, restricting all JavaScript precludes much useful functionality, such as Ajax or dynamically manipulating page content without reloading.

Instead, FBML interprets the contents of developer-provided <script> trees and other page elements with these constraints in mind. On top of that, Facebook provides JavaScript libraries to make common scenarios easy yet controlled. Together, these modifications constitute Facebook's Platform JavaScript emulation suite, called FBJS, which makes applications dynamic yet safe by:

- Rewriting FBML attributes to enforce virtual document scope
- Deferring active script content until a user initiates action on the page or element
- Providing Facebook libraries to implement common script scenarios in a controlled way

Clearly, not all container sites implementing their own platforms need these modifications, but FBJS demonstrates the kinds of solutions needed to work around a new web architecture like this. We present only the solutions as general ideas here; much of FBJS involves incremental improvements incorporated into FBML and extensive proprietary JavaScript libraries.

First, JavaScript generally has access to the entire Document Object Model (DOM) tree of the document that contains it. Yet in a platform canvas page, Facebook includes many of its own elements, which developers are not allowed to change. The solution? Prefix developer-provided HTML elements and JavaScript symbols with the ID of the app itself (e.g., app1234567). In this way, attempting to call this disallowed `alert()` in developer JavaScript will call the undefined function `app1234567_alert`, and only portions of the document's HTML that the developer provided himself can be accessed by something such as JavaScript's `document.getElementById`.

As an example of the kinds of transforms FBJS needs to make on provided FBML (including `<script>` elements), we create a simple FBML page implementing AJAX functionality in Example 6-30.

EXAMPLE 6-30. An FBML page using FBJS

```
These links demonstrate the Ajax object:
<br /><a href="#" onclick="do_ajax(Ajax.RAW); return false;">AJAX Time!</a><br />
<div>
<span id="ajax1"></span>
</div>

<script>
function do_ajax(type) {
  var ajax = new Ajax(); // FBJS Ajax library.
  ajax.responseType = type;
  switch (type) {
  <!-- note FBJS's Ajax object also implements AJAX.JSON and AJAX.FBML, omitted
  for brevity -->
    case Ajax.RAW: ajax.ondone = function(data) {
      document.getElementById('ajax1').setTextValue(data);
    };
    break;
  };

 ajax.post('http://www.fettermansbooks.com/testajax.php?t='+type);

}
</script>
```

FBML with our FBJS modifications transforms this input to the HTML in Example 6-31. The NOTE comments in this example refer to each kind of transform required, and are not part of the actual output.

EXAMPLE 6-31. Example HTML and JavaScript output

```
<!-- NOTE 1 -->
<script type="text/javascript" src="http://static.ak.fbcdn.net/
.../js/fbml.js"></script>

<!-- Application's HTML -->
These links demonstrate the Ajax object:
<br>
<!-- NOTE 2 -->
<a href="#" onclick="fbjs_sandbox.instances.a1234567.bootstrap();
 return fbjs_dom.eventHandler.call(
[fbjs_dom.get_instance(this,1234567),function(a1234567_event) {
a1234567_do_ajax(a1234567_Ajax.RAW);
return false;
}
,1234567],new fbjs_event(event));return true">
AJAX Time!</a>
<br>

<div>

<span id="app1234567_ajax1" fbcontext="b7f9b437d9f7"></span><!-- NOTE 3 -->
</div>

<!-- Facebook-generated FBJS bootstrapping -->
<script type="text/javascript">
var app=new fbjs_sandbox(1234567);
app.validation_vars={ <!-- Omitted for clarity -->};
app.context='b7f9b437d9f7';
app.contextd=<!-- Omitted for clarity -->;
app.data={"user":8055,"installed":false,"loggedin":true};
app.bootstrap();
</script>

<!-- Application's script -->

<script type="text/javascript">
function a1234567_do_ajax(a1234567_type) { <!-- NOTE 3 -->

var a1234567_ajax = new a1234567_Ajax();<!-- NOTE 3 -->
 a1234567_ajax.responseType = a1234567_type;
 switch (a1234567_type) {
  case a1234567_Ajax.RAW:
a1234567_ajax.ondone = function(a1234567_data) {
a1234567_document.getElementById('ajax1').setTextValue(a1234567_data);
};
break;
};

<!-- NOTE 4 -->
a1234567_ajax.post('http://www.fettermansbooks.com/testajax.php?t='+a1234567_type);
}
</script>
```

The following are explanations of the NOTEs in the code:

NOTE 1

Facebook needs to include its own specialized JavaScript, including the definition of `fbjs_sandbox`, in order to render developer script.

NOTE 2

Remember the `$rewrite_attrs` element from the earlier FBML initialization flow? FBML rewrites attributes in this list to Facebook-specific functionality; this is really part of FBJS. So `onclick` here would activate other elements in the page that would be inactive until a user action took place.

NOTE 3

Notice how elements within both the HTML and the script are prefixed with the application ID of the application. This means a developer call to `alert()` would become a call to `app1234567_alert()`. If Facebook's backend JavaScript allowed this method in this context, this would be routed ultimately to `alert()`. If not, this would be an undefined call. Similarly, this prefixing effectively namespaces the DOM tree, so changes to parts of the document are limited to those parts defined by the developer. Similar sandboxing techniques allow developers to contribute limited-scope CSS as well.

NOTE 4

Facebook provides specialized JavaScript objects such as `Ajax` and `Dialog`, designed to enable (and often improve) common scenarios. For example, requests made through the `Ajax()` object are actually able to obtain FBML as results, so they are redirected through a proxy on the Facebook domain, where Facebook does online FBML-to-HTML transformation.

Enabling FBJS requires changes to FBML, specialized JavaScript, and server-side elements such as the AJAX proxy to work around the limitations of the application web architecture, but the results are powerful. Developers then enjoy most of the capabilities of JavaScript (and even improved capabilities, such as FBML-enabled AJAX), and the Platform enforces the application content to deliver the controlled experiences users expect on Facebook, through entirely technical means.

Service Improvement Summary

Solving some of the remaining problems created by our new conception of the social n-tier, we've again improved our service architecture with the newly added COOKIE and FBJS items in Figure 6-6.

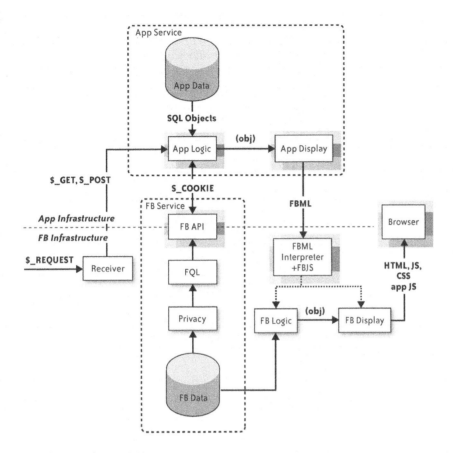

FIGURE 6-6. Facebook Platform services

With developers' social applications becoming more of an integrated service consumed by Facebook rather than an external site consumed by a browser, we've had to recreate or reengineer some of the functionality of that browser (through Platform Cookies, FBJS, etc.). These are two examples of the significant modifications required when trying to change or reinvent the idea of an "application." The Facebook Platform includes additional architectural tweaks along this line that are not detailed here, including the Data Store API and the browser-side web service client.

Summation

Facebook's user-contributed social information effectively motivates the utility of most any page on *http://facebook.com*. However, this data is so universal that some of its best uses appear when it is integrated with the stacks of outside developers' applications, made possible through data technologies such as Facebook Platform's web services, data query services, and FBML.

Starting from simple internal APIs that get a user's friends or profile information, the whole range of improvements we've detailed in this chapter show how to reconcile continually expanding methods of data access with the expectations of the container site, especially the requirements of data privacy and site experience integrity. Each new change to the data architecture presents new problems in the web architecture, which are resolved through even more powerful improvements to the data access pattern.

Though we've focused entirely on the potentials and constraints of applications built using Facebook's social data platform, new data services like these need not be limited to social information. As users contribute and consume more information that is useful across many container sites (data such as collections, reviews, location information, personal scheduling, collaboration, etc.), platform providers of all kinds can benefit from applying the ideas behind the unique data and web architecture of the Facebook Platform.

PART III

第 3 部分

Systems Architecture
系统架构

Chapter 7　Xen and the Beauty of Virtualization
第 7 章　XEN 与虚拟化之美

Chapter 8　Guardian: A Fault-Tolerant Operating System Environment
第 8 章　Guardian：一个容错操作系统环境

Chapter 9　JPC: An x86 PC Emulator in Pure Java
第 9 章　JPC：纯 Java 的 X86 PC 模拟器

Chapter 10　The Strength of Metacircular Virtual Machines: Jikes RVM
第 10 章　元循环虚拟机的力量：Jikes RVM

Principles and properties		Structures	
✓	Versatility	✓	Module
	Conceptual integrity	✓	Dependency
✓	Independently changeable		Process
	Automatic propagation		Data access
	Buildability		
✓	Growth accommodation		
	Entropy resistance		

CHAPTER SEVEN

Xen and the Beauty of Virtualization

Derek Murray
Keir Fraser

Introduction

XEN IS A VIRTUALIZATION PLATFORM THAT HAS GROWN FROM AN ACADEMIC research effort to become a major open source project. It enables its users to run several operating systems on a single physical machine, with particular emphasis on performance, isolation, and security.

The Xen project has had great impact in a variety of fields: from software to hardware, academic research to commercial development. A large part of its success is due to it being released as open source, under the GNU General Public License (GPL). However, the developers did not simply sit down one day and decide to write an open source hypervisor. It began as part of a larger—and even more ambitious—research project called *Xenoservers*. This project provided the motivation for developing Xen, so we'll use it here to explain the need for virtualization.

Making Xen open source not only made it available to a vast range of users, but also allowed it to enjoy a symbiotic relationship with other open source projects. The unique thing about Xen is that, when it was first released, it employed *paravirtualization* to run commodity operating systems such as Linux. Paravirtualization involves making changes to the operating systems that run on top of Xen, which both improves performance and simplifies Xen itself. However, paravirtualization only goes so far, and it is only with hardware support from processor vendors that Xen can run unmodified operating systems, such as Microsoft Windows. One of the new frontiers in processor development is the addition of new features to support virtual machines and remove some of the performance penalties.

The architecture of Xen is slowly evolving as new features are added and new hardware becomes available. However, the same basic structure has persisted from the original prototype through to the present version. In this chapter, we trace how Xen's architecture has matured from its early days as a research project, through three major versions, and to the present day.

Xenoservers

Work on Xen began at the University of Cambridge in April 2002. It was initially developed as part of the Xenoservers project, which aimed to create a "global distributed computing infrastructure."

Around the same time, *grid computing* was being advanced as the best way to make use of computing resources that are scattered throughout the world. The original grid proposal cast computer time as a utility, like electricity, which could be obtained from a grid—or network—of collaborating computers. However, subsequent implementations concentrated on *virtual organizations*: groups of companies and institutions that established possibly complicated relationships of trust, which are enforced by heavyweight public-key cryptography for authentication and authorization.

Xenoservers approached the problem from the opposite direction. Instead of forging trust relationships with service providers, the customer chooses a resource on the open market through a broker known as a *XenoCorp*. The XenoCorp stores a list of *xenoservers*—computers offered for lease by third parties—and matches customers with servers, collecting and passing on payment for the utility. Crucially, there is *mutual distrust* between the customer and the provider: the customer cannot harm the provider's machine, and the provider cannot tamper with the customer's job.

▶ 从另一个角度看，对输入参数的不信任也是设计的一部分。这里对distrust的判断，其实也体现了在架构设计中对风险的考虑。

TRUST

It might sound counterintuitive that *distrust* is a useful architectural feature. However, the main goal of security in this context is to prevent other individuals from accessing or interfering with your sensitive data. A *trusted* system, then, is one that is allowed access to your data. When distrust is built into the architecture, the number of trusted components is minimized, and this therefore provides security by default.

Enter virtualization. Instead of giving the customer an account on the server, the provider gives him a fresh *virtual machine* to use as he pleases. The customer then can run any operating system and any applications (see Figure 7-1). Virtualization software ensures that these are isolated from the rest of the machine (which may be leased out to more customers). The *hypervisor*, on which the virtual machines run, contains two main parts: a *reference monitor*, which makes sure that no virtual machine can access another virtual machine's resources (especially its data), and a *scheduler*, which ensures that each virtual machine gets a fair share of the CPU.

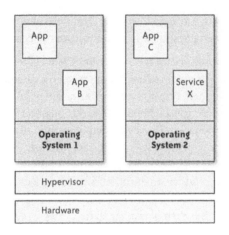

FIGURE 7-1. Virtual machine architecture

COULDN'T YOU JUST USE AN OPERATING SYSTEM?

Time-sharing operating systems have existed since the early 1960s, and enable several mutually distrusting users to run processes at the same time. Would it not be sufficient to give each user an account on, say, a Unix-based machine?

This would certainly let users share the computational resources. However, it is unsatisfactory because the user has much less flexibility and performance isolation.

In terms of flexibility, the user could only run software that is compatible with the machine's operating system; there is no way to run a different operating system, or to change the operating system. Indeed, it would be impossible for the user (without administrative support) to install software that requires root permissions.

As for performance isolation, it is difficult for an operating system kernel, which is an extremely complicated piece of software, to account for all the resources that are being used by a particular user. An example is the *fork bomb*, where a user starts an exponentially growing number of processes. This rapidly consumes all of the processor's resources and causes denial of service to other users. Hence multiuser systems typically require some amount of trust or etiquette between users, so that attacks like these are not carried out.

In Figure 7-1 we see two virtual machines running on top of a hypervisor. The first virtual machine runs Operating System 1 (e.g., Microsoft Windows) and two applications; the second virtual machine runs Operating System 2 (e.g., Linux), an application, and a service.

It turns out that virtualization is useful for other things as well. For example, in many data centers, a dedicated server is used for each application, such as a database or a web server, but each server uses only a fraction of its processor, memory, and bandwidth resources. Obviously, it would be better if these were consolidated onto fewer physical machines, saving power, space, and component maintenance. However, simply running the applications on the same operating system can give poor results. The various applications, when run together, might cause unpredictable poor performance. Worse, there is a risk of correlated failures, which occur when one application crashes and causes the others to crash as well. By placing each application in a virtual machine and then running them on top of a hypervisor, the hypervisor can protect the applications and ensure that each gets a fair share of the server's resources.

The idea of using virtualization for utility computing has gained currency in recent years. One of the best-known utility computing services is Amazon's EC2, which allows customers to create virtual machines that run on servers in Amazon's data centers. The customer is then charged for the processor time and network bandwidth that his virtual machine uses. In fact, the servers run Xen as their virtualization software, making it even closer to the Xenoservers vision (although it admits only a single service provider).

Virtualization has also had an influence on grid computing. Globus, the de facto standard middleware for grid computing, now supports *virtual workspaces*, which marry virtual machines with existing grid security and resource management protocols. An added benefit is that a virtual workspace—like any virtual machine—can be *migrated* to another physical location if conditions change.

The key advantage of virtual machines in the Xenoservers model is that they can be used to run popular commodity operating systems and existing applications. In practice, this means running on the dominant x86 architecture, which presents the hypervisor developer with several challenges.

The Challenges of Virtualization

At a high level, operating system virtualization is used to multiplex several virtual machines onto a single physical machine. The virtual machines run operating systems; the physical machine can run operating systems. So what is the difference between a virtual machine and the physical machine?

Hardware is the most obvious difference. On a physical machine, the operating system has direct control of all attached hardware: network cards, hard drives, the graphics card, the mouse and keyboard. However, the virtual machines cannot have direct access to this hardware, or else they will undermine the isolation between each virtual machine. For example, a virtual machine (or VM) might not want other VMs to see what it stores in its secondary storage, or to read its network packets. Moreover, it would be difficult to ensure fair use in this scheme. You could have one device of each type for each virtual machine, but this would negate the cost and power savings of virtualization. The solution is to give each virtual machine a set of *virtual hardware*, which provides the same functionality as real hardware, but which is then multiplexed on the physical devices.

A more subtle difference arises when an operating system runs in a virtual machine. Traditionally, the operating system kernel is the most privileged software running on a computer, which allows it to execute certain instructions that user programs cannot. Under virtualization, the hypervisor is most privileged, and operating system kernels run at a lower privilege level. If the operating system now tries to execute these instructions, they will fail, but the way in which they fail is crucial. If they cause an error, which the hypervisor then traps, the hypervisor can correctly emulate the instruction and return control to the virtual machine. On the x86, however, there are some instructions that behave differently in lower privilege levels—for example, by failing silently without a trap to the hypervisor. This is bad news for virtualization because it stops operating systems from working properly in virtual machines. Obviously it is necessary to change these instructions, and the prevailing technique (at least, before Xen) was to scan the operating system code at runtime, looking for certain instructions and replacing them with code that calls the hypervisor directly.

Indeed, before Xen, most virtualization software aimed to make virtual hardware look exactly like physical hardware. So the virtual devices behaved like physical devices, emulating the same protocols, while the code rewriting ensured that the operating system would run without modifications. Although this gives perfect compatibility, it comes at a heavy cost in performance. When Xen was released it showed that by abandoning perfect compatibility, performance increased dramatically.

Paravirtualization

The idea of *paravirtualization* is to remove all the features of an architecture (such as the x86) that are difficult or expensive to virtualize, and to replace these with *paravirtual* operations that communicate directly with the virtualization layer. The technique was first used in Denali, a virtual machine monitor that hosts the specially written Ilwaco guest operating system. Xen went one step further by running paravirtualized versions of commodity operating systems.*

Paravirtualizing an operating system involves rewriting all of its code that is incompatible with the paravirtualized architecture. Performance improves because the changes are made in advance, by developers, rather than at runtime. To demonstrate the power of paravirtualization, the Xen team first required an operating system that they could change. Fortunately, Linux was available, open source, and widely used. Only 2,995 lines in the Linux kernel were modified or added to make it run on Xen: this represents less than 2% of the x86 Linux codebase. With paravirtualization (as with virtualization), all of the existing user applications can continue to be used without modification, so the overall modifications are not too invasive.

To achieve paravirtualization, you must either write the operating system yourself (the Denali approach), modify an existing open source operating system (such as Linux or BSD), or convince the developers of a proprietary operating system that paravirtualizing their code is worthwhile. The original research version of Xen achieved impressive, near-native performance on a number of benchmarks running on the modified version of Linux. This performance, and the fact that Xen was released as open source software, has brought us to a point where proprietary operating system developers are paravirtualizing parts of their code, so that they will run more efficiently on hypervisors such as Xen. Even more encouragingly, the paravirtual operations developed for Xen and other hypervisors have been standardized in the latest version of the Linux kernel. By incorporating Xen (and other hypervisor) support in the standard kernel, the uptake of virtualization becomes even easier.

How does paravirtualization work? The full details are too involved to cover here, but the "Further Reading" section at the end of this chapter includes papers that cover the techniques in depth. Here, we'll look at two examples of paravirtualization: for virtual memory and for virtual devices.

The first step in paravirtualizing an operating system is to make it aware that it is not the most privileged software running on the computer; that distinction is awarded to the hypervisor. Most processors have at least two modes: *supervisor* and *user* mode. Normally, the operating system kernel would run in supervisor mode, but this is reserved for Xen, so it must be modified to run in user mode.† However, when running in user mode, several operations are illegal.

---

* Of course, it could be argued that VM/370—IBM's operating system from the 1960s and the progenitor of virtualization—was the first paravirtualized operating system. However, since IBM designed the instruction set, operating system, and virtual machine monitor, this approach faced different challenges to modern paravirtualization.

† The x86 architecture has four privilege levels, or *rings*, with 0 being the most privileged and 3 the least privileged. In its 32-bit incarnation, Xen ran in ring 0, paravirtualized kernels ran in ring 1, and user applications ran, as normal, in ring 3. However, on the 64-bit version, paravirtualized kernels run in ring 3, due to a difference in the memory segmentation hardware.

This is crucial for protecting processes from one another in a regular operating system. Therefore the kernel must ask the hypervisor to carry out these operations on its behalf, using a mechanism called a *hypercall*. A hypercall is similar to a system call (from a user process to the kernel), except that it is used for communication between a kernel and the hypervisor, and it typically implements lower-level operations.

Virtual memory is used to ensure that processes cannot interfere with the data or code of other processes. Each process is given a virtual *address space*, which ensures that that process can access only its allocated memory. The kernel is responsible for creating the virtual address space, by maintaining *page tables*, which map virtual addresses to the physical addresses that identify the actual location of data on the memory chips. When it is running in a virtual machine, the kernel does not have carte blanche to manage these tables, as it could conceivably make a mapping to memory that belongs to another virtual machine. Therefore, Xen must validate all updates to the page tables, and the kernel must inform the hypervisor when it wants to change any page table. This could be very inefficient if the hypervisor were involved in every page table update (for example, when a new process starts and its page tables are first built). However, it turns out that these cases are relatively rare, and Xen can amortize the cost of going to the hypervisor by batching the update requests or "unhooking" the page table while it is being updated.

Look at Figure 7-2. Each virtual machine has a share of the total physical memory.[‡] However, it might not be contiguous, and, in most cases, it will not start at physical address zero. Therefore, each virtual machine kernel deals with two types of addresses: physical (or *machine*) and *pseudophysical*. The physical addresses correspond to the actual location of data on the memory chips, whereas the pseudophysical addresses provide the virtual machine with the illusion of a contiguous physical address space that starts at zero. Pseudophysical addresses may be useful for certain algorithms and subsystems that rely on this assumption and would otherwise need to be paravirtualized.

To be of any practical use, it must be possible to interact with a virtual machine. At a bare minimum, the virtual machine needs a disk (more properly known as a *block device*), and a network card.[§] Since most operating systems include support for at least one block device and network card, it might seem tempting for the hypervisor to emulate these devices so that the original drivers could be used. However, the software implementation would struggle to emulate the performance of the real devices, and the emulated device models may have to go through contortions (such as implementing hardware protocols) that are unnecessary and inefficient when providing a device's function in software.

[‡] Note that Xen does not support overcommitting physical memory, so there is no swapping of virtual machines. However, the memory footprint of a virtual machine can be altered using a process called *ballooning*.

[§] You might think that a mouse, keyboard, and video output would be necessary for interactivity, but these can be provided by a remote desktop client such as VNC. Nevertheless, recent versions of Xen have included support for these virtual devices.

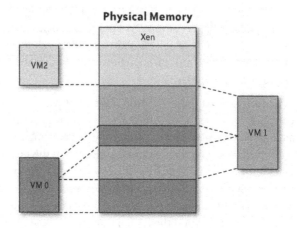

FIGURE 7-2. Virtual machine memory layout

HOW ELSE COULD YOU DO IT?

The standard means of virtualizing virtual memory (when you cannot change the operating system) is to use *shadow page tables*. With these, the guest deals with pseudophysical addresses (i.e., contiguous and beginning at 0) in place of physical addresses.

The guest maintains its own page tables against this address space. However, they cannot be used by the hardware, because they don't correspond to real physical addresses. Therefore, the hypervisor monitors updates to these guest page tables, and uses them to construct a shadow page table, which translates virtual addresses to real physical addresses.

This method clearly incurs some overhead, but it is necessary when you cannot modify the operating system. Xen uses a variant of this method for hardware virtualized guests, as described later in this chapter.

Since Xen is not constrained by having to support unmodified operating systems, it is free to introduce virtual block and network drivers. Both operate in a similar manner: they comprise a frontend driver in the guest virtual machine and a backend driver in the virtualization software. The two devices communicate using a ring buffer, which is a high-performance mechanism for transferring large volumes of data between virtual machines. This results in a flexible layered architecture (Figure 7-3): the frontend implements the operating system's network or block device interface so that it appears to the operating system as a regular hardware device, and the backend connects the virtual device to real hardware. A virtual block device might be connected to a file containing a disk image or a real disk partition; a virtual network device might be attached to a software network bridge, which is itself attached to a real network card. The ring buffer abstraction ensures that the frontend and backend are totally decoupled. One backend can support frontends from Linux, BSD, or Windows, whereas the same frontend can be used with various backends, so features such as copy-on-write, encryption, and compression can be added transparently to the guest. Like the Internet Protocol, the Xen split device model can operate on a vast array of hardware, and it supports a multitude of higher-level clients, as shown in Figure 7-3.

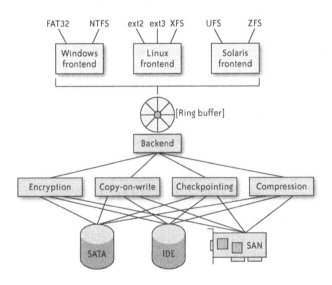

▶ Ring Buffer 就像是一个桥（Bridge 模式），通过引入抽象的接口隔离了前端，使其可以更好地扩展支持各种操作系统的网络（Network）或块设备（Block Device）。

FIGURE 7-3. Hourglass architecture of a split block device

Paravirtualization encompasses far more than these examples. For example, the meaning of time changes when an operating system can be switched out from the CPU, and Xen introduces a virtual time concept to ensure that operating systems still behave as expected. There are many more virtual devices, and the paravirtualization of memory makes use of many more optimizations to ensure efficient performance. For more details on these, see the "Further Reading" section at the end of the chapter.

The Changing Shape of Xen

The traditional representation of a Xen-based system shows several virtual machines (known in Xen as *domains*) sitting on top of the hypervisor, which itself sits directly on the hardware (Figure 7-4). When it boots up, the hypervisor launches a special domain, known as *domain zero*. Domain zero has special privileges that allow it to manage the rest of the system, and it is analogous to a root or administrator process in a regular operating system. Figure 7-4 shows a typical Xen-based system, with domain zero and several *guest domains* (known as *DomUs* in Xen jargon) running on top of the Xen hypervisor.

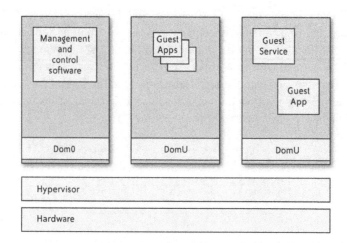

FIGURE 7-4. Xen system architecture

HOSTED VIRTUALIZATION

Xen is an example of native virtualization (also known as Type 1 virtualization). The alternative approach is to run a hypervisor on top of a *host* operating system. In this case, each virtual machine effectively becomes a process in the host operating system. The host operating system is responsible for the management functions that domain zero performs on Xen. The hosted hypervisor and management software are like a regular application, which sits on top of (and might plug into) a commodity operating system; see Figure 7-5.

Hosted hypervisors are commonly used in the "workstation" versions of other virtualization products, such as VMWare Workstation, Parallels Workstation, and Microsoft Virtual PC. The main advantage of this approach is that installing a hosted hypervisor is as simple as installing a new application, whereas installing a native hypervisor such as Xen is more akin to installing a new operating system. Therefore hosted virtualization is better suited for nonexpert users.

On the other hand, the advantage of a native hypervisor is that it can achieve better performance, because the native hypervisor is a far thinner layer of software than the combined host operating system and hypervisor. Hosted virtual machines are scheduled at the mercy of the host operating system, which can lead to performance degradation if other applications are running alongside the hypervisor. By contrast, because domain zero is scheduled like a regular virtual machine, applications running there do not have an impact on the performance of the other virtual machines.

Hosted virtual machines are typically used for desktop virtualization: they allow a user running, say, Mac OS X to run Linux in a window on her desktop. This is useful for running applications that are not available for the host operating system, and the performance hit is less noticeable when using interactive applications. Native virtualization is more suited to a server setting, where both raw performance and predictability are critical.

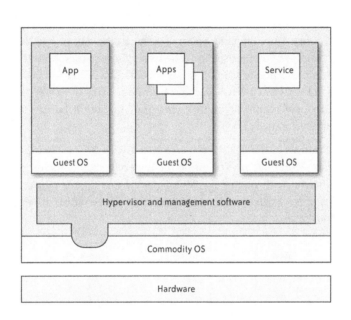

FIGURE 7-5. *Hosted virtualization system architecture*

When designing the Xen architecture, a primary concern was separating policy from mechanism wherever this was feasible. The hypervisor was designed to be a thin layer that managed the low-level hardware, acting as the reference monitor and scheduler and multiplexing access to hardware devices. However, since the hypervisor runs at the highest privilege level (and a bug here could compromise the whole system), the higher-level management is delegated to domain zero.

For example, when creating a new virtual machine, the bulk of the work is done in domain zero. From the hypervisor's point of view, a new domain is allocated, along with a portion of physical memory, some of that memory is mapped (in order to load the operating system), and the domain is unpaused. Domain zero takes care of admission control, setting up virtual devices, and building the memory image for the new domain. This split was particularly useful in the development process, as it is much easier to debug the management software in domain zero than the hypervisor. Moreover, it allows support for different operating systems to be

▶ 注意 Xen 1.0 与 Xen 2.0 之间的演化。对比图7-6，虽然Xen 1.0的架构显得更统一，且将Dom0 与 DomU 同等对待，引入了共同的抽象，却在性能上产生了不必要的损耗；Xen 2.0通过将Native drivers 移到Dom0，并与Hypervisor分离，使Hypervisor得到了简化。同时，各个Domain 之间的职责也变得更加清晰，例如Dom0就可以视为driver domain。当然，正如后文所说，如果将过多的功能都搬移到Dom0，则可能使它成为整个系统的单一故障点。

事实上，2.0版本的设计将native device driver 从 hypervisor 中剥离出去，是符合职责分离原则的，但带来的结果是导致Dom0变得更复杂。在后文介绍的3.0版本中，设计者意识到了这个问题，又进一步对Dom0进行了分解。观察Xen 几个版本的架构演化，可以很好地理解Evolutionary Design 的思想。

added in domain zero rather than in the hypervisor, where additional complexity is generally undesirable.

Earlier we noted how Xen benefited from the availability of an open source operating system, which provided a testbed for paravirtualization. A second benefit of using Linux is its vast range of support for different hardware devices. Xen is able to support almost any device for which a Linux driver exists, as it reuses the Linux driver code. Xen has always reused Linux drivers in order to support a variety of hardware. However, between versions 1.0 and 2.0, the nature of this reuse changed significantly.

In Xen 1.0, all virtual machines (including domain zero) accessed hardware through the virtual devices, as described in the previous section. The hypervisor was responsible for multiplexing these accesses onto real hardware, and therefore it contained ported versions of the Linux hardware drivers and the virtual driver backends. Although this simplified the virtual machines, it placed a lot of complexity in the hypervisor and put the burden of supporting new drivers on the Xen development team.

Figure 7-6 shows changes to the device architecture between Xen 1.0 and 2.0. In version 1.0, the virtual backends were implemented inside the hypervisor: all domains, including domain zero, accessed the hardware through these devices. In version 2.0, the hypervisor was slimmed down, and domain zero was given access to the hardware with native drivers. Therefore, the backend drivers moved to domain zero.

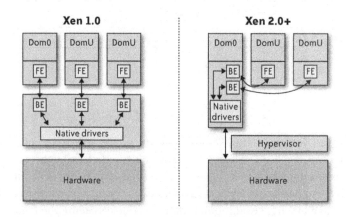

FIGURE 7-6. Changes to the device architecture between Xen 1.0 and 2.0

In the development of Xen 2.0, the device architecture was completely redesigned: the native device drivers and virtual backends were moved out of the hypervisor and into domain zero.‖ The frontend and backend drivers now communicate using *device channels*, which

---

‖ In fact, the architecture allows any authorized virtual machine to access the hardware, and therefore act as a *driver domain*.

enable efficient and secure communication between domains. Thanks to device channels, Xen's virtual devices achieve near-native performance. Their performance rests on two design principles: copyless transfer and asynchronous notification.

Look at Figure 7-7. This diagram shows how a split device is used. The guest provides the frontend driver with a page of memory, either containing data to be written or to hold data that is read in (1). The frontend driver places a request in the next available slot in the shared ring-buffer, which contains a reference to the provided page (2), and tells the hypervisor to notify the driver domain that a request is pending (3). The backend wakes up and maps the provided page into its address space (4) so that the hardware can interact with it using DMA (5). Finally, the backend notifies the frontend that the request has completed (6), and the frontend notifies the guest application (7).

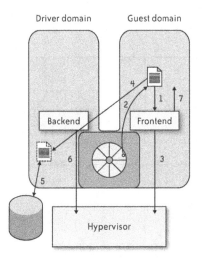

FIGURE 7-7. Anatomy of a split device

Copying data using the CPU is expensive, which is why techniques such as *Direct Memory Access* (DMA) have been developed to transfer device data directly to and from memory without CPU involvement. However, when the data has to move between address spaces, Xen must take special measures to avoid the copy. Xen supports a shared memory mechanism called *grant tables*, whereby each virtual machine maintains a table that defines which of its pages can be accessed by other virtual machines. An index in this table is called a *grant reference*, which, when given to another virtual machine, acts as a *capability*. The hypervisor ensures that only the intended recipient can map the grant reference, which in turn maintains memory isolation. The device channel itself is used to send grant references, which are then used to map buffers for sending or receiving data.

When a new request or response is made, the sender must notify the receiver. This would traditionally use a synchronous notification—akin to a function call—whereby the sender waits until it knows that the notification was received. As Figure 7-8 shows, this mode of operation leads to poor performance, especially when only a single processor is available. Xen instead uses *event channels* to send asynchronous notifications. Event channels implement virtual interrupts, but a virtual interrupt is serviced only when the target domain is next scheduled. Therefore, the requestor can generate multiple requests, raising the event channel each time, before the target domain is scheduled to act upon them. Then, when the target domain is scheduled, it can process several requests and send responses, again asynchronously.

Look at Figure 7-8. With synchronous notification, the frontend has to wait for the backend to complete its work before it can make the next request. This means waiting for the backend domain to be scheduled, and then for the frontend domain to be rescheduled. By contrast, with asynchronous notification the frontend can send as many requests as possible while it is scheduled, and the backend can send as many responses as possible. This leads to much-improved throughput.

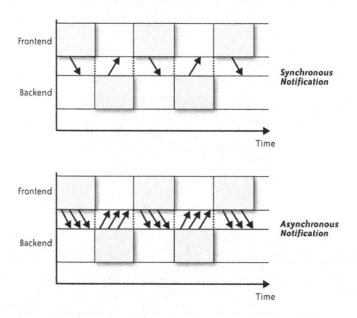

FIGURE 7-8. The advantages of asynchronous notification

Of course, if you move too much functionality into domain zero, it becomes a single point of failure. This is especially true of device failures, which can bring down the whole operating system (and with it, the entire virtualized system). Hence Xen allows for *driver domains*, to which domain zero can delegate the control of one or more devices. These are simply implemented by placing the backend driver in the driver domain and granting some I/O privileges to the domain. Then, if a driver should fail, the failure would be isolated within the driver domain, which can be restarted without harming the system or the client domain.

This model has been applied to other parts of domain zero. The latest versions of Xen include *stub domains*, which provide device support for "hardware virtualized" domains (described in the following section). Moving this code into isolated domains allows better performance isolation, improves robustness, and—somewhat surprisingly—improves raw performance. As development continues, more features may be moved out of domain zero, especially where doing so might improve security.

Changing Hardware, Changing Xen

Up to this point, our discussion has concentrated on paravirtualization. However, between Xen versions 2.0 and 3.0, Intel and AMD introduced distinct but similar support in their processors for *hardware virtual machines*. It became possible to run unmodified operating systems, including Microsoft Windows or native Linux, in virtual machines. So did this spell the end for paravirtualization?

First of all, let's look at how hardware virtual machines are implemented. Both Intel and AMD introduced a new mode (*nonroot mode* on Intel and *guest mode* on AMD) in which attempting to execute a privileged operation, even at the highest (virtual) privilege level, generates an exception that notifies the hypervisor. Therefore it is no longer necessary to scan the code and replace these instructions (either at runtime or in advance through paravirtualization). The hypervisor can use shadow page tables to provide the virtual machine with an illusion of contiguous memory, and it can trap I/O operations in order to emulate physical devices.

Xen added support for hardware virtual machines in version 3.0. The transition was aided greatly by open source development. Since Xen is an open source project, it was possible for developers from Intel and AMD to contribute low-level code that supports the new processors. Furthermore, thanks to its GPL status, Xen could incorporate code from other open source projects. For example, the new hardware virtual machines required an emulated BIOS and emulated hardware devices; implementing either of these would require a huge development effort. Fortunately, Xen could call on the open source BIOS from the Bochs project and emulated devices from QEMU.

EMULATION VERSUS VIRTUALIZATION

The latest version of Xen includes code from Bochs and QEMU, which are both *emulators*. What is the difference between emulation and virtualization, and how can the two combine?

Bochs provides an open source implementation, in software, of the x86 family of processors, as well as the supporting hardware. QEMU emulates several architectures, including the x86. Both can be used to run unmodified x86 operating systems and applications. Moreover, because they include a full implementation of the hardware—including the CPU—they can run on hardware that uses an incompatible instruction set.

Virtualized and emulated systems differ in how each instruction is executed. In a virtualized system, applications and most of the operating system run directly on the processor, whereas in an emulated system, the emulator must simulate or translate each instruction in order to execute it. Therefore, an emulator introduces more overhead than a virtual machine monitor for the same platform.[#]

However, even though they use parts of Bochs and QEMU, Xen's hardware virtual machines are virtualized and not emulated. The Bochs code provides the BIOS, which supports the boot process, and QEMU provides emulated drivers for a range of common devices. However, these pieces of code are only invoked at startup and when an I/O operation is attempted. The majority of other instructions run directly on the CPU.

At this point, you might wonder what happened to the much-vaunted advantages of paravirtualization. Surely all these emulated devices, shadow page tables, and additional exceptions would lead to poor performance? It's often true that a naïvely hardware-virtualized operating system performs worse than a paravirtualized operating system, but there are two mitigating factors.

First, the processor vendors are continually developing new features that optimize virtualization. Just as a *memory management unit* (MMU) lets programmers deal with virtual rather than physical addresses, an *IOMMU* does the same for input and output devices. An IOMMU can be used to give a virtual machine (whether hardware-virtualized or paravirtualized) safe, direct access to a piece of hardware (Figure 7-9). The normal problem with giving a virtual machine direct access to hardware is that many devices can perform DMA, and therefore without an IOMMU, it can read or overwrite other virtual machines' memory. The IOMMU can be used to ensure that while a particular virtual machine is in control, only memory belonging to that virtual machine is available for DMA.

Figure 7-9 illustrates a simplified DMA request from a virtual machine (DomIO) using an IOMMU. The hardware driver uses pseudophysical (virtual machine–specific) addresses when communicating with the device (1). The device makes DMA requests using these addresses (2), and the IOMMU (using I/O page tables, which are configured by the hypervisor) converts these to use physical addresses (3). The IOMMU also stops any attempts by the virtual machine to access memory that it does not own.

[#] KQEMU is a Linux kernel module that enables user-mode code—and some kernel-mode code—to run directly on the CPU. Where the host and target platforms are the same, this provides a huge speed-up. The result is a hybrid of emulation and virtualization.

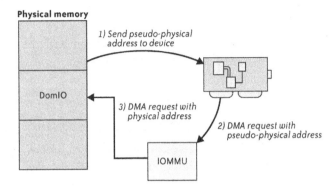

FIGURE 7-9. *Direct device access using an IOMMU*

Enhancing the memory management hardware can also remove the need for shadow page tables.* Both AMD and Intel have technology (respectively, Rapid Virtualization Indexing and Enhanced Page Tables) which perform the translation between pseudophysical addresses and physical addresses. Therefore there is no need for the hypervisor to create shadow page tables, as the whole translation occurs in hardware.

Of course, a far cheaper solution is to take the lessons learned from paravirtualization and apply them to unmodified guest operating systems. Although it is not possible to change core parts of the operating system, we can add device drivers, and moreover, Xen can modify the virtual hardware on which the operating system runs. To this end, the emulated hardware provides a *Xen platform device*, which appears as a PCI device to unmodified guest operating systems and provides access to the virtual platform. It is then possible to write frontend devices for the unmodified operating systems, which operate in the same way as frontends in paravirtualized operating systems. By doing this, we achieve I/O performance in hardware virtual machines that is comparable to the paravirtualized case.

When we introduced paravirtualization earlier in this chapter, we said that the only ways to get a commodity operating system running as a paravirtualized guest would be by doing it ourselves or by convincing the developers of a proprietary operating system that they should do it. As a testament to the success of paravirtualization, Microsoft has included *enlightenments* in Windows Server 2008, which improve the performance of memory management when running in a virtual machine. These enlightenments are equivalent to paravirtualized operations, as they rely on hypercalls to inform the hypervisor of the current operation.

---

* It should be noted that Xen's shadow page table implementation is highly optimized, and achieves competitive performance, but still has some overhead when compared with paravirtualized page tables.

Lessons Learned

Looking back, there are two main lessons that can be drawn from Xen: the importance of paravirtualization, and the benefits of open source development.

Paravirtualization

Foremost is the success of paravirtualization. A famous quote reminds us:

> **Any problem in computer science can be solved with another layer of indirection.** *But that usually will create another problem.*
>
> —David Wheeler

Virtualization is simply a form of indirection, and even though modern computers have hardware support for virtualization, naïve reliance on this support leads to poor performance. The same problems arise when you make naïve use of any type of virtualization.

For example, virtual memory uses a hard disk to provide the illusion of a vast amount of available memory. However, if you write a program that tries to use all of it as if it were real, physical memory, the performance will be atrocious. In this case you could imagine "paravirtualizing" that program to make it aware of the physical limits, changing the algorithms and data structures used to make it run efficiently in combination with the virtual memory system.

In the context of operating systems, Xen has shown that paravirtualization—whether it be adding a virtual driver, changing the operating system wholesale, or adding enlightenments to improve performance in select areas—is an important technique for improving performance when running in a virtual environment.

Open Source Development

Perhaps the boldest decision taken during Xen's development was choosing to make it available as open source software when other hypervisors were available only as proprietary software.

This decision has definitely benefited Xen because of the sheer amount of software that it has been able to harness: from the Linux kernel and the QEMU machine emulator, down to the tiny program that draws the Xen logo at boot time.[†] Without this software, the Xen project would have involved a huge amount of reimplementation. By including software from these other projects, Xen benefits when the software is updated, and the other projects benefit from patches submitted by Xen developers.

Xen, which started out as the part-time project of a single research student at the University of Cambridge, has grown to include over 100 contributors from around the world. Some of

---

† Figlet: *http://www.figlet.org*

the largest contributions have come from Intel and AMD, who provided much of the code to support hardware virtual machines. This enabled Xen to be one of the first hypervisors to support these processor extensions.

What's more, because Xen is freely available, several other projects have adopted it. Major Linux distributions such as Debian, Red Hat, SUSE, and Ubuntu now include Xen packages and have contributed code back into the project, along with useful tools for using Xen. Some contributors have taken on the effort of porting Xen to other architectures and porting other operating systems to run directly on the hypervisor. Xen has been used to run paravirtualized OpenSolaris, FreeBSD, and NetBSD, among others. Xen now runs on the Itanium architecture, and work is underway to port it to the ARM processor. The latter is particularly exciting because it will enable Xen to run on "nontraditional" devices, such as mobile phones.

As we look to the future, some of the most interesting uses of Xen are in the research community. Xen appeared at the Symposium on Operating Systems Principles (SOSP) in 2003, and has formed the basis of a variety of research, both within and outside of its original research group. One of the earliest papers written about Xen was from Clarkson University, where a group of researchers repeated the results in the SOSP paper. The authors remarked that open source software improves computer science because it enables repeated research and, in turn, strengthens any claims made about performance or other characteristics. More recent research work has led directly to interesting new features in Xen. One particular example is *live migration*, which enables a virtual machine to be moved between physical computers with only a negligible period of downtime. This was detailed in a paper in 2005, and was added to Xen in version 2.0.

Further Reading

This chapter could only scratch the surface of the Xen project, and the relevant research papers are the best source of further details.

These first two papers describe the architecture of Xen 1.0 and 2.0, respectively:

> Barham, Paul, et al. "Xen and the art of virtualization," *Proceedings of the 19th ACM Symposium on Operating System Principles,* October, 2003.
>
> Fraser, Keir, et al. "Safe hardware access with the Xen virtual machine monitor," *Proceedings of the 1st OASIS Workshop*, October, 2004.

The following papers describe some of the new chipset and processor technology that has been developed to aid virtualization:

> Ben-Yehuda, Muli, et al. "Using IOMMUs for virtualization in Linux and Xen," *Proceedings of the 2006 Ottawa Linux Symposium,* July, 2006.
>
> Dong, Yaozu, et al. "Extending Xen with Intel virtualization technology," *Intel® Technology Journal,* August, 2006.

Finally, Xen is under active development and continually evolving. The best way to keep abreast of new developments is to download the source code and participate in the mailing lists. Both can be found at *http://www.xen.org/*.

▶ 从学习架构设计，尤其针对系统架构领域的角度看，本章介绍的Xen确有值得借鉴之处。但是在运用场景下，不可否认Xen已经受到了KVM（Kernel-based Virtual Machine）强有力的挑战。由于早期的Xen并没有很好地支持Linux，导致目前许多虚拟化商用平台放弃了Xen，而选择了KVM。此外，本文也介绍了Xen的Paravirtualization技术，不过在新的Xen版本中，实现的则是基于硬件支持的完全虚拟化。阅读时，需要明确这一点。作为经典知识，自有借鉴之处，但也需要及时更新最近的技术内容。（注：此处为该章的点评。）

Principles and properties	Structures
✓ Versatility	✓ Module
✓ Conceptual integrity	✓ Dependency
✓ Independently changeable	✓ Process
✓ Automatic propagation	✓ Data access
Buildability	
✓ Growth accommodation	
✓ Entropy resistance	

CHAPTER EIGHT

Guardian: A Fault-Tolerant Operating System Environment

Greg Lehey

▶ 系统的容错性一直是架构设计的一大挑战，尤其是对于分布式系统而言，难度岂止倍增。多数情况下，要满足容错性，就是要为可能出现问题的资源（包括本文提到的操作系统、数据库、存储介质、应用服务器、消息队列等）增加冗余，避免"单一故障点"。

ARCHITECTURE IS NOTHING NEW. REAL BUILDING ARCHITECTURE has been around for thousands of years, and some of the most beautiful examples of building architecture are also thousands of years old. Computers haven't been around that long, of course, but here too there have been many examples of beautiful architectures in the past. As with buildings, the style doesn't always persist, and in this chapter I describe one such architecture and consider why it had so little impact.

Guardian is the operating system for Tandem's fault-tolerant "NonStop" series of computers. It was designed in parallel with the hardware to provide fault tolerance with minimal overhead cost.

This chapter describes the original Tandem machine, designed between 1974 and 1976 and shipped between 1976 and 1982. It was originally called "Tandem/16," but after the introduction of its successor, "NonStop II," it was retrospectively renamed "NonStop I." Tandem frequently used the term "T/16" both for the system and later for the architecture.

I worked with Tandem hardware full-time from 1977 until 1991. Working with the Tandem machine was both exhilarating and unusual. In this chapter, I'd like to bring back to life some of the feeling that programmers had about the machine. The T/16 was a fault-tolerant machine, but that wasn't its only characteristic, and in this discussion I mention many aspects that don't directly contribute to fault tolerance—in fact, a couple detract from it! So prepare for a voyage into the past, back to about 1980, starting with one of Tandem's marketing slogans.

Tandem/16: Some Day All Computers Will Be Built Like This

Tandem describes the machines as single computers with multiple processors, but from the perspective of the 21st century, they're more like a network of computers operating as a single machine. In particular, each processor works almost completely independently from the others, and the system can recover from the failure of any single component, including processors. The biggest difference from conventional networked processors is that the entire system runs from a single kernel image.

Hardware

Tandem's hardware is designed to have no potential for a "single point of failure": any one component of the system, hardware *or* software, can fail without causing the entire system to fail. Beyond this, it is designed for *graceful degradation*. In most cases, the system as a whole can continue running despite multiple failures, though this depends greatly on the nature of the individual failure.

The first implication of this architecture is that there must be at least two of each component in case one should fail. In particular, this means that the system requires at least two CPUs.

But how should the CPUs be connected? The traditional method, then as now, is for the CPUs to communicate via shared memory. At Tandem we call this *tightly coupled multiprocessors*. But if the processors share memory, that memory could be a single point of failure.

Theoretically, it is possible to duplicate memory (a later Tandem architecture actually did that), but it's very expensive, and it creates significant timing problems. Instead, at the hardware level, Tandem chose a pair of high-speed parallel buses, the "interprocessor bus" or *IPB*, sometimes also referred to as *Dynabus*, which transfer data between the individual CPUs. This architecture is sometimes called *loosely coupled multiprocessors*.

There's more to a computer than the CPU, of course. In particular, the I/O system and data storage are of great importance. The basic approach here is also duplication of hardware; we'll look at it further down.

The resultant architecture looks something like Figure 8-1, the so-called *Mackie diagram*, named after Dave Mackie, a vice president of Tandem.

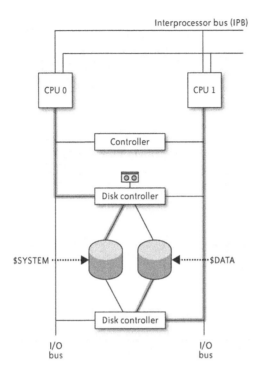

FIGURE 8-1. Mackie diagram

This could easily have led to at least doubling the cost of a system, as is the case with "hot standby" systems, where one component is only present in order to wait for the failure of its partner. Tandem chose a different approach for the more expensive components, such as CPUs. In the T/16, each CPU is active, and instead the operating system processes provide the hot standby function.

Diagnosis

The operating system needs to find out when a component fails. In many cases, there's not much doubt: if it fails catastrophically, it stops responding altogether. But in many cases, a failed component continues to run but generates incorrect results.

Tandem's solution to this problem is neither particularly elegant nor efficient. The software is designed to be paranoid, and at the first suggestion that something has gone wrong, the operating system stops the CPU—there's another to take over the load. If a disk controller returns an invalid status, it is taken offline—there's another to continue processing without interruption. But if the failure is subtle, it could go undetected, and on rare occasions this results in data corruption.

> 这里的watchdog,也就是我们常说的"看门狗",或者所谓"心跳侦测",在诊断场景中是极为常见的。不过,在解决了侦测故障这一问题的同时,其实watchdog自身亦要保障其健康。同时,在多数架构设计中,当故障出现时,除了提供这些自动侦测的手段外,还需要引入人工干预。这需要侦测程序实现告警通知。我曾经参与一个电信行业的告警程序开发,针对告警信号进行分解,并提供不同的处理策略。只有将自动与手动方式结合起来,才能有效地处理紧急情况,保护系统的正常运行。

It's not enough for a CPU to fail, of course; other CPUs have to find out that it has failed. The solution here is a watchdog. Each CPU broadcasts a message, the so-called "I'm alive" message, over both buses every 1.2 seconds. If a CPU misses two consecutive "I'm alive" messages from another CPU, it assumes that that CPU had failed. If the CPUs share resources (processes or I/O), the CPU that detects the failure then takes over the resources.

Repair

It's not enough to take a defective component offline; to maintain both fault tolerance and performance, it needs to be brought back online ("up") as quickly as possible, and of course without taking any other components offline ("down").

How this happens depends on the component and the nature of the failure. If the operating system has crashed in one CPU (possibly deliberately), it can be rebooted ("reloaded") online. The standard way to boot a system is to first boot one processor from disk and then boot all other processors across the IPB. Failed processors are also rebooted via the IPB.

If, on the other hand, the hardware is defective, it needs to be replaced. All system components are *hot-pluggable*: they can be removed and replaced in a running system with power up. If a CPU fails because of a hardware problem, the appropriate board is replaced, and then the CPU is rebooted across the bus as before.

Mechanical Layout

The system is designed to have as few boards as possible, so all boards are very large, about 50 cm square. All boards use low power Schottky TTL logic.

The CPU consists of two boards, the processor and the MEMPPU. The MEMPPU contains the interface to memory, including virtual memory logic, and the interface to the I/O bus. The T/16 can have up to 512 kW (1 MB) of semiconductor memory or 256 kW of core memory. Memory boards come in three sizes: 32 kW core, and 96 kW and 192 kW semiconductor memory. This means that there is no way of getting exactly 1 MB of semiconductor memory with fully populated boards. Core memory has word parity protection, whereas semiconductor memory has ECC protection, which can correct a single-bit error and detect a double-bit error.

Processor cabinets are about 6 feet high and house four CPUs with semiconductor memory or four CPUs with core memory. The processors are located at the top of the cabinet, with the I/O controllers located in a second rack directly below. Below that are fans, and at the bottom of the cabinet there are batteries to maintain memory contents during power failures.

Most configurations have a second cabinet with a tape drive. The disk drives are freestanding 14-inch units. There is also a system console, a DEC LA-36 printing terminal.

Processor Architecture

The CPU is a custom TTL design that shows significant similarities to the Hewlett-Packard 3000. It has virtual memory with a 2 kB page size, a stack-based instruction set, and fixed-width 16-bit instructions. Raw processor speed is about 0.8 MIPS per processor, giving 13 MIPS in a fully equipped 16-processor system.

Memory Addressing

The T/16 is a 16-bit machine, and the address space is limited to 16 bits in width. Even in the late 1970s, this was beginning to become a problem, and Tandem addressed it by providing a total of four address spaces at any one time:

User code
> This address space contains the executable code. It is read-only and shared between all processes that use it. Due to the architecture (separate memory for each CPU), the code can be shared only on a specific CPU.

User data
> The data space for user processes.

System code
> The code for the kernel.

System data
> The kernel data space.

With one exception, only one data space and one code space is accessible at any one time. They are specified in the *Environment Register*, which contains a number of flags describing the current CPU state, as shown in Figure 8-2.

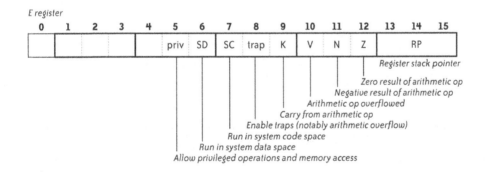

FIGURE 8-2. E register

The SD bit determines the data space, and the SC bit determines the code space. The SG-relative addressing mode is an exception to this rule: it always addresses system data.

In keeping with the aim of reliability and data integrity, the `trap` bit in the E register enables, among other things, traps on arithmetic overflow. There are "logical" equivalents of the arithmetic instructions that do not set the condition codes.

The CPU has a hardware stack addressed by two registers, the *S register* or stack pointer, and the *L register*, which points to the current stack frame. The L register is a relatively new idea: it points to the base of the current frame. Unlike the S register, it does not change during the execution of a procedure.* The stack is limited by addressing considerations to the first 32 kB of the current data space, and unlike some other machines, it grows upward.†

In addition to the hardware stack, there is a *register stack* of eight 16-bit words. The registers are numbered R0 to R7, but the instruction set uses them as a circular stack, where the top of stack is defined by the RP bits of the E register. In the following example, RP is set to 3, making R3 the top of stack, referred to as the A register; see Figure 8-3.

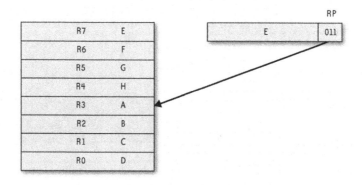

FIGURE 8-3. Register stack

Assuming that the register stack is "empty" at the beginning, a typical instruction sequence might be:

```
LOAD    var^a           -- push var^a on stack (R0)
LOAD    var^b           -- push var^b on stack (R1)
ADD                     -- add A and B (R1 and R0), storing result in R0 (A)
STOR    var^c           -- save A to var^c
```

Instructions are all 16 bits wide, which does not leave much space for an address field: it is only 9 bits wide. To work around this problem, Tandem bases addressing on offsets from a series of registers; see Figure 8-4.

---

* This is the same thing as the base pointer register used in most 21st-century processors.

† Stacks were quite a new idea in the 1970s. Like its predecessor, the HP 3000, Tandem's support for stacks went significantly beyond that of systems such as DEC's PDP-11, the most significant other stack-based machine of the time.

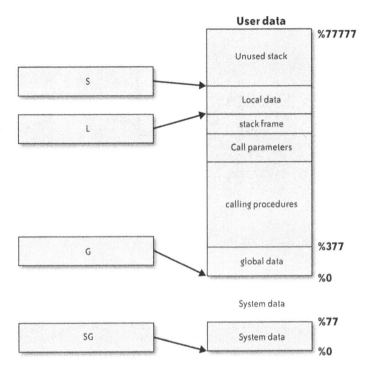

FIGURE 8-4. Memory addressing

Only the following memory areas can be addressed directly (in other words, without indirect addressing):

- The first 256 words of the current data space, referred to as "G" (*global*) mode. These are frequently used for indirect pointers.
- The first 128 words of positive offset from the L register, called L+. These are the local variables of the current procedure invocation, which would be called *automatic* variables in C.
- The first 64 words of system data ("SG+" mode). System calls run in user data space, so the CPU needs some means for privileged procedures to access system data. They are not accessible, even read-only, to unprivileged procedures.
- The first 32 words below the current value of the L register. This includes the caller stack frame (3 words) and up to 29 words of parameters passed to the procedure.
- The first 32 words below the top of the stack (S- addressing). These are used for *subprocedures*, procedures defined inside another procedure, which are called without leaving a stack frame. This address mode thus handles both the local variables and the parameters for a subprocedure.

These address modes are encoded in the first few bits of the address field of the instruction; see Figure 8-5.

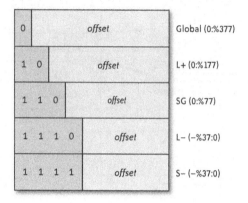

FIGURE 8-5. Address format

The % symbol represents octal numbers, like %377 (decimal 255, or hexadecimal 7F). Tandem does not use hexadecimal.

The instruction format also provides single-level indirection: if the I bit is set in the instruction, the retrieved data word is taken as the address of the final operand, which has to be in the same address space. The address space and the data word are both 16 bits wide, so there is no possibility for multilevel indirection.

One problem with this implementation is that the unit of data is a 16-bit word, not a byte. The instruction set also provides "byte instructions" with a different addressing method: the low-order bit of the address specifies the byte in the word, and the remainder of the address are the low-order 15 bits of the word address. For data accesses, this limits byte addressability to the first 32 kB of the data space; for code access, it limits the access to the same half of the address space as the current instruction. This has given rise to the restriction that a procedure cannot span the 32 kB boundary in the code space.

There are also two instructions, LWP (*load word from program*) and LBP (*load byte from program*), that can access data in the current code space.

Procedure Calls

Tandem's programming model owes much to the Algol and Pascal world, so it reserves the word *function* for functions that return a value, and uses the word *procedure* for those that do not. Two instructions are provided to call a procedure: PCAL for procedures in the current code space, and SCAL for procedures in the system code space. SCAL fulfills the function of a *system call* in other architectures.

All calls are indirect via a *Procedure Entry Point Table*, or *PEP*, which occupies up to the first 512 words of each code space. The last 9 bits of the PCAL or SCAL instruction are an index in this table.

This approach has dangers and advantages: the kernel uses exactly the same function call methods as user code, which simplifies coding conventions and allows code to be moved between kernel and user space. On the other hand, at least in theory, the SCAL instruction enables any user program to call any kernel function.

The system protects access to sensitive procedures based on the priv bit in the E register. It distinguishes between three kinds of procedures:

- Nonprivileged procedures, which can be called from any procedure, regardless of whether they are privileged.
- Privileged procedures, which can be called only from other privileged procedures.
- *Callable* procedures, which can be called from any procedure, but which set the priv bit once called. They provide the link between the privileged and the nonprivileged procedures.

The distinction between privileged, nonprivileged, and callable procedures is dependent on their position in the PEP. Thus it is possible to have nonprivileged library procedures in the system PEP, sometimes called the *SEP*. The table has the structure shown in Figure 8-6.

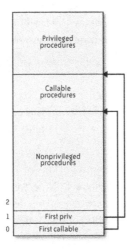

FIGURE 8-6. Procedure entry point table

Action of the PCAL and SCAL Instructions

The PCAL instruction performs the following actions:

- If the priv bit in the E register is not set (meaning that the calling procedure is nonprivileged), check the "first priv" value (word 1 of the code space). If the offset in the instruction is greater or equal, the procedure is trying to call a privileged procedure. Generate a protection trap.
- If the priv bit in the E register is not set, check the "first callable" value (word 0 of the code space). If the offset in the instruction is greater or equal, set the priv bit in the E register.
- Push the current value of the P register (program counter) onto the stack.
- Push the *old* value of the E register onto the stack.
- Push the current L register value onto the stack.
- Copy the S register (stack pointer) to the L register.
- Set the RP field of the E register to 7 (empty).
- Load the contents of the PEP word addressed by the instruction into the P register.

The SCAL instruction works in exactly the same way, except that it also sets the SC bit in the E register, thus ensuring that execution continues in kernel space. The data space does *not* change.

The PCAL and SCAL instructions are very similar, and the programmer normally does not need to distinguish between them. That is done by the system at execution time. Thus library procedures can be moved between user code and system code with no recompilation.

The Interprocessor Bus

All communication between CPUs goes via the *interprocessor bus*, or *IPB*. There are in fact two buses, called X and Y (see Figure 8-1), in case one fails. Unlike other components, both buses are used in parallel when they're up.

Data is passed across the bus in fixed-length packets of 16 words. The bus is fast enough to saturate memory on both CPUs, so the client CPU performs it synchronously in the *dispatcher* (scheduler) using the SEND instruction. The destination (server) CPU reserves buffer space for a single transfer at boot time. On completion of the transfer, the destination CPU receives a *bus receive* interrupt and handles the packet.

Input/Output

Each processor has a single I/O bus with up to 32 controllers. All controllers are dual-ported and connected to two different CPUs. At any one time, only one CPU has access to any specific controller. This relationship between CPU and controller is called *ownership*: the controlling

CPU "owns" the controller. The backup path is not used until the primary path fails or the system operator manually switches to it (a so-called *primary switch*).

Disks are a particularly sensitive issue because many components could fail. It could be a disk itself, the physical connection (cable) to the disk, the disk controller, the I/O bus, or the CPU to which it is connected. As a result, in addition to the dual-ported controllers, each disk is physically duplicated—at least in theory—and it is also dual-ported and connected to two different controllers, both connected to the same two CPUs The restriction remains that only one CPU can access each controller at any one time, but it is possible for one of the CPUs to own one of the controllers and the other CPU to own the other controller. This is also desirable from a performance point of view.

Figure 8-1 shows a typical configuration: as the gray highlighted paths indicate, the I/O process for the system disk $SYSTEM accesses it via CPU 0 and the first disk controller, while the I/O process for another disk $DATA, connected to the same two controllers, accesses the disk via CPU 1 and the second disk controller. CPU 0 "owns" the first controller, while CPU 1 "owns" the second controller. If CPU 0 were to fail, the backup I/O process for $SYSTEM in CPU 1 would take over and take ownership of the controller, and then continue processing. If the second disk controller were to fail, the I/O process for $DATA would not be able to use the first disk controller, since it is owned by CPU 0, so the I/O process would first do a primary switch, after which the primary I/O process would be running in CPU 0. It would then access $DATA by the same path as $SYSTEM.

That's the theory, anyway. In practice, disks and drives are expensive, and many people run at least some of their disks in degraded mode, without duplicating the drive hardware. This works as well as you would expect, but of course there is no longer any further fault tolerance: effectively, one of the disks has already failed.

Process Structure

Guardian is a microkernel system: apart from the low-level interrupt handlers (a single procedure, IOINTERRUPT) and some very low-level code, all the system services are performed by system processes which run in system code and data space.

The more important processes are:

- The *system monitor*, PID 0 in each CPU, which is responsible for starting and stopping other processes and for miscellaneous tasks such as returning status information, generating hardware error messages, and maintaining the time of day.
- The *memory manager*, PID 1 in each CPU, which is responsible for I/O for the virtual memory system.
- The I/O processes, which are responsible for controlling I/O devices. All access to I/O devices from anywhere in the system goes via its dedicated I/O process. The I/O controllers

are connected to two CPUs, so each device is controlled by a pair of I/O processes running in those CPUs: a *primary* process that performs the work and a *backup* process that tracks the state of the primary process and waits to fail or hand over control to it voluntarily ("primary switch").

The main issue in the choice of primary CPU is the CPU load, which needs to be balanced manually. For example, if you have six devices connected between CPUs 2 and 3, you would probably put the primary process of three of them in CPU 2, and the primary process of the other three in CPU 3.

Process Pairs

▶ Pairs的概念非常重要，其实就是前面提到的"冗余（Redundant）"。

The concept of process pairs is not limited to I/O processes. It is one of the cornerstones of the fault-tolerant approach. To understand the way they work, we need to understand the way messages are passed in the system.

Message System

▶ 可以想象任何节点之间皆可以互联，且互联路径不止一条的网络。这样的网络虽然结构变得复杂了，但却是健壮稳定的，具有极强的容错性。在实现容错性方面，Guardian 与类似 HDFS 之类的分布式存储系统不一样，由于它不需要保持数据的同步，因而实现相对简单，不需要记录发生故障时的 checkpoint，而只需保障消息的传递是正确的，即文中提及的 interprocess communication。

As we've seen, the biggest difference between the T/16 and conventional computers is the lack of any single required component. Any one part of the system can fail without bringing down the system. This makes it more like a network than a conventional shared memory multiprocessor machine.

This has far-reaching implications for the operating system design. A disk could be connected to any 2 of 16 CPUs. How do the others access it? Modern networks use file systems such as NFS or CIFS, which run on top of the network protocols, to handle this special case. But on the T/16 it isn't a special case; it is the norm.

File systems aren't the only thing that require this kind of communication: interprocess communication of all kinds requires it, too.

Tandem's solution to this issue is the *message system*, which runs at a very low level in the operating system. It is not directly accessible to user programs.

The message system transmits data between processes, and in many ways it resembles the later TCP or UDP. The initiator of the message is called the *requestor*, and the object is called the *server*.‡

All communication between processes, even on the same CPU, goes via the message system. The following data structures implement the communication:

- Each message is associated with two *Link Control Blocks*, or *LCBs*, one for the requestor and one for the server. These small data objects are designed to fit in a single IPB packet. If more data is needed than would fit in the LCB, a separate buffer is attached.

‡ These names correspond closely in function to the modern terms *client* and *server*.

- To initiate a transfer, the requestor calls the procedure link. This procedure sends the message to the server process and queues the LCB on its *message queue*. At this point the server process has not been involved, but the dispatcher awakes the process with an LREQ (*link request*) event.

  On the requestor side, the call to link returns immediately with information to identify the request; the requestor does not need to wait for the server to process the request.

- At some time the server process sees the LREQ event and calls listen, which removes the first LCB from the message queue.

- If a buffer is associated with the LCB and includes data to be passed to the server, the server calls readlink to read in the data.

- The server then performs whatever processing is necessary, and next calls writelink to reply to the message, again possibly with a data buffer. This wakes the requestor on LDONE.

- The requestor sees the LDONE event, examines the results, and terminates the exchange by calling breaklink, which frees the associated resources.

Only other parts of the kernel use the message system directly; the *file system* uses it to communicate with the I/O devices and other processes. Interprocess communication is handled almost identically to I/O, and it also is used for maintaining fault-tolerant *process pairs*.

This approach is inherently asynchronous and multithreaded: after calling link, the requestor continues its operations. Many requestors can send requests to the same server, even when it's not actively processing requests. The server does not need to respond to the link request immediately. When it replies, the requestor does not need to acknowledge the reply immediately. Instead, in each case the process is woken on an event that it can process when it is ready.

Process Pairs, Revisited

One of the requirements of fault tolerance is that a single failure must not bring the system down. We've seen that the I/O processes solve this by using process pairs, and it's clear that this is a general way to handle the failure of a CPU. Guardian therefore provides for creation of user-level process pairs.

All process pairs run as a *primary* and a *backup* process. The primary process performs the processing, while the backup process is in a "hot standby" state. From time to time the primary process updates the memory image of the backup process, a process called *checkpointing*. If the primary fails or voluntarily gives up control, the backup process continues from the state of the last checkpoint. A number of procedures implement checkpointing, which is performed by the message system:

- The backup process calls checkmonitor to wait for checkpoint messages from the primary process. It stays in checkmonitor until the primary process goes away or relinquishes control.

During this time, the only use of the CPU is message system traffic to update its data space and calls to open and close to update file information.

- The primary process calls checkpoint to copy portions of its data space and file information to the backup process. It is up to the programmer to decide which data and files to checkpoint, and when.
- The primary process calls checkopen to checkpoint information about file opens. This effectively results in a call to open from the backup process. The I/O process recognizes that this is a backup open and treats it as equivalent to the primary open.
- The primary process calls checkclose to checkpoint information about file closes. This effectively results in a call to close from the backup process.
- The primary process may call checkswitch to voluntarily release control of the process pair. When this happens, the primary and backup processes reverse their roles.

When the backup process returns from checkmonitor, it has become the new primary process. It returns to the location of the old primary's last call to checkpoint, not to the location from which it was called. It then carries on processing from this point.

In general, the life of a process pair can look like Table 8-1.

TABLE 8-1. Life of a process pair

Primary	Backup
Perform initialization	
call newprocess to create backup process	
	Perform initialization
	call checkmonitor to receive checkpoint data
call checkpoint	Wait in checkmonitor
call checkopen	call open from checkmonitor
Processing	Wait in checkmonitor
call checkpoint	Wait in checkmonitor
Processing	Wait in checkmonitor
Voluntary switch: call checkswitch	Take over
call checkmonitor to receive checkpoint data	Processing
Wait in checkmonitor	call checkpoint
Wait in checkmonitor	Processing
Wait in checkmonitor	call checkpoint
Wait in checkmonitor	*CPU fails*
Take over	*(gone)*
Processing	

Synchronization

This approach proves very reliable, and it can deliver reliability superior to that of a pure lockstep approach. In some classes of program error, notably race conditions, a process that is running in lock-step will run into exactly the same program error and crash as well. A more loosely coupled approach can often avoid the exact same situation and continue functioning.

A couple of issues are not immediately obvious:

- Checkpointing is CPU-intensive. How often should a process checkpoint? What data should be checkpointed? This decision is left to the programmer. If he does it wrong and forgets to checkpoint important data, or does it at the wrong time, the memory image in the backup process will be inconsistent, and it may malfunction.
- If the primary process performs externally visible actions, such as I/O, after performing a checkpoint but before failing, the backup process will repeat them after takeover. This could result in data corruption.

In practice, the issue of incorrect checkpointing has not proved to be a problem, but duplicate I/O most certainly is a problem. The system solves this problem by associating a sequence number called a *sync ID* with each I/O request. The I/O process keeps track of the requests, and if it receives a duplicate request, it simply returns the completion status of the first call to the request.

Networking: EXPAND and FOX

The message system of the T/16 is effectively a self-contained network. That puts Guardian in a good position to provide wide-area networking by effectively extending the message system to the whole world. The implementation is called *EXPAND*.

From a programmer's point of view, EXPAND is almost completely seamless. Up to 255 systems can be connected.

System names

Each system has a name starting with a backslash, such as \ESSG or \FOXII, along with a node number. The node numbers are much less obvious than modern IP addresses: from the programmer's perspective, they are necessary almost only for encoding file names, which we'll see later.

EXPAND is an extension of the message system, so most of the details are hidden from the programmer. The only issues are the difference in speed and access requirements.

FOX

Considering purely practical constraints, it is difficult to build a system with more than 16 CPUs; in particular, hardware constraints limit the length of the interprocessor bus to a few meters, so a realistic limit is 16 CPUs. Beyond that, Tandem supplies a fast fiber-optic connection capable of connecting up to 14 systems together in a kind of local area cluster. In most respects it is a higher-speed version of EXPAND.

File System

Tandem uses the term *file system* to mean the access to system resources that can supply data ("read") or accept it ("write"). Apart from disk files, the file system also handles devices, such as terminals, printers, and tape units, and processes (interprocess communication).

File Naming

There is a common naming convention for devices, disk files, and processes, but unfortunately it is complicated by many exceptions. Processes can have names, but only I/O processes and paired processes *must* have a name. In all cases, the file "name" is 24 characters long and consists of three 8-byte components. Only the first component is required; the other two are used only for disk files and named processes.

Unnamed processes use only the first 8 bytes of the name. Unpaired system processes, such as the monitor or memory manager, have the format shown in Figure 8-7.

FIGURE 8-7. Name format for unpaired system processes

Unpaired user processes have the format shown in Figure 8-8.

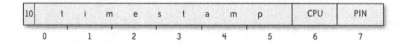

FIGURE 8-8. Name format for unpaired user processes

The combination *CPU* and *PIN* together forms the process ID, or *PID*. The PIN is the *process identification number* within the CPU. This limits each CPU to 256 processes.

Real names start with a $ sign. Devices use only the first 8 bytes, and disk files use all three components. The individual components look like the names of the disk, the directory, and the file, though in fact there is only one directory per disk volume. Processes can also use the other two components for passing information to the process.

Typical names are shown in Table 8-2.

TABLE 8-2. Typical file names

$TAPE	Tape drive
$LP	Printer
$SPLS	Spooler process
$TERM15	Terminal device
$SYSTEM	System disk
$SYSTEM SYSTEM LOGFILE	System log file on disk $SYSTEM
$SPLS #DEFAULT	Default spooler print queue
$RECEIVE	Incoming message queue, for interprocess communication

If a component is less than 8 bytes long, it is padded with ASCII spaces. Externally, names are represented in ASCII with periods, for example, $SYSTEM.SYSTEM.LOGFILE and $SPLS.#DEFAULT.

There are still further quirks in the naming. Process subnames must start with a hash mark (#), and user process names (but not device names, which are really I/O process names) have the PID at the end of the first component; see Figure 8-9.

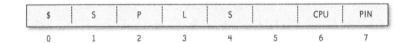

FIGURE 8-9. Name format for named user processes

The PID in this example is the PID of the primary process. It limits the length of user process names to six characters, including the initial $.

As if that wasn't enough, there is a separate set of names for designating processes, disk files, or devices on remote systems. In this case, the initial $ sign is replaced by a \ symbol, and the second byte of the name is the system number, shifting the rest of the name one byte to the right. This limits the length of process names to five characters if they are to be network-visible. So from another system, the spooler process we saw earlier might have the external name \ESSG.$SPLS and have the internal format shown in Figure 8-10.

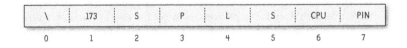

FIGURE 8-10. Name format for network-visible processes

The number 173 is the node number of system \ESSG.

> 通常我们说的Blocking IO、Non-blocking IO 与 I/O multiplexing从本质上讲都是Synchronous IO，二者有着本质的不同。理论上，Synchronous IO是指用户线程发起I/O请求后需要等待或者轮询内核I/O完成后再继续；而Asynchronous IO则是指用户线程发起I/O请求直接退出，当内核I/O操作完成后会通知用户线程来调用回调函数。显然，后者的响应性会更好。

Asynchronous I/O

One of the important features of the file system interface is the strong emphasis on asynchronous I/O. We've seen that the message system is intrinsically asynchronous in nature, so this is relatively simple to implement.

Processes can choose synchronous or asynchronous ("no wait") I/O at the time they open a file. When a file is opened no-wait, an I/O request will return immediately, and only errors that are immediately apparent will be reported—for example, if the file descriptor isn't open. At a later time the user calls `awaitio` to check the status of the request. This gives rise to a programming style where a process issues a number of no-wait requests, then goes into a central loop to call `awaitio` and handle the completion of the requests, typically issuing a new request.

Interprocess Communication

At a file system level, interprocess communication is a relatively direct interface to the message system. This causes a problem: the message system is asymmetrical. The requestor sends a message and may receive a reply. There's nothing that corresponds to a file system `read` command. On the server side, the server reads a message and replies to it; there's nothing that corresponds to a `write` command.

The file system provides `read` and `write` procedures, but `read` only works with I/O processes, which map them to message system requests. `read` doesn't work for the interprocess communication level, and in practice `write` also is not used much. Instead, the requestor uses a procedure called `writeread` to first write a message to the server and then get a reply from it. Either the message or the reply can be null (zero length).

These messages find their way to the server's message queue. At a file system level, the message queue is a pseudofile called `$RECEIVE`. The server opens `$RECEIVE` and normally uses the procedure `readupdate` to read a message. At a later point it can reply with the procedure `reply`.

System Messages

The system uses `$RECEIVE` to pass messages to processes. One of the most important is the *startup message*, which passes parameters to a newly started process. The following example is written in TAL, Tandem's low-level system programming language (though the name stands for "Tandem Application Language"). TAL is derived from HP's SPL, and it is similar to Pascal and Algol. One of the more unusual characteristics is the use of the caret (^) character in identifiers; the underscore (_) character is not allowed. This example should be close enough to C to be intelligible. It shows a process that starts a child server process and then communicates with it.

The first piece shows the parent process (requestor):

```
call newprocess (program^file^name,,,,,, process^name); -- start the server process
call open (process^name, process^fd);              -- open process
```

```
    call writeread (process^fd, startup^message, 66); -- write startup message
    while 1 do
      begin
      read data from terminal
      call writeread (process^fd,
                      data, data^length,         -- write data
                      reply, max^reply,          -- read data back
                      @reply^length);            -- return real reply length
      if reply^length > 0
        write data back to terminal
      end;
```

The following shows the child process (server):

```
    call open (receive, receive^fd);
    do
      call read (receive^fd, startup^message, 66);
    until startup^message = -1;              -- first word of startup message is -1.
    while 1 do
      begin
      call readupdate (receive^fd, message, read^count, count^read);
      process message received, replacing buffer contents
      call reply (message, reply^length);
      end;
```

The first messages that the child receives are system messages: the parent open of the child sends an open message to the child, and then the first call to writeread sends the startup message. The child process handles these messages and replies to them. It can use the open message to keep track of requestors or receive information passed in the last 16 bytes of the file name. Only then does the process receive the normal message traffic from the parent. At this point, other processes can also communicate with the child. Similarly, when a requestor closes the server, the server receives a close system message.

Device I/O

It's important to remember that device I/O, including disk file I/O, is handled by I/O processes, so "opening a device" is really opening the I/O process. Still, I/O to devices and files is implemented in a slightly different manner, though the file system procedures are the same. In particular, the typical procedures used to access files are the more conventional read and write, and normally disk I/O is not no-wait.

Security

In keeping with the time, the T/16 is not an overly secure system. In practice, this hasn't caused any serious problems, but one issue is worth mentioning: the transition from nonprivileged to privileged procedures is based on the position of the procedure entry point in the PEP table and the value of the priv bit in the E register. Early on, exploits became apparent. If you could get a privileged procedure to return a value via a pointer and get it to overwrite the saved E register on the stack in such a way that the priv bit was set, the process would remain privileged on return from that procedure. It is the responsibility of callable procedures to check their pointer parameters to ensure that they don't have any addressing exceptions, and that they return values only to the user environment. A bug in the procedure `setlooptimer`, which sets a watchdog timer and optionally returns the old value, made it possible to become the SUPER.SUPER (the root user, with ID 255,255, or –1):

```
proc make^me^super main;
begin
int .TOS = 'S';                         -- top of stack address

call setlooptimer (%2017);              -- set a timer value
call setlooptimer (0, @TOS [4]);        -- reset, return old value to saved E reg
pcb [mypid.<8:15>].pcbprocaid := -1;    -- dick in my PCB and make me super
end;
```

The second call to `setlooptimer` returns the old value `%2017` to the saved E register contents on stack, in particular setting the `priv` bit, which leaves the process in privileged state. Theoretically this value could have been decremented to `%2016`, but this would not make any difference (this is the saved RP field, which is not restored). The program then uses SG-relative addressing to modify the user information in its own process control block (PCB). `mypid` is a function returning the current process's PID, and the last 8 bits (`<8:15>`) were the PIN, which is used as an index in the PCB table.

This bug was quickly fixed, of course, but it showed a weakness in the approach: it is up to the programmer to check the parameters passed to callable procedures. Throughout the life of the architecture, such problems have reoccurred.

File Access Security

Tandem's approach to file access security is similar to that of Unix, but users can belong only to a single group, which is part of the username. Thus my username, SUPPORT.GREG, also written numerically as 20,102, indicates that I belong to the SUPPORT group (20) only, and that within that group my user ID is 102. Each of these fields is 8 bits long, so the complete user ID fits in a word. If I wanted to be a member of another group, I would need another user ID, possibly with a different number—for example, SUPER.GREG with user ID 255,17.

Each file has a number of bits describing what access the owner, the group, or all users have to the file. Unlike Unix, however, the bits are organized differently: the four permissions are *read*, *write*, *execute*, and *purge*. *Purge* is the Tandem name for *delete*, and it's necessary because directories don't have their own security settings.

For each of these access modes, there is a choice of who is allowed to use them:

- *Owner* means only the owner of the file.
- *Group* means anybody in the same group.
- *All* means anybody.

All of these relate only to the same system as the one in which the file is located. A second set of modes was introduced with networking to regulate access from users on other systems:

- *User* means only a user with the same user and group number as the owner of the file.
- *Class* means anybody with the same group number as the owner of the file.
- *Network* means anybody, anywhere.

There is no security whatsoever for devices, and user processes have to roll their own. The former is a particular disadvantage in a networked environment. At a security seminar in early 1989, I was able to demonstrate stealing the SUPER.SUPER (root) password on system \TSII, which was in the middle of the management area in Cupertino, simply by putting a fake prompt on the system console. I was in Düsseldorf (Germany) at the time.

Folklore

Coming back into the present, early 21st century, it's easy to forget the sheer fun of working with the computer. Tandem was a fun company, and it looked after its employees. One Friday in late 1974, early in the development of the system, the founders finally got the software to work on the hardware; up to this point the software had been developed on simulators. You can imagine the excitement. The story goes that one of the VPs went out and brought in a crate of beer, and they all sat around the crate, celebrating the event and discussing the future. One thing they decided was that the crate of beer should be a weekly event, and the Tandem Beer Bust was born. It really did continue into the 1990s, during which time it became increasingly politically incorrect and was finally canceled.

Tandem gave rise to lots of slogans and word plays, of course—the name "Tandem" itself was one. In those days we had T-shirts with slogans such as "So nice, so nice, we do it twice," "There's no stopping us," and "Tandem users do it with mirrors." And, of course, the standard answer when anybody came up with an excess of just about anything: "It's there in case one fails."

This last slogan was more than just wordplay. It sat deep in our thought processes. In May 1977, on returning from five weeks of initial Tandem training, I was faced with the sad discovery that our cat had run away. After establishing that she wasn't going to return, we went out and got…two new cats. It wasn't until much later that I realized that this was the result of successful brainwashing. Even today I have a phobia of rebooting a computer unless it's absolutely unavoidable.

The Downside

The T/16 was a remarkably successful machine for its intended purpose—at one time over 80% of all ATMs in the U.S. were controlled by Tandem systems—but of course there were disadvantages as well. Some, like the higher cost in comparison with conventional systems, are inevitable. Others were not so obvious to the designers.

Performance

Tandem was justifiably proud of the near-linear scaling of performance when hardware was added. Horst and Chou (1985), which refers to a later system, the TXP, shows how a FOX cluster can scale linearly from 2 to 32 processors.

Bartlett (1982) shows the downside: the performance of the message system limited the speed even of small systems. A single message with no attached data takes over 2 ms to transmit, and messages with 2,000 bytes of data in each direction take between 4.6 ms (same CPU) and 7.0 ms (different CPUs). This is the overhead for a single I/O operation, and even in its day it was slow. The delay between sequential I/O requests to a disk file was long enough that they would not occur until the data had passed the disk head, meaning that only one request could be satisfied per disk revolution. A program that sequentially reads 2 kB from disk and processes it (for example, the equivalent of *grep*) would get a throughput of only 120 kB/s. Smaller I/O sizes, such as 512 bytes, could limit the throughput to floppy disk speeds.

Hardware Limitations

As the name "Tandem/16" suggests, the designers had a 16-bit mindset. That is fairly typical for the mid-1970s, but the writing was already on the wall that "real" computers had a 32-bit word. In the course of time, successor machines addressed a number of the issues. In 1981, Tandem introduced the *NonStop II* system with an upward-compatible instruction set and fewer hardware limitations. Over the next 10 years, a number of compatible but faster machines were introduced. None were extremely fast, but they were fast enough for online transaction processing. In addition, the operating system was rewritten to address the more immediate problems, and over the course of time additional improvements were made. The changes included:

- Introduction of a 31-bit address mode to give user processes "unlimited" memory space. This mode used byte addresses, but it didn't remove the limitations on stack size and code spanning the 32 kB boundary, since the old instruction formats remained.

- Increase in the number of hardware virtual memory maps. The T/16 had only four, for the code and data spaces. The TNS/II, as it was called, had a total of 16 memory maps, which meant that the processor could directly address up to 2 MB without involving the memory manager. One of these maps was used as a kind of translation lookaside buffer to handle the 31-bit extended addresses.

- Guardian II, the new version of Guardian that came with the TNS/II, also featured system library and user library spaces, which increased the total space available to processes to 384 kB. Still later, the number of library spaces was increased from 2 (system and user)

to up to 62 (31 each for system and user) by segment switching. Only a single user library and system library map could be active at any one time.
- Message queue size proved to be a problem. The monitor processes sent status messages at regular intervals to every process that wanted them. If the process didn't read the messages, large numbers of resources (LCBs and message buffers) could be used for duplicate messages. To address this problem, Guardian II introduced a *messenger* process that kept a single copy of these status messages and sent them to a process when it called listen.

Missed Opportunities

The T/16 was a revolutionary machine, but it also offered an environment that few other machines of the day had. Ultimately, though, it was the small things that got in the way. For example, device independence is one of the most enduring aims of operating systems, and Tandem went a long way toward this goal. Ultimately, though, they missed their full potential because of naming issues and almost gratuitous incompatibilities. Why was it not possible to use read in interprocess communication? Why did process names have to differ in format from device names? Why did they need a # character in the ninth byte?

Split Brain

A more serious issue was with the basic way of detecting errors. It worked fine as long as only one component failed, and usually quite well if two failed. But what if both interprocessor buses failed? Even in a two-CPU system, the results could be catastrophic. Each CPU would assume that the other had failed and take over the I/O devices—not once, but continually. Such circumstances did occur, fortunately very rarely, and they often resulted in complete corruptions of the data on disks shared between the two CPUs.

Posterity

From 1990 on, a number of factors contributed to a decline in Tandem's sales:

- Computer hardware in general was becoming more reliable, which narrowed Tandem's edge.
- Computer hardware was becoming *much* faster, highlighting some of the basic performance limitations of the architecture.

In the 1990s, the T/16 processor architecture was replaced by a MIPS-based solution, though much of the remaining architecture remained in place. On the other hand, the difference in performance was big enough that as late as 2000, Tandem was still using the MIPS processors to emulate the T/16 instructions. One of the reasons was that most Tandem system-level software was still written in TAL, which was closely coupled to the T/16 architecture. Moves to migrate the codebase to C were rejected because of the cost involved.

For such a revolutionary system, the Tandem/16 has made a surprisingly small impression on the industry and design of modern machines. Much of the functionality is now more readily available—mirrored disks, network file systems, the client-server model, or hot-pluggable hardware—but it's difficult to see anything that suggests that they happened by following Tandem's lead. This may be because the T/16 was so different from most systems, and of course the purely commercial environment in which it was developed didn't help either.

Further Reading

Hewlett-Packard has a number of papers on its website; start looking at the Tandem Technical reports at *http://www.hpl.hp.com/techreports/tandem/*. In particular:

Bartlett, Joel. "A NonStop Kernel," June 1981. *http://www.hpl.hp.com/techreports/tandem/TR-81.4.html?jumpid=reg_R1002_USEN*. (Gives more information about the operating system environment.)

Bartlett, Joel, et al. "Fault tolerance in Tandem computer systems," May 1990. *http://www.hpl.hp.com/techreports/tandem/TR-90.5.html*. (Describes the hardware in more detail.)

Gray, Jim. "The cost of messages," March 1988. *http://www.hpl.hp.com/techreports/tandem/TR-88.4.html*. (Describes some of the performance issues from a theoretical point of view.)

Horst, Robert, and Tim Chou. "The hardware architecture and linear expansion of Tandem nonstop systems," April 1985. *http://www.hpl.hp.com/techreports/tandem/TR-85.3.html*.

Principles and properties		Structures	
Versatility		Module	✓
Conceptual integrity		Dependency	
Independently changeable		Process	✓
Automatic propagation	✓	Data access	
Buildability			
Growth accommodation	✓		
Entropy resistance	✓		

CHAPTER NINE

JPC: An x86 PC Emulator in Pure Java

Rhys Newman
Christopher Dennis

"EMULATORS ARE SLOW AND JAVA IS SLOW; THUS THE COMBINATION could only mean computation at a snail's pace." As this conventional wisdom would suggest, the first JPC prototype ran 10,000 times slower than a real machine.

Nevertheless, a pure Java x86 PC emulator is a compelling idea—imagine booting Linux and Windows inside a secure Java Sandbox while remaining fast enough to be practical. Not that this task was ever likely to be easy, as it required replicating the internals of one of the most complex pieces of machinery humankind has produced. Navigating the task of reproducing the physical x86 PC design, layered on top of the Java Virtual Machine, and then fitting the result inside the security restrictions of the Java Applet sandbox has been an often difficult journey of discovery.

On the way we have experienced computing challenges seldom encountered by modern software engineers, but which offer timely reminders of the fundamentals usually taken for granted. Now we have a beautiful architecture that shows that pure Java emulation of x86 hardware is possible, and also fast enough to be practical after all.

Introduction

With the increasing processor speed and network performance enjoyed by even domestic computer users, more and more things that would have been considered impractical only a few years ago are becoming commonplace. A decade ago when a small technology company called VMWare started up in California, the idea of running a completely virtual computer as software within a physical computer was viewed as rather esoteric. After all, if you have a computer, why slow it down by adding a virtualization layer only to run what you'd be running anyway? The software you used needed the full power of the hardware to run, and as you could simply buy more machines to do more work if needed, what would be the point?

A decade later we all see the benefits of virtual machines. Hardware is so fast that modern machines can run many virtual machines without a major impact on overall performance, and the importance of software services is so significant that the security and reliability benefits of isolating them completely in virtual machines are clear.

However, pure virtualization has its problems, as it relies on some degree of hardware support[*] to function and is therefore exposed to instabilities caused by such close links to the physical machine. Emulators, by contrast, are virtual computers built entirely in software, and therefore have no specific requirements on the underlying hardware. This means the emulated machine is completely separated from the real hardware. Its presence on the system neither suffers nor causes any additional instabilities than the normal application software would. As running application software is the raison d'être of a computer in the first place, an emulator will always be the most stable and secure means to create virtual machines.

As with virtualization a decade ago, current critiques of emulation focus on the speed penalty incurred, which is often significant. However, history shows that speed issues are resolved by technological progress, but given the ever increasing complexity of modern hardware and software stacks, ever more difficult issues arise from subtle interactions between hardware, operating systems, and application software. Separating systems at a very low level in a very robust and secure way while still sharing the physical resource will therefore become increasingly necessary but increasingly difficult. Emulation offers the required degree of robustness, security, and flexibility, and so will become an increasingly compelling option.

There are a number of emulators currently available, and the most notable examples for emulating an x86 PC are Bochs and QEMU. Both have been developed over a number of years and are sufficiently accurate to boot modern operating systems and run application software. However, both are written in native code (C/C++) and need recompilation if they are to run on a new underlying hardware architecture/OS stack (i.e., a new type of host system). Furthermore, for very high-security applications, there is always the concern that the emulator has been tampered with for nefarious purposes, or to let guest code do nefarious things, or that

---

[*] At a bare minimum you need hardware that is the same as that being "virtualized." Thus products such as Xen and VMWare enable virtual x86 PCs to be created on x86 hardware only.

it contains a bug that permits either of these. For example, QEMU uses dynamic binary translation to achieve acceptable speed, and if this process were compromised or a fault were exploited, the software could become unstable or breach security safeguards.

If users will accept the speed penalty of emulation when security can be absolutely guaranteed, then why not build an emulator on the most widely deployed and secure virtual machine, the Java VM (JVM)? The JVM has been tested for over 10 years as a secure means of running code, and users are often content to let unvetted code downloaded from the Internet execute within the Applet Sandbox, the security container supplied by the JVM. This is because the JVM guards against fundamental programmatic errors, such as accessing arrays in memory beyond their valid size and reading data from unallocated memory. Further, a JVM with a security manager installed to enforce a sandbox can veto any sensitive operation attempted by guest software.

JPC is just this: an x86 PC emulator written entirely in Java. There is no native code in JPC; it emulates all the standard hardware components of an x86 PC while remaining entirely inside the Java Applet Sandbox. Thus within these security restrictions the x86 operating systems and software running inside JPC are totally isolated from the underlying hardware, even to the point that said hardware does not need to be an x86 PC.[†] From the host's point of view, JPC represents just another Java application/applet, and thus the host can be confident that the code (whatever it may be) is safe to run. JPC can boot DOS and modern Linux, giving the guest software and OS complete unfettered access to all the hardware of the virtual machine, including root/admin access, all while staying inside the Java sandbox.

To break out of JPC, an attacker would have to find a bug in JPC's code coinciding with a bug in the JVM, which could enable a sensitive action on the host computer that was also within the powers of the user running the JVM. This represents a breach of three completely independent layers of security. Each layer is typically built by completely different companies, and because each layer is needed in so many different circumstances, their security is under constant independent testing and review. Note also the coincidence requirement in breaking the layers. It is not sufficient to find a bug in JPC and then move on to the task of breaking out of the JVM; the hacker needs to find a bug in JPC that directly (and already) connects to a suitable security bug within the JVM being used. As JPC is open source, it is a relatively simple task to review the code and build a "clean" version, and in high security applications a security-hardened JVM could be used. JPC thus presents an impregnable barrier to malicious x86 code, and indeed provides the most secure, convenient, and safe way to examine malicious x86 code in action.

Like virtualization, as hardware speed increases, the applications of emulation can expand from the security area to mission-critical systems that need to ensure robustness. No matter how the emulated machine crashes (or how hard), the host machine is unaffected and can continue

---

† JPC has been adapted to run inside a J2ME environment and can then boot original and unmodified DOS on an ARM11-based mobile phone!

running other emulated instances. This technique avoids possible issues with even the most carefully thought through features available in modern hardware to support virtualization.[‡]

In the rest of this chapter, we describe the process used to prototype and develop JPC, and we show how a number of incremental improvements to design resulted in the JPC available today. We then outline more applications and implications of the unique combination of technologies that JPC represents.

Proof of Concept

> ▶ 在进行系统架构设计的过程中，"Proof of Concept（简称PoC）"这个环节非常重要，它能够有效地规避技术风险，确定架构目标是否正确。在这个过程中，通常会开展一些技术预研（Spike），甚至通过编写一些Sample来验证。

The x86 PC has been around for over 30 years and has evolved through many different generations of hardware. At each stage, backward-compatibility has been maintained so that even today, an original 8086 program is likely to run on a new PC. Although this has had undoubted benefits and has contributed to the unparalleled success of the platform, it does mean the architecture is packed with extra complexity as new technologies are incorporated, in order to avoid breaking existing code. If a PC were built today from scratch, many aspects of the hardware would be substantially different, and almost certainly a lot simpler.

Nevertheless, this x86 platform is ubiquitous, with over 1 billion in the world today and over 200 million more being manufactured each year. Consequently, the most widely useful emulator will be one that targets the x86 PC architecture.

However, this is not an easy task. Just some of the hardware components that must be emulated in software include the x86 processor, hard disk (and its controller), keyboard and mouse drivers, VGA graphics card, DMA controller, PCI bus, PCI host bridge, interval timer, real-time clock, interrupt controller, and PCI ISA bridge. Each device has its own specification sheet, which must be read and translated into software. The x86 processor manual runs to 1,500 pages, and in all there are approximately 2,000 pages of technical manuals. The x86 instruction set is large, with up to 65,000 possible instructions that could be issued in a program's code, four different protection levels to set on memory pages, four different processor "modes," and in each mode the instructions can (and do) have different effects.

So before embarking on this mammoth task, it is important to assess whether the outcome will be worth it. Both the Bochs and QEMU projects have achieved this feat, and some reassurance can be gained by examining how they approached the problems. However, these projects are C/C++ programs and therefore give only limited assistance, as JPC has to stay within the pure Java environment with the extra design restrictions and performance considerations that implies.

Some simplification is possible by selecting a simple set of hardware to emulate. As the processor emulation will be the major bottleneck in the system, there is little point in emulating

---

[‡] For example, the hardware-supported x86 CPU virtualization (Intel VT, AMD Pacifica) has security vulnerabilities due to the shared L1/L2 cache of multicore chips.

a complex but fast hard disk controller (or drive). A simple and reliable hard disk emulation will easily suffice, and the same holds true for all other hardware components. Even when looking at the processor, you can choose a Pentium II as the target, and there is no need to emulate instruction set extensions that are present in the latest chips. Because modern software, including operating systems, typically work perfectly well without such instructions to ensure backward compatibility, this decision is not significantly limiting.

Nevertheless, emulation speed remains the major challenge. The obvious way to get the best performance is to use some form of dynamic binary translation to move from a laborious step-by-step software emulation to a compiled mode where the raw speed of the underlying hardware is more efficiently exploited. This technique is used in many different guises; the modern x86 processor breaks the x86 instructions on first use into smaller microcodes, which it then caches for quick repeated execution. Just-In-Time-compiled Java environments similarly compile blocks of bytecodes into native code to improve the speed of execution for code that is repeated many times. The effectiveness of such techniques is due to the fact that in almost all software, 10% of the code represents 90% of the execution time.§ Finding simple ways of improving the speed of repeating this 10% of "hot" code can increase the overall speed dramatically. Also note that applying this technique does not imply compiling all the code; rather, selective optimization can result in a massive performance gain without the unacceptable delay that would ensue if all code were optimized.

The fact that virtually all software, whether compiled into native x86 from C/C++ or Java bytecode, is subjected to various amounts of dynamic binary translation by modern computers prior to execution shows the power and effectiveness of this technique. Thus, when approaching the task of creating JPC, there is hope for reasonable speed if similar techniques can be applied.

By gaining reassurance from existing emulators that this work is plausible, and by selecting a simple PC architecture as the initial target, we reduced the original scope to an achievable level. We then needed to assess the potential of dynamic binary translation to improve speed and ensure that a realistic performance could be achieved, even in principle. If the best outcome for JPC with all programming tricks applied was to run at 1%, then the project would not have been worth it.

▶ 这是性能调优的关键。首先需要搞清楚性能瓶颈发生在哪里，然后再对症下药。例如在我参与的一个项目中，我们发现程序员编写的代码对Spring的application-context.xml进行了重复加载，导致性能下降。在另外一个项目中，我们发现可以并行处理的几个步骤却采用了串行执行。整体而言，代码中如果涉及访问外部资源、对象频繁创建、多线程处理、过多的抽象、远程调用及对象的序列化等都可能导致性能问题。

Potential Processor Performance Tests

In order to evaluate the potential performance level achievable using various emulation tactics, a "Toy" processor was invented as a simple model. The Toy processor has 13 instructions,

---

§ This apocryphal rule of thumb was actually verified using JPC during the boot sequence of DOS and when playing numerous DOS games. With the total control of an emulator, it is easy to compile such statistics on program execution. See also Donald E. Knuth's "An empirical study of FORTRAN programs." (*Software Practice and Experience*, 1: 105-133. Wiley, 1971.)

2 registers, and 128 bytes of RAM. A simple program equivalent to the following C code was written in Toy assembly language for speed tests:

```
for (int i = 0; i < 10; i++)
    for (int j = 0; j < 50; j++)
        memory[51 + j] += 4;
```

The memory is initially all zero, and the output of the program is the memory state at the end (the first 50 bytes of memory are reserved for the program code).

A Java emulation of the Toy architecture was built, and 100,000 sequential runs of this program on the Sun HotSpot VM took 8000 ms. On the same hardware, a GCC-compiled C program took 86 ms (compiled with all static optimizations). Not surprisingly, naive emulation suffered a performance penalty of two orders of magnitude. But this simple emulation was indeed very simple, reading the next assembly instruction each time from memory, selecting the operation to carry out via a switch statement, and looping around until complete.

A better version would be to eliminate this lookup-dispatch process. This represents the commonplace trick of inlining; a C compiler inlines often-used code at compile time and the HotSpot VM does this at runtime in response to real execution telemetry. This done, the emulator then took 800 ms, a 10 times speed improvement.

Now, when using lookup-dispatch, the instruction pointer must be updated after each instruction so that the processor knows where to fetch the next one. However, with the code inlined, the instruction pointer needs to be updated only when the whole inline block has been executed. Removing these interleaved increments also helps HotSpot optimize the code; it can focus on the key operations without worrying about the need to keep this instruction pointer register always consistent with progress. With the instruction pointer updates shifted to the end of the inlined program sections, the execution speed reduced to 250 ms, an additional improvement of 3.2 times.

Nevertheless, the assembly code, translated by a simplistic automatic algorithm, resulted in many unnecessary movements of data between memory and registers. In several places a result was stored into the byte array of memory and then immediately read out again. A simple flow control analysis meant that these inefficiencies can be detected automatically and removed, resulting in a execution speed of 80 ms. This is as fast as the optimized native program!

Thus, with suitable runtime compilation, not much of it particularly complicated, a Java emulator can run this Toy native code at 100% native hardware speed. Clearly, expecting 100% when the vastly more complex hardware of the x86 PC is examined would be unrealistic, but these tests did suggest that practically useful speed might be attainable.

So, despite what might have been initial skepticism for the whole concept of a pure Java x86 emulator, there was actually sufficient evidence that such technology could be built. The next sections detail how the architectural aspects of the x86 PC hardware design were exploited and mapped onto the JVM to make an efficient emulation. A number of critically important software design decisions, based purely on how the JVM behaves, are also outlined where they had a significant performance benefit.

The PC Architecture

The modern PC is a very complicated beast. Its hardware has been optimized and iterated over many times to produce a highly effective and generalized computing platform. However, it also carries with it legacy components and functionality designed to maintain its backward compatibility. In some ways this is a blessing. The basic architecture of an IA-32 machine has not changed since its advent in 1985 with the 386. In fact, in terms of system architecture, the 386 itself was not much of a departure from its x86 predecessors.

Although Figure 9-1 is in some ways grossly simplified, with some text changes and some duplication of boxes, this could easily pass as an architectural diagram for JPC itself.

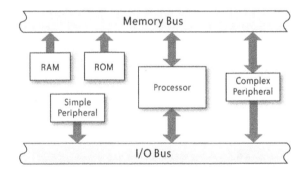

FIGURE 9-1. Basic architecture of a modern PC

▶ 正如一位计算机科学家提过:"软件中的任何问题都可以通过引入一层间接来解决；任何性能问题都可以通过减少间接来解决。"所以，我们在设计时真的需要针对实际的项目情况对二者进行设计上的权衡。

Designing the bulk of JPC was a relatively simple matter of systems analysis, and mapping from the original system to JPC for the bulk of the emulation is almost a 1:1 correspondence between the hardware specs and the Java class. For example, a serial port in JPC for example is represented by a single class, `SerialPort`, that implements `HardwareComponent` and `IOPortCapable`. This simplistic approach gives rise to a design that is easy to understand and navigate, and on the whole, objects within the architecture are loosely coupled to each other. This gives JPC the benefit of being very flexible, so just as in a real machine, virtual devices can be "plugged in" to the PCI bus, and components can be interchanged to build virtual machines of wide-ranging specifications.

▶ 在设计上，我们需要去关注 hot spot（即所谓 "热点"）。一方面热点往往是需求变化较为频繁的地方，另一方面则是指被频繁调用的代码，这可能成为性能瓶颈。

The only reason to depart from this path is when clarity of design and modularity are in direct competition with performance. This occurs in two key places in JPC: once at the concentration of computation (the processor) and once at the concentration of bandwidth (the memory system). These hot spots have two effects on the project:

- The codebase becomes lopsided. Simple hardware devices such as the IDE interface are simple translations of the specification documents, and correspond to two classes: `IDEChannel` and `PIIX3IDEInterface`. The processor, a more complicated device, comparatively has a huge amount of code related to it. In total, it is represented by eight distinct packages and over 50 classes.

- As developers, we find that we need to become as schizophrenic as the codebase is. Crudely speaking, when you are working on hardware emulation outside the memory or processor systems, you are aiming for ultimate code clarity and modular design. When working within the processor or memory system, you are aiming for ultimate performance.

The hard hat that must be worn while working in the sensitive parts of the architecture is one of pessimistic inventiveness. In order to gain as much performance as possible, we continually have to be prepared to experiment. But even small changes to the codebase must be viewed with suspicion until they have been proven to have, at a minimum, no detrimental effect on performance.

The requirement for maximum performance at the bottlenecks of the emulation is what makes JPC an interesting project to work on and with. The remainder of this chapter concentrates on how we achieved what we believe is a maintainable and logical design, without compromising the performance of the system.

Java Performance Tips

The First Rule of Optimization: Don't do it.

The Second Rule of Optimization (for experts only): Don't do it yet.

—*Michael A. Jackson*

Like all performance tips, the following are guidelines and not rules. Code that is well designed and cleanly coded is almost always infinitely preferable to "optimized" code. Invoke these guidelines only when either a positive effect will be seen on the design or that last drop of performance is really necessary.

Tip #1: Object creation is bad

Excessive object instantiation (especially of short-lived objects) will cause poor performance. This is because object churn causes frequent young generation garbage collections, and young generation garbage-collection algorithms are mostly of the "stop-the-world" type.

Tip #2: Static is good

If a method can be made static, then make it so. Static methods are not virtual, and so are not dispatched dynamically. Advanced VMs can inline such methods much more easily and readily than instance methods.

Tip #3: Table switch good, lookup switch bad

    Switch statements whose labels are a reasonably compact set are faster than those whose values are more disparate. This is because Java has two bytecodes for switches: `tableswitch` and `lookupswitch`. Table switches are performed using an indirect call, with the switch value providing the offset into a function table. Lookup switches are much slower because they perform a map lookup to find a matching value:function pair.

Tip #4: Small methods are good methods

    Small chunks of code are nice, as just-in-time environments generally see code on a method granularity. A large method that contains a "hot" area, may be compiled in its entirety. The resultant larger native code may cause more code-cache misses, which is bad for performance.

Tip #5: Exceptions are exceptional

    Exceptions should be used for exceptional conditions, not just for errors. Using exceptions for flow control in unusual circumstances provides a hint to the VM to optimize for the nonexception path, giving you optimal performance.

Tip #6: Use decorator patterns with care

    The decorator pattern is nice from a design point of view, but the extra indirection can be costly. Remember that it is permitted to remove decorators as well as add them. This removal may be considered an "exceptional occurrence" and can be implemented with a specialized exception throw.

Tip #7: `instanceof` *is faster on classes*

    Performing `instanceof` on a class is far quicker than performing it on an interface. Java's single inheritance model means that on a class, `instanceof` is simply one subtraction and one array lookup; on an interface, it is an array search.

Tip #8: Use `synchronized` *minimally*

    Keep synchronized blocks to a minimum; they can cause unnecessary overhead. Consider replacing them with either atomic or volatile references if possible.

Tip #9: Beware external libraries

    Avoid using external libraries that are overkill for your purposes. If the task is simple and critical, then seriously consider coding it internally; a tailor-made solution is likely to be better suited to the task, resulting in better performance and fewer external dependencies.

Four in Four: It Just Won't Go

Many of the problems that we have come across while developing and designing JPC have been those associated with overhead. When trying to emulate a full computing environment inside itself, the only things that prevent you from extracting 100% of the native performance are the overheads of the emulation. Some of these overheads are time related such as in the processor emulation, whereas others are space related.

The most obvious place where spatial overheads cause a problem is in the address space: the 4 GB memory space (32-bit addresses) of a virtual computer won't fit inside the 4 GB (or less) available in real (host) hardware. Even with large amounts of host memory, we can't just declare `byte[] memory = new byte[4 * 1024 * 1024 * 1024];`. Somehow we must shrink our emulated address space to fit inside a single process on the host machine, and ideally with plenty of room to spare!

To save space, we first observe that the 4 GB address space is invariably not full. The typical machine will not exceed 2 GB of physical RAM, and we can get away with significantly less than this in most circumstances. So we can crush our 4 GB down quite quickly by observing that not all of it will be occupied by physical RAM.

The first step in designing our emulated physical address space has its origin in a little peek at the future. If we look up the road we will see that one of the features of the IA-32 memory management unit will help guide our structure for the address space. In protected mode, the memory management unit of the CPU carves the address space into indivisible chunks that are 4 KB wide (known as pages). So the obvious thing to do is to chunk our memory on the same scale.

Splitting our address space into 4 KB chunks means our address space no longer stores the data directly. Instead, the data is stored in atomic memory units, which are represented as various subclasses of `Memory`. The address space then holds references to these objects. The resultant structure and memory accesses are shown in Figure 9-2.

NOTE
To optimize `instanceof` lookups, we design the inheritance chain for `Memory` objects without using interfaces.

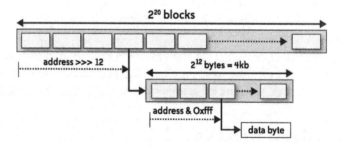

FIGURE 9-2. Physical address space block structure

This structure has a set of 2^{20} blocks, and each block will require a 32-bit reference to hold it. If we hold these in an array (the most obvious choice), we have a memory overhead of 4 MB, which is not significant for most instances.

TIP #7: INSTANCEOF IS FASTER ON CLASSES

Performing instanceof on a class is far quicker than performing it on an interface. Java's single inheritance model means that on a class, instanceof is simply one subtraction and one array lookup; on an interface, it is an array search.

Where this overhead is a problem, we can make further optimizations. Observe that memory in the physical address space falls into three distinct categories:

RAM

Physical RAM is mapped from the zero address upward. It is frequently accessed and low latency.

ROM

ROM chips can exist at any address. They are infrequently accessed and low latency.

I/O

Memory-mapped I/O can exist at any address. It is fairly frequently accessed, but is generally higher latency than RAM.

For addresses that fall within the RAM of the real machine, we use a one-stage lookup. This ensures that accesses to RAM are as low latency as possible. For accesses to other addresses, those occupied by ROM chips and memory-mapped I/O, we use a two-stage lookup, as in Figure 9-3.

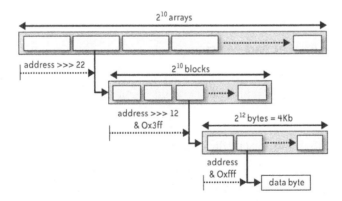

FIGURE 9-3. Physical address space with a two-stage lookup

Now a memory "get" from RAM has three stages:

```
return addressSpace.get(address);
    return blocks[i >>> 12].get(address & 0xfff);
        return memory[address];
```

And one from a higher address has four:

```
return addressSpace.get(address);
    return blocks[i >>> 22][(i >>> 12) & 0x3ff].get(address & 0xfff);
        return memory[address];
```

This two-layer optimization has saved us memory while avoiding the production of a bottleneck in every RAM memory access. Each call and layer of indirection in a memory "get" performs a function. This is indirection the way it should be used—not for the sake of interfacing, but to achieve the finest balance of performance and footprint.

Lazy initialization is also used in JPC wherever there is chance of storage never being used. Thus a new JPC instance has a physical address space with mappings to Memory objects that occupy no space. When a 4 KB section of RAM is read from or written to for the first time, the object is fully initialized, as in Example 9-1.

EXAMPLE 9-1. Lazy initialization

```
public byte getByte(int offset)
{
    try {
        return buffer[offset];
    } catch (NullPointerException e) {
        buffer = new byte[size];
        return buffer[offset];
    }
}
```

The Perils of Protected Mode

The arrival of protected mode brings a whole new system of memory management into play, with another layer of complexity added on top of the physical address space. In protected mode, paging can be enabled, which allows the rearrangement of the constituent 4 KB blocks of the physical address space. This rearrangement is controlled by a sequence of tables held in memory that can be dynamically modified by the code running on the machine. Figure 9-4 shows the path followed on a complete page translation.

In principle, every memory access on the machine will require a full lookup sequence through this paging structure to find the physical address that the given linear address maps to. As this process is so convoluted and costly, a real machine will cache the result of these lookups in translation look-aside buffers (TLBs). In addition to these added layers of indirection, there are extra protection features with memory paging. Each mapped page can be given a user or supervisor and a read or read/write status. Code that attempts to access pages without sufficient privileges, or tries to access pages that simply do not exist, results in the raising of a processor exception in a manner analogous to a software interrupt.

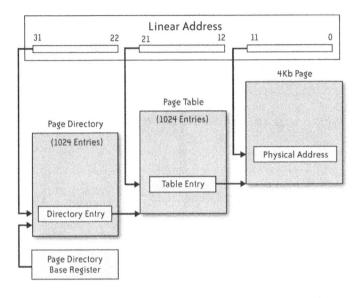

FIGURE 9-4. Paging mechanism of the IA32 architecture (4 KB pages only)

To optimize such a structure, it is clear that our first tactic should be to adopt the approach of the real processor; in other words, we need to have some form of lookup caching. To devise this, we make two key observations:

- The remapping of paging occurs on a granularity of 4 KB. Conveniently, and not by coincidence, this is also the granularity we chose for our physical address space.
- When a protected mode process accesses the memory for read or write, it merely sees a remapping of these 4 KB blocks (while some of the blocks cause processor exceptions). The physical address space is just one possible ordering of the original Memory objects (where all the objects are by coincidence in address order) but is otherwise no more significant than any other ordering.

From these observations we see that the most natural form for our cache (i.e., our TLBs) is a duplication of the physical address space structure. The memory pages are mapped according to a new ordering as determined by the page table lookups. On the first request for an address within a given page, a full traversal of the table structure is performed in order to find the matching physical address space memory object. A reference to this object is then deposited at the correct location within the linear address space, and from then on it is cached until we choose to clear it.

To solve the problem of read/write and user/supervisor, we make a tactical decision to sacrifice some memory overhead for speed. We produce a linear address space for each combination:

read-user, read-supervisor, write-user, and write-supervisor. Memory "gets" use the read indices and "sets" use the write indices. Thus, a switch between user mode and supervisor mode just requires changing two references in the memory system, as shown in Figure 9-5.

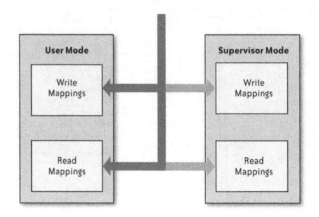

FIGURE 9-5. Protection level switching in the linear address space

NOTE

This fits in nicely with the Linux kernel's approach to paging. In Linux, every user-mode process has the kernel pages mapped in its linear address space. In hardware this prevents a context switch on transferring to kernel code for routine maintenance operations or system calls. Our switches to kernel space are just the flipping of array references, which is also a low-cost process.

This still leaves us with the problem of page faults and protection violations. Our attitude toward dealing with these conditions is, in some ways, an indication of an acceptance of the true nature of exceptions, both in the low-level processor sense and in the Java language. This attitude can be summed up as follows:

> An Exception is an exception, and an Error is an error.

More verbosely: an Exception should represent rare or exceptional conditions, but not necessarily fatal ones; an Error should represent fatal or certainly seriously undesirable situations.

Considering the extent to which exception handling is implemented in the Java language, it is in some ways curious that many programmers are reluctant to use the exception-handling mechanism to their advantage. Exceptions represent the most efficient and cleanest way for software to handle rare but correctable conditions. Throwing an exception will bias the execution cost by effectively optimizing for the common path, and concurrently will increase the cost involved in the inevitable exceptional situation. It is obvious from this discussion that the throwing of a page fault or protection violation is an exceptional but normal occurrence that should be mapped to an instance of Exception.

TIP #5: EXCEPTIONS ARE EXCEPTIONAL

Exceptions should be used for exceptional conditions, not just for errors. Using exceptions for flow control in unusual circumstances provides a hint to the VM to optimize for the nonexception path, giving you optimal performance.

In handling page faults and exceptions, we have been slightly cavalier in throwing out two of the usual conventions on the use of exceptions:

- `ProcessorException`, which is the class that represents all Intel processor exception types, extends `RuntimeException`.
- Page faults, and to a lesser extent protection violations, are exceptional conditions, but they are common enough that in order to refrain from sacrificing performance, we throw them using static exception instances.

Both of these decisions were made for good reasons, but were born in large part out of the peculiarities of the JPC project.

Choosing to extend `RuntimeException` was mainly a code beautification decision. Traditionally, runtime exceptions are reserved for exceptions that cannot or should not be handled. Most texts recommend that if thrown, they are allowed to propagate up and cause the termination of the throwing thread. We know that the distinction between runtime and checked exceptions is source "candy" in the same vein as generics, autoboxing, or the for-each loop. Not that we mean to denigrate these constructs, but by observing that the classification of an exception has no effect on the compiled code, we can safely choose the most convenient classification without risking any performance penalty. In our case, `ProcessorException` is a runtime exception in order to avoid having to declare `throws ProcessorException` on a multitude of methods.

▶在Java项目中，我个人也倾向于使用runtime exception。checked exception 会导致异常对接口的污染，从而影响接口的可扩展性。当然，亦有另外一种声音，认为Java的cheekeel exception 可以有效地保证代码的健壮性。两种类型的异常各有其适用场景，关键在于你对代码质量的追求。

Throwing statically initialized exception instances (effectively singleton exceptions) for page faults and protection violations yields benefits at both ends of the exception-handling chain. First, we save ourselves the cost of instantiating a new object every time we want to throw. This is avoids the cost of reconstructing the stack trace of the calling thread (far from trivial in a modern JVM). Second, we save ourselves the cost of determining the nature of the thrown type, as with a limited set of static exceptions determining the thrown type, we only need to perform a sequence of reference comparisons.

Fighting A Losing Battle

Nowhere in the IA-32 architecture does its enduring popularity show more than in the instruction set. What was once a simple accumulator-based architecture in the days of the 8080 has grown through the years into a vast and complicated array of instructions. IA-32 has become a RISC-like chip with numerous bolt-on extras and a bewildering array of addressing modes.

When approaching such a landscape as a Java developer, it is very tempting to revert to type and start writing classes as if the more structure you code, the simpler the problem will get. This approach would be fine, if not ideal, were we developing a disassembler. In such a system with so much object creation, there is also inevitably a large amount of garbage collection.

This results in a double speed penalty. Not only do we suffer the overhead of large amounts of object allocation, but we also suffer from frequent garbage collections. In a modern generational garbage-collected environment (of which the Sun JVMs are an example), small, short-lived objects are created in the young generation and almost all young-generation collection algorithms are stop-the-world. So a decoder with large amounts of object churn will suffer poor performance not only from unnecessary object allocation, but also from a large number of very short GC pauses while the collector cleans up all the object churn.

For this reason it was quite important to reduce object churn within the decoder that drives the interpreted execution. In the real mode decoder, this minimalist approach results in a 6,500-line class with just 42 instances of the "new" keyword:

- 4 × `new boolean[]` at class load time (`static final`)
- 3 × `new Operation()` for a rotating buffer
- 2 × `new int[]` for the expanding buffer in `Operation`
- 33 × `new IllegalStateException()` on exception paths

Once an instance of the decoder is created, the only necessary object construction is for expansion of the `int[]` buffers in `Operation`. Once the buffers have expanded, there is no object construction and no garbage collection, and therefore there is pause-less decoding.

The design of the decoder illustrates what we consider one of the important tenets of programming (and in this case, of Java):

> Just because you can, it doesn't mean you should.

In this case, just because a JVM can do automated garbage collection, it doesn't mean you are forced to exercise it. In a performance-critical section of code, approach object instantiation with caution. Beware of silent object instantiation done on your behalf with classes such as `Iterator`, `String`, and `varargs` calls.

TIP #1: OBJECT CREATION IS BAD

Excessive object instantiation (especially of short-lived objects) will cause poor performance. This is because object churn causes frequent young generation garbage collections, and young generation garbage-collection algorithms are mostly of the "stop-the-world" type.

Microcoding: Less Is More or More Is Less

So now we have a GC-less decoder for parsing the IA-32 instruction stream, but we have not discussed what such a decoder should decode to. The IA-32 architecture is not a fixed-length instruction system; instructions range in length from a single byte to a maximum of 15 bytes. Much of the complexity in the set is down to the plethora of memory-addressing modes that can be used for any given operand of an instruction.

The initial highest level of factorization splits each operation into four stages:

Input operands
    Loading the operation's data from registers or memory.

Operation
    Data processing on the input operands.

Output operands
    Saving the operation's results out to registers or memory.

Flag operations
    Adjusting the flag register bits to represent the result of the operation.

This factorization into operands and operations allows us to separate the simplicity of the operation from the complexity of its operands. An operation such as add eax,[es:ecx*4+ebx+8] is initially factorized into five operations:

```
load eax
load [es:ecx*4+ebx+8]
add
store eax
updateflags
```

It is immediately clear that load [es:ecx*4+ebx+8] is a far from simple operation, and easily could be factorized into a number of smaller elements. In fact, for this addressing format alone there are:

- Six possible memory segments
- Eight possible index registers
- Eight possible base registers

> 对于追求高性能的产品，尽可能早地针对性能进行benchmark测试非常有必要。因为它甚至可能影响到整个系统的架构设计。例如是否支持水平伸缩、是否需要引入缓存、存储的数据结构与介质等。另外，在选择架构方案时，还必须根据项目的情况进行判断，例如，如果是读远远多于写，且写操作并不要求实时，就可以引入CQRS，为查询和命令提供不同的架构范式，并利用消息队列完成命令的异步调用。再比如说针对大数据处理，也可以考虑引入Streaming而非MapReduce满足实时处理的需求。这些决策都是难以改变的设计，尽量在开发前确定。这与本原则并不冲突，因为在了解了性能诉求的前提下，提前对架构进行性能方面的考量，并非一种Premature optimization。

This results in 384 possible combinations for this subset of the addressing modes alone. Evidently some more factorizing is required on these address calculations. This type of memory access is therefore broken down further, until we get:

```
load eax
memoryreset
load segment es
inc address ecx*4
inc address ebx
inc address imm 8
load [segment:address]
add
store eax
updateflags
```

To produce an emulated instruction set that performs acceptably, we have had to balance two sets of priorities:

- First, we must balance decode time against execution time in order to optimize overall execution speed. We must remember that in the interpreted processor, our main target is low-latency initial execution. Commonly executed code should get handed on to later optimization stages. Our initial aim here is to get the code out the door without blocking the whole emulation. Later optimizations can be done asynchronously, and time spent here holds up everything. So we are looking for a relatively simple set that can be decoded quickly.

- Second, we have to balance instruction set size against "compiled" code length. A small set will naturally produce verbose code, and a larger set should be more compact. We say "should" because a large set can still require longer sections if it is badly chosen. An interpreter for a smaller set will be smaller in code and footprint, and therefore will execute each of its operations much faster, but conversely it has a larger set to execute. So we are looking for a reasonable balance of set size against code length in order to get a near-optimal performance out of the interpreter.

In finding the optimal point for both of these trade-offs, it is important to keep Hoare's Dictum[‖] in the back of your mind:

> **Premature optimization is the root of all evil.**
>
> —C. A. R. Hoare

The precise sweet spot for lots of these optimizations will be system-dependent. In a Java environment, the system includes not only the physical hardware but also the JVM. With the added factor that the Java component of the environment is invariably just-in-time compiled, small scoped performance benchmarks are notoriously unreliable. To this end we refrained from overusing such benchmarks to guide our coding choices, and instead relied on first

---

[‖] http://en.wikiquote.org/wiki/C._A._R._Hoare

principles and trusted benchmarks only when large shifts in performance occurred. In a just-in-time compiled environment, small changes in these micro benchmarks are, at best, not repeatable on a separate system. At worst, they are so highly dependent on the benchmarked scenario that they are not even reliably repeatable on the same system.

> **NOTE**
> One important feature of the microcode set is that the integer values to which the constants are set are sequential. The core of the interpreter is a switch statement on this set of constants, and we need to ensure that this switch runs as fast as possible.

With these factors in mind, we concluded after several experiments that a set with 750 or so codes for complete integer and floating-point emulation represented a good trade-off. This gives us an approximate factor of 10 conversion from x86 operation to microcodes. Although this set may seem large, it decodes quickly and the operations are reasonably atomic. This makes them good candidates to feed into the later optimizing stages.

TIP #3: TABLE SWITCH GOOD, LOOKUP SWITCH BAD

Switch statements whose labels are a reasonably compact set are faster than those whose values are more disparate. This is because Java has two bytecodes for switches: tableswitch and lookupswitch. Table switches are performed using an indirect call, with the switch value providing the offset into a function table. Lookup switches are much slower because they perform a map lookup to find a matching value:function pair.

Hijacking the JVM

The refrain of "Java is slow" haunts Java developers to this day. The bulk of this comment derives from the experiences of non-Java developers with early JVMs back in the mid-to-late 1990s. Since that time those of us who work with Java know that it has moved in leaps and bounds. The driving force behind this improvement is also the key to speeding up JPC: the environment of a conventional Java process can be partitioned very simply between program regions and data regions.

In Figure 9-6 we can see that the data region is then further split between static data, which can be known at compile time, and dynamic data, which cannot. The bulk of the static data in a Java environment are the class bytes loaded from the classpath. Although the class bytes are loaded as data, it is clear that they actually represent code, and they will get interpreted by the JVM at runtime. So it is obvious that we would like to maneuver these class bytes onto the other side of the diagram.

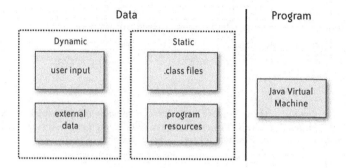

FIGURE 9-6. *Program and data regions in a Java process*

In a "Just-In-Time" compiled environment such as Sun HotSpot, the commonly used sections of bytecode are translated or dynamically compiled into the native instruction set of the host machine. This moves the class bytes from the data region into the code region. These classes then execute as native code that accelerate the program to native speed. See Figure 9-7.

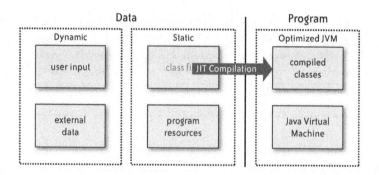

FIGURE 9-7. *Just-In-Time compilation in a Java environment*

In JPC we take advantage of the fact that not all class data has to be known at JVM startup. In fact, in Java-speak, "static data" would be better referred to as "final data." When a class is loaded, its class bytes are fixed and cannot be changed (let's ignore the JVM TI[#] for the sake of convenience). This allows us to define new classes at runtime, a concept that will be immediately familiar to those who work with plug-in architectures, applets, or J2EE web containers.

[#] For those of us that like mucking around with the naughty bits of the JVM, the Tool Interface (*http://java.sun.com/javase/6/docs/technotes/guides/jvmti/*) can do some very interesting things, including class file redefinition.

We then repeat the just-in-time compilation trick performed by the JVM, but do so at the JPC level. So for JPC we have a second-tier program-data information divide within the confines of the Java Runtime Environment. Our compilation now has two stages:

1. IA-32 machine code is compiled into bytecode on demand within JPC. These x86 blocks thus become valid Java class files that can be loaded as such by the JVM.
2. Classes are compiled by the JVM into native code. As the JVM does not distinguish between the original "static" classes that comprise the handwritten code and the dynamic classes built automatically, both types are optimized to get the best native performance possible.

After these two stages of compilation, our original IA-32 machine code will have been translated into the host machine's architecture (see Figure 9-8). With a nice dollop of good fortune, the new instruction count will not be significantly larger than the original, which means performance will not be significantly slower than native.

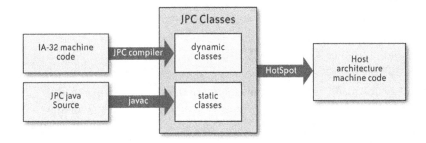

FIGURE 9-8. *The three compilers in the JPC architecture*

So now we know how to get more speed out of our emulator, but we have slightly glossed over the details. These details (our new problems) now fall into two distinct areas: how do we perform this compilation, and how should we load the resultant classes?

Compiling: How to Reinvent the Wheel

The compiler we describe here is not the most optimal example, but we are in a slightly unusual situation. Both *javac* and the JPC compiler are only first-stage compilers, meaning they simply feed their output into a second stage, be this a bytecode interpreter or a just-in-time compiler. We know that *javac* does very little to optimize its output; bytecodes output by *javac* are simply a translation of the input Java code. See Example 9-2.

EXAMPLE 9-2. Javac's nonoptimal compiling

```
public int function()         public int function();
{                              0:   iconst_1
    boolean a = true;          1:   istore_1
    if (a)                     2:   iload_1
        return 1;              3:   ifeq      8
    else                       6:   iconst_1
        return 0;              7:   ireturn
}                              8:   iconst_0
                               9:   ireturn
```

This minimal optimization is easily justified because most of the optimization work is being left as an exercise for the latter stages of compilation. In JPC not only do we have this reasoning, but we also have the burden of extra time pressure:

- We want minimal overhead compiling in order to avoid stealing CPU cycles from the other emulation threads.
- We want minimal latency compiling so that the interpreted class is replaced as soon as possible. A high-latency compiler may find that the code is no longer needed by the time it has finished the compile.

Simple code generation

The compiling task in JPC is now a simple matter of translating the microcodes of a single interpreted basic block into a set of Java bytecodes. In the first approximation we will assume that the basic block cannot throw exceptions. This means that a basic block is a strictly defined basic block and has exactly one entry point and one exit point. Each variable modified by a given basic block can now be expressed as a function of the set of input registers and memory state. Internally in JPC, we collectively represent these functions as a single directed acyclic graph.

Graph source
  Sources represent input data in the form of register values or instruction immediates.

Graph sink
  Sinks represent output data in the form of register values and memory or I/O port writes. For each state variable that the block affects, there will be one sink.

Graph edge
  Edges represent variable value propagation within the graph.

Graph node
  Nodes represent operations performed on incoming edges whose results propagate along outgoing edges. In JPC, the operations are a single-state modification component of an interpreted microcode. Hence one microcode may map to multiple graph nodes if it affects multiple variables.

Converting this graph representation of the interpreted basic blocks into Java bytecode is a simple depth-first traversal of the graph from each sink in turn. Each node in the graph takes the topmost elements of the stack as input and then leaves its result at the top, ready for processing by any child nodes. See Figure 9-9.

> **NOTE**
> Classically optimal traversal order for the graph requires that at each node, the ascendants should be evaluated in depth order. The node with the longest path to the farthest source is evaluated first, and that with the shortest path last. On a register-based target, this will produce the shortest code because it eliminates the majority of the register juggling. On a stack-based machine such as the JVM, the same decision-making process will give rise to code with a minimal stack depth. In JPC we neglect such niceties and rely on the JVM to iron out the difference, the extra complication of tracking all the node depths is simply not worth the effort.

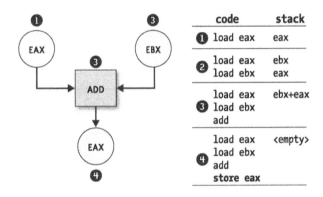

FIGURE 9-9. *Representing x86 operations as a directed acyclic graph*

This sink-by-sink depth-first parsing causes natural optimization of the graph, as shown in Figure 9-10. Orphaned sections of the graph that correspond to unused code cannot be accessed via the sinks and so will be automatically removed, as the parse will never reach them. Reused code sections will find themselves evaluated multiple times in one parse of the graph. We can cache their result in a local variable and simply load the result on subsequent visits to the node, thus saving any code duplication.

The code associated with the node in a tree is then represented by a single static function compiled from source as part of the JPC codebase. The code we generate is then just a sequence of pushes of processor variables and immediates onto the stack, followed by a sequence of `invokestatic` calls for each node, and finally a sequence of pops back into the processor object.

▶ 这一技巧需斟酌，至少应根据场景而定。从执行效率的角度考虑，自然 static method 的效果更高；但是，若从可扩展性以及可测试性的角度看，我们恰恰需要避免定义 static method，因为静态方法不具备多态能力，也很难被 mock。故而，任何设计原则和设计技巧都需要结合当前上下文来综合判定，不可武断。

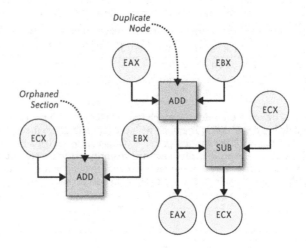

FIGURE 9-10. Directed acyclic graph features

TIP #2: STATIC IS GOOD

If a method can be made static, then make it so. Static methods are not virtual, and so are not dispatched dynamically. Advanced VMs can inline such methods much more easily and readily than instance methods.

Handling exceptions

Now that we can handle the compilation of basic blocks, we must restore some of the ugliness to the situation. As we discussed previously, exceptions are not always errors. Page faults and protection violations are thrown with abandon as a matter of course. When an exception is thrown, the processor state must be consistent up to the last successful execution of an operation. This obviously has rather severe implications for the compiler. Having mapped exceptions in IA-32 to Java exceptions, we know that the only practical solution to this problem is to catch the exception and, once inside the exception handler, ensure that the state is consistent with the last successful operation.

The behavior of an exception path within any given basic block is just like that of the basic block itself. It has one entry point and one exit point, and the only difference from the main path is the exit point. As the exception path is not so very different from the basic block itself, the most natural way to represent it is with its own directed acyclic graph. The graph for the exception path will share the same node set as the basic block, but will have a distinct set of sinks mapped to its own distinct exit point.

The code for the exception handler is then produced by traversing the chosen paths graph. The nodes shared in common with the main path are reset back to their state at the exact point mid main path traversal when the exception is thrown. This means that all cached and calculated values from the main path can be reused in the exception handler, thus avoiding repeating any work.

Bytecode manipulation

Converting our compiled sections of bytecode into loadable classes is something for which there are a number of established and well-engineered solutions. Apache's Byte Code Engineering Library (BCEL) considers itself to conveniently "analyze, create, and manipulate (binary) Java class files."[*] ASM is an "all purpose Java bytecode manipulation and analysis framework."[†]

Unfortunately, all we want to do is modify a single method (always the same one) in a single skeleton class. We can only generate a small subset of the possible set of bytecode sequences, and we don't need to provide any analysis tools. It would appear then that both BCEL and ASM are overkill for our needs. Instead, we developed a custom bytecode manipulation library with very limited capabilities that exactly matched what we needed. For example, our stack-depth algorithm is tuned to rapidly assess the maximum stack depth of our methods (so that they can pass verification). Although this algorithm does not work for general class compiling, it is sufficient and more efficient for our purposes.

TIP #9: BEWARE EXTERNAL LIBRARIES

Avoid using external libraries that are overkill for your purposes. If the task is simple and critical, then seriously consider coding it internally; a tailor-made solution is likely to be better suited to the task, resulting in better performance and less external dependencies.

Class Loading and Unloading, but on a Big Scale

So now we have classes, and they need to be loaded. The obvious first question is, "How many?" Figure 9-11 gives an illustration of the scale of the problem. Loading 100,000 classes into a single JVM can prove to be a slight challenge.

---

[*] http://jakarta.apache.org/bcel

[†] http://asm.objectweb.org/

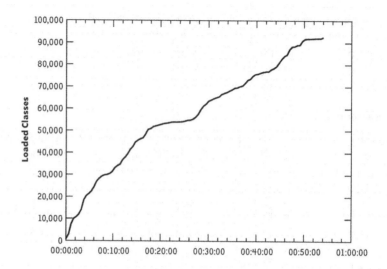

FIGURE 9-11. Class count during a modern GNU/Linux boot

So we need to find the space for all these classes. Class file storage in the JVM occurs in a special memory area outside the normal object heap known as the "Permanent Generation" space. In a typical Sun JVM, the permanent generation starts out at 16 MB and can grow to at most 64 MB. Obviously 100,000 classes are not going to fit into a 64 MB heap. There are two ways we can solve this problem.

The first is with the command *java -XX:MaxPermSize=128m*. Clearly, solution one is to increase the size of the permanent generation. Crude though this may be, it does help to solve the problem. Unfortunately, it's no solution on its own, because all we have done is delay the inevitable. Eventually we will load enough classes to fill the new space, and we can't just keep adding more room.

The second half of the solution involves reducing the number of loaded classes. In fact, shouldn't the garbage collector be clearing away all the unused classes? Classes are really no different from heap objects as far as garbage collection goes. A class can be garbage collected (unloaded) only if no live references to the class are held. Of course every instance of a class holds a strong reference to the `Class` object of its type. So for a class to be collected, there must first be no live instances of the class, and then no additional references to the class. Therefore, perhaps we can solve the problem if we define our custom classloader as shown in Example 9-3.

EXAMPLE 9-3. Simple no-holding ClassLoader

```
public class CustomClassLoader extends Classloader
{
    public Class createClass(String name, byte[] classBytes)
    {
        return defineClass(name, classBytes, 0, classBytes.length);
```

```
    }

    @Override
    protected Class findClass(String name) throws ClassNotFoundException
    {
        throw new ClassNotFoundException(name);
    }
}
```

The `ClassLoader` in Example 9-3 holds no references to the classes it defines. Each class is a singleton with no relations, and so `findClass` is safe to throw `ClassNotFoundException` and all is right with the world. Once the singleton instance of each class becomes a GC candidate, so will the class itself, and both can be collected. This works great, and the classes all get loaded. However, something goes wrong. For some unknown reason, no classes get unloaded, ever. It appears that some object somewhere is holding onto references to our classes.

Let's look down the list of calls on defining a new class:

```
java.lang.ClassLoader: defineClass(...)
java.lang.ClassLoader: defineClass1(...)
ClassLoader.c: Java_java_lang_ClassLoader_defineClass1(...)
vm/prims/jvm.cpp: JVM_DefineClassWithSource(...)
vm/prims/jvm.cpp: jvm_define_class_common(...)
vm/memory/systemDictionary.cpp: SystemDictionary::resolve_from_stream(...)
vm/memory/systemDictionary.cpp: SystemDictionary::define_instance_class(...)
```

This is what happens when you're not satisfied with the answer "because it does." We get down this far into the guts of the JVM, and we find this little snippet of code:

```
// Register class just loaded with class loader (placed in Vector)
// Note we do this before updating the dictionary, as this can
// fail with an OutOfMemoryError (if it does, we will *not* put this
// class in the dictionary and will not update the class hierarchy).
if (k->class_loader() != NULL) {
  methodHandle m(THREAD, Universe::loader_addClass_method());
  JavaValue result(T_VOID);
  JavaCallArguments args(class_loader_h);
  args.push_oop(Handle(THREAD, k->java_mirror()));
  JavaCalls::call(&result, m, &args, CHECK);
}
```

which is a call back up to the Java level into the classloader instance that is doing the loading:

```
// The classes loaded by this class loader.  The only purpose of this table
// is to keep the classes from being GC'ed until the loader is GC'ed.
private Vector classes = new Vector();

// Invoked by the VM to record every loaded class with this loader.
void addClass(Class c) {
    classes.addElement(c);
}
```

So now we know which naughty little object is holding a reference to all of our classes and keeping them from being garbage collected. Unfortunately, there is little we can do about this.

Well, little that we can do without violating one of the absolute maxims and declaring a class in the java.lang package. So we now know that our superclass is going to helpfully hold references for us. What does this mean for the class unloading?

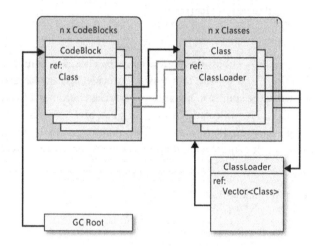

FIGURE 9-12. The GC root path for classes

In the GC root path in Figure 9-12, we can see that until all the instances of all the classes loaded by a classloader are GC candidates, all the classes will remain loaded. So one active codeblock can stop n classes from being unloaded. That's no good at all.

There is one simple way to mitigate this problem. No one said that we had to let n rise uncontrollably. If we limit the number of classes loaded by any one loader, then we decrease the chances of classes being held hostage. In JPC we have a custom classloader that will only load up to 10 classes by default. Loading the 10th class triggers the construction of a new loader, which will be used for the next 10, and so on. This method means that any one class can only hold up to 10 others hostage; see Example 9-4.

EXAMPLE 9-4. ClassLoader implementation in JPC

```
private static void newClassLoader()
{
    currentClassLoader = new CustomClassLoader();
}

private static class CustomClassLoader extends ClassLoader
{
    private int classesCount;

    public CustomClassLoader()
    {
        super(CustomClassLoader.class.getClassLoader());
    }
```

```java
    public Class createClass(String name, byte[] b)
    {
        if (++classesCount == CLASSES_PER_LOADER)
            newClassLoader();

        return defineClass(name, b, 0, b.length);
    }
    protected Class findClass(String name) throws ClassNotFoundException
    {
        throw new ClassNotFoundException(name);
    }
}
```

HOTSPOT CODE CACHE

In JPC, when running complex jobs that load such large numbers of classes, there is another memory limit that will get hit. In Sun HotSpot JVMs, the just-in-time compiled code is stored in a nonheap area called the code cache. Not only does JPC produce a large number of classes, it is also unusual in that a high fraction of these are candidates for HotSpot. This means that the HotSpot caches get filled rapidly. This just means that any boost to the permanent generation size normally is also accompanied by an increase in the size of the code cache.

Codeblock replacement

We now have a compiled, loaded, and instantiated custom code block instance. Somehow we have to get this block in place where it's needed. How this is done is also closely related to how the blocks are initially scheduled for execution; see Example 9-5.

EXAMPLE 9-5. Decorator pattern for compilation scheduling

```java
public class CodeBlockDecorator implements CodeBlock
{
    private CodeBlock target;

    public int execute()
    {
        makeSchedulingDecision();
        target.execute();
    }

    public void replaceTarget(CodeBlock replacement)
    {
        target = replacement;
    }
}
```

Example 9-5 shows how the code block decorator intercepts calls for execution and is then able to make a decision as to whether to queue the block for compilation. In addition to this, we also have a method that can replace the decorator's target with a different block instance. Once the block has been compiled, the decorator is useless, and so ideally we would like to replace it. This replacement is actually achieved quite easily. By replacing the original interpreted block with a block like Example 9-6, we can propagate the notice about the new block back up the call stack. Once the exception reaches the right level, we can replace the reference to the decorator with a direct reference to the compiled block.

EXAMPLE 9-6. Block replacing CodeBlock

```
public class CodeBlockReplacer implements CodeBlock
{
    private CodeBlock target;

    public int execute()
    {
        throw new CodeBlockReplacementException(target);
    }

    static class CodeBlockReplacementException extends RuntimeException
    {
        private CodeBlock replacement;

        public CodeBlockReplacementException(CodeBlock compiled)
        {
            replacement = compiled;
        }

        public CodeBlock getReplacementBlock()
        {
            return replacement;
        }
    }
}
```

TIP #6: USE DECORATOR PATTERNS WITH CARE

The decorator pattern is nice from a design point of view, but the extra indirection can be costly. Remember that it is permitted to remove decorators as well as add them. This removal may be considered an "exceptional occurrence" and can be implemented with a specialized exception throw.

Ultimate Flexibility

Armed with all of this trickery, we now have a highly optimized emulation system that can be improved and extended without the need for major architectural revision. A better compiler can be plugged into the backend of the system, and other components can be adapted and replaced with different implementations to suit a wide range of purposes. For example:

- The data that forms the virtual hard disk could actually be served (on demand) by any server anywhere in the world.
- The user interaction of the emulated system (virtual screen, keyboard, and mouse) could be via a remote system.
- JPC can run x86 software on any standard Java 2 virtual machine, and thus the underlying hardware can be chosen independently from the choice of operating system and software. In addition, the complete state of the virtual machine can be saved and the emulated machine "frozen" in time. It can then be resumed at a later date or on a different physical machine without any of the hosted software being aware of any change.

Flexible Data Anywhere

With JPC, your disk image can be carried with you on a memory stick, together with a complete JVM and JPC code. You can then plug this into any computer and "boot" your machine up to do all your private email and other work, and when you finish and unplug, you've left nothing on the host hardware.

Alternatively, your hard disk image could reside on a server on the Internet, and you could access your own machine from anywhere in the world simply by loading a local JPC and pointing it to your server. Together with suitable authentication and transport security, this becomes a powerful tool for the mobile workforce. JPC's natural fit in the Java space means that almost any device can be used as the end portal for remote access, from Java-enabled browsers to mobile devices.

For sensitive work in a highly secure environment, data security and integrity can be enforced at the most basic hardware level by working on local JPC instances whose hard drives are located on a secure server close by. Each worker gets full control of the virtual hardware they're working on, thus enabling them to work effectively. However, there is no way for them to extract data from the system, even if they try to hack the physical security of the computer in front of them: the local machine knows practically nothing of the application running inside the JVM, inside JPC, inside the guest operating system, and so on.

Even when employee trust is not an issue, working on virtual machines, especially one as flexible as JPC, means the physical hardware becomes completely replaceable at a moment's notice. Thus for disaster recovery and ultimate backups, where the entire state of the machine is backed up, not just the hard disk data, emulators such as JPC offer powerful advantages where instant failover is important (even over the WAN).

Because of JPC's hardware agnosticism, these stories apply equally well to non-x86 hardware, the ideal solution for thin clients, and users get their favorite x86 environment in all cases.

Flexible Auditing and Support

The screen, keyboard, and mouse of a running JPC instance could be viewed and overridden remotely by a suitably authorized system. Fraud could be effectively monitored by remotely collecting keystrokes and screenshots taken of suspicious activity to form an evidence base. Given the low-level hardware access made possible by JPC, this feature cannot be subverted by the technically able person who attempts to detect and remove monitoring software from within the emulated operating system. Even a user given administrator rights (on the guest system) would not be able to escape the monitoring, no matter how knowledgeable.

An auditing JPC system could simply record activity, scanning and flagging actions in real time, or go one step further and prevent certain actions. For example, in collaboration with suitable server software, a monitored instance could scan the video output for inappropriate images and then obscure them (or replace them with other content) at the virtual video card level. Such low-level monitoring means users could not subvert content protection systems by installing alternative viewing software.

Remote assistance could be far more effective if a help desk could literally see exactly what the entire screen was doing, and directly interact with the keyboard and mouse at the virtual hardware level. This would be possible even when JPC ran operating systems for which remote access was never implemented—for example, DOS, whose use is ongoing in many industries and countries around the world.

Flexible Computing Anywhere

Rather than run the main emulation on a local resource and merely import data from remote resources as necessary, the core emulation could be carried out on a central "JPC" server. Because JPC needs only a standard JVM to operate, this central JPC server could be based on hardware that is completely different from a normal x86 PC. There are some candidates already that could achieve this, and JPC has already been demonstrated on a 96 core Azul compute appliance. Other possibilities include Sun's Niagara-based servers and systems built from mobile phone technology (JPC has already booted DOS on a Nokia N95, an ARM11-based system).

But why centralize a server to run all these JPC instances? Presumably any resource on the Internet could run a JPC instance on behalf of anyone else, with the screen output and user input being piped via the network to the virtual machine owner. In this way of working, the world is viewed as "N" users with "M" machines, and there is no fixed mapping of hardware ownership relationships between users in the former group and machines in the latter. If a machine is idle, any one of the users can use it, remotely launching a JPC instance to work on their personal disk image data. If the idle machine is suddenly needed for other purposes, the JPC instance can be "frozen" and the state moved to another idle physical resource.

Although this latter modus operandi is perhaps more difficult to imagine for interactive users, for whom the freezing and resuming would take too long over the Internet to be convenient, it makes a lot of sense for users who want to run many simultaneous virtual machines in parallel and without much interaction. This is the experience of users who currently use large "batch" farms to run massively parallel tasks, such as rendering frames of an animated movie, searching for drugs via molecular simulation, optimizing engineering design problems, and pricing complex financial instruments.

Ultimate Security

Allowing unvetted code to run on your machine is fraught with danger, and this danger is getting worse. There is a rapidly growing list of malicious software ("malware") on the Internet, variously known as "trojans," "keyloggers," "hostageware," "spamware," and "viruses." You could fall victim to data loss, identity theft, and fraud, and worst of all, might become implicated in a criminal offense if you did not exercise caution when running software downloaded from an unknown or unverified source.

For every security hole patched by the makers of the popular operating systems and Internet browsers, it seems two more grow in its place. Knowing this, how can you ever run code that might genuinely enhance your browsing experience or provide useful services?

Java code, when run in the Java Applet Sandbox, has provided this level of reassurance for over a decade. Add the extra independent layer of security represented by JPC and you have a double-insulated sandbox in which to run unvetted code. The JPC website (*http://www-jpc .physics.ox.ac.uk*) demonstrates how JPC can boot DOS and run a number of classic games inside a standard applet as part of a web page; in other words, they show an unvetted x86 (DOS) executable running in a completely secure container on any machine.

There is one major downside to running JPC within an applet sandbox: the security restrictions do not allow JPC to create classloaders, and therefore the dynamic compilation that gives JPC much of its speed is disabled. The good news is that by using the inherent flexibility of the JPC design, this can be circumvented without compromising security.

Java code in an applet sandbox can load classes from the network on demand as long as they come from the same server as the applet code originally did. Exploiting this, we have built a remote compiler that compiles classes on demand from JPC instances running in applets, sending them back to these instances when asked by the JVMs responsible for running them. The local JVM merely regards these classes as static resources that just happen to be needed rather later than the rest of the classes inside JPC, whereas in fact these classes have been compiled as needed by the JPC applet instances.

In this way, we get the speed of compiled classes even within the standard Java Applet Sandbox, and so users can be reassured that no matter what JPC is running, the JVM is enforcing the basic restrictions on executing code that make the activity safe in a dangerous world.

A nice side effect of the remote compiler is that as time goes on and many JPC instances make use of it, it builds up a library of compiled classes that it can share among the JPC instances. Thus the compiler server rapidly becomes a simple web server for previously compiled classes. Moreover, by counting the number of times each compiled class is requested, the server knows which classes are most used, and therefore which code to spend more time optimizing. Although this latter feature has not yet been implemented, we believe such focused optimization could offset the execution speed penalty suffered by JPC applet clients introduced by the network delay in loading classes. Thus, JPC applet clients might perform as well as normal JPC application clients where compilation is carried out locally.‡

It Feels Better the Second Time Around

> Everybody knows that it is always possible to do a thing better the second time.
>
> —*Henry Ford*, My Life and Work

Developing in an academic environment comes with its own challenges, which are somewhat different from those in a commercial setting. In an academic environment, performance targets are mostly self imposed, which is both a blessing and a curse. Discipline is required on the part of the developers to keep the project on track and to prevent its focus from shifting. However, a free environment also allows ideas to be rapidly developed and tested to confirm or disprove their benefits. For the most creative and ambitious projects, this type of atmosphere is critical to eventual success.

The fact that the architecture of JPC has progressed as far as it has with such a small team of developers§ is the result of an overriding attitude toward coding. As Figure 9-13 shows, in the life of the project over 500,000 lines of code have been written. Of this, only 85,000 survive to this day. There have been numerous rewrites of various sections of the emulation, including one complete purge and rewrite from scratch.

---

‡ Indeed, when the compiler server and JPC Applet are connected via a decent 100 MBit LAN, there is virtually no perceived performance penalty due to network issues. The advantage of having compilation carried out elsewhere frees up CPU resources locally, and this seems to balance the network delay.

§ The average team size during the major 30-month development phase was 2.5 programmers, not all Java experts.

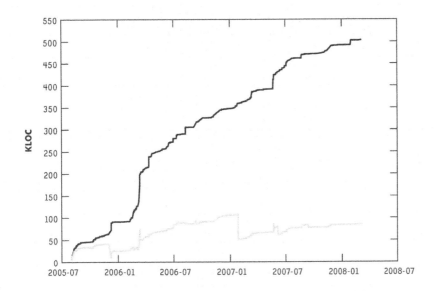

FIGURE 9-13. Accumulated code in JPC

A continuous cycle of purge, rewrite, and refine is difficult to achieve, although it is easier in an academic environment. Provided you don't get too emotionally attached to your code, the process of code deletion can be cathartic, not only for the codebase, but also for the developer's attitude toward it. Henry Ford was right: almost invariably your second attempt will be better. As long as this series of iterative improvements has a termination condition, all will be well.

So, the road to a "Beautiful Architecture" would be a four-step plan:

1. Take a large and complex problem in its entirety, and find a set of credible simpler stages that would enable a complete end-to-end prototype system to be built. Each stage represents a system simpler and less capable than the final goal, but each stage can be tested within its design limitations as part of a complete system prototype rather than as individual prototypes of small parts of the final design (i.e., a more traditional unit test).

2. Before building each part of each stage, be clear what aspect is being developed and why. Ideally, the bottlenecks in each stage can be easily identified and improvements to these will be the main targets for the next stage or stages. Look hard to find ways to prove the methodology in principle before embarking on large chunks of work—even for each part of each stage.

▶ 这四个步骤暗暗契合了风险驱动设计（Risk Driven Design，RDD）的设计思路，首先识别风险，并对风险排定优先级；然后寻找解决该风险的最适宜（注意，不是最好）的解决方案，并运用之；最后，对风险进行评估，判断风险是否已得到解决。RDD的一个要求是进行恰如其分的架构设计，不可无限期地追求方案的完美。

3. Code each stage to completion and system test the whole prototype, resisting the temptation to push too quickly beyond the stage design parameters. Make sure to get full system tests of each stage to inform the design of the next.

4. Iterate the design and return to step 2. At no stage should you be afraid to rewrite complete components from scratch.

In an academic environment, where commercial pressure is practically absent, it is easy to apply the knife to the code in stage 4. It takes bravery to be as cruel as necessary in commercial settings, but we contend that lack of nerve at this type of juncture is a major underlying, and often misidentified, cause of project failure, especially for the most innovative and challenging ones.

To get a beautiful architecture with truly unexpected benefits, stay true to your beliefs and have no fear.

Principles and properties		Structures	
	Versatility	✓	Module
	Conceptual integrity		Dependency
✓	Independently changeable	✓	Process
	Automatic propagation		Data access
✓	Buildability		
	Growth accommodation		
	Entropy resistance		

CHAPTER TEN

The Strength of Metacircular Virtual Machines: Jikes RVM

Ian Rogers
Dave Grove

RUNNING CODE IN A MANAGED RUNTIME ENVIRONMENT IS THE PREVALENT CHOICE for today's developers. In fact, a large fraction of all developed code is for a managed runtime environment. However, although runtime environments are increasingly popular, the majority are written in a different language than the one the runtime environment supports. In the case of Java Virtual Machines, which act as a runtime environment for Java applications, the programming languages C and C++ are most commonly used to implement the runtime environment itself.

In this chapter, we present an overview of a mature virtual machine called Jikes RVM, which is written in Java to run Java applications. Not only is the runtime system written in Java, but all other components of the architecture are written in Java. These components include adaptive and optimizing compilation systems, threading, exception handling, and garbage collection. We present an overview of these systems here, and will explain why having a singular vision of language, runtime, and implementation leads to systems that are inherently more compelling and potentially more optimal.

Background

The issue of how to develop a new programming language is a staple of computer science that has formalisms such as the T-Diagram (Aho 1986). Figure 10-1 shows a T-Diagram that depicts using a C compiler, which runs and creates PowerPC machine code, to compile a Pascal compiler written in C that creates PowerPC machine code, producing a compiler that runs with PowerPC machine code, creating PowerPC machine code.

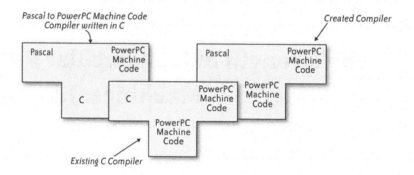

FIGURE 10-1. A T-Diagram showing the creation of a Pascal-to-PowerPC machine-code compiler

Unlike traditional programming languages, which compile down to the machine code for the computer upon which the program was intended to run, most modern languages can be compiled to an architecture-neutral machine code. In the case of Java, this is known as Java bytecode. A neutral machine code allows applications to be ported to any environment where the runtime is present. So, Java can be run anywhere a Java virtual machine is present.

Modern languages aim to help the programmer by designing out potential programming language pitfalls. The most prevalent feature is to have memory safety, by limiting what a programmer may do with data types and allowing only automatic garbage collection to release memory. Another feature is the ability to throw exceptions.

Self-hosting is seen as an important principle for programming languages. Self-hosting means that the programming language should allow enough expression that the programming language can be written in its own programming language. For example, a Pascal compiler written in Pascal is self-hosting, whereas a Pascal compiler written in C is not. Self-hosting allows the programming language developer to use the features of the programming language for which they are responsible. Critically, self-hosting creates a virtuous cycle in which language implementers desire to utilize advanced and/or expressive language features in performance-critical parts of the language implementation, and therefore often discover innovative ways to efficiently implement said language features.

Although making the compiler self-hosting is seen as important, many runtime environments are not written in the language in which they typically run. For example, a runtime written in C or C++ may run Java applications. If the runtime were to have a bug relating to memory safety, it could crash the Java application, even though the Java application itself has memory safety. Removing bugs is an important reason to have a self-hosting runtime.

As computer systems are better understood and evolve, the requirements of a programming language change. For example, the programming languages of C and C++ have no standard library for utilizing multiple processors using threading (although popular extensions such as POSIX threads and OpenMP do exist). Modern languages will have such features designed into the language and standard library. Allowing the runtime to take advantage of better libraries and abstractions is another important reason to have a self-hosting runtime.

Finally, whenever a runtime and the application it is running must communicate with each other, there is a layer for the communication. One job of this communication layer can be to marshal objects, changing the format in one programming language to that of the other. In the case of objects, the communication layer also needs to remember not to garbage collect any objects that may be in use from outside of the managed runtime. Such communication layers are not necessary, or at least not necessary in as many situations, when the runtime is self-hosting.

We hope we've provided compelling reasons for making a self-hosting runtime. In this chapter, we present an overview of such a runtime, Jikes RVM, which is written in Java and runs Java applications. Self-hosting runtime environments are known as *metacircular* (Abelson et al. 1985). Jikes RVM is not unique in being metacircular, and indeed it draws inspiration from similar runtime systems, such as Lisp (McCarthy et al. 1962) and the Squeak virtual machine for Smalltalk (Ingalls et al. 1997). Being a metacircular virtual machine written in Java allows the use of excellent tools, development environments, and libraries. As Java lacks credibility in certain communities, we will first address some of the myths that may lead people to believe that a metacircular Java runtime has inherent flaws.

Myths Surrounding Runtime Environments

There is still much active debate on how best to create applications for different environments. Factors such as the resources available where the application will be run, the productivity of the developers challenged with creating the application, and the maturity of the development environment come into play. If the application is implemented in the same way, performance and memory requirements are a feature of the development environment. Next, we look to overturn the most common myths about managed environments.

As Runtime Compilers Must Be Fast, They Must Be Simple

A misconception about runtime environments is that they are interested purely in *just-in-time* (JIT) compilation. JIT compilation must create code quickly because it will be put to use as soon as it is ready. Although this simple execution model was used in many early JVMs and in prototype runtime environments, most modern production virtual machines rely on some form of *selective optimization*. In selective optimization, online profiling is used to identify a subset of the executing methods to compile with an aggressive optimizing compiler; the remainder of the methods are either interpreted or compiled by a very fast nonoptimizing compiler immediately before execution. It is this selectivity that enables the use of sophisticated optimizing compilers at runtime.

Unlimited Analysis in a Static Compiler Must Mean Better Performance

As runtime environments will be heavily used by many applications, optimizing them to extract greater performance across all applications makes sense. However, a range of optimizations cannot be performed if the runtime is not created as part of a dynamic environment:

Online profiling
> Aspects of the runtime vary at runtime—for example, the average size of pieces of data, or particular coding styles that may be based on differing design patterns. Online profiling allows timely use of this information to reduce overheads, such as branch prediction, and also allow more advanced optimizations, such as value speculation. An example from Java of value speculation could be to predict that the majority of stream output operations may be occurring to the `java.lang.System.out` file stream. Value speculation is an extension to partial evaluation, which we will consider further in the later section "Partial evaluation."

Variance in the underlying system
> The range of systems an application runs on is increasingly becoming more varied. The abilities of different processors, the amount of memory, the number of different processors, the power requirements, and the load of the system that the runtime is executing upon are all important in knowing how the runtime should best adapt.

Intraprocedural analysis
> Intraprocedural analysis is an important tool for an optimizing compiler, allowing optimization across method boundaries. Although off-line analysis can be unlimited, often this can result in so much data that the compiler cannot determine which data is important. As runtime feedback is more timely, it can better guide intraprocedural and other compiler optimizations.

Runtime Analysis Uses a Lot of Resources

Having a runtime environment has an overhead for the memory required by the environment. Similar requirements exist for the standard libraries that conventional applications use. On top of these, the runtime environment must keep information that can help guide its future compilation and execution. These memory requirements are modest, and through timely and memory-efficient sampling, the runtime environment can gain the most benefit with little cost.

Dynamic Class Loading Inhibits Performance

Many modern runtime environments, including Java, have the ability to be extended dynamically. This can be useful in an environment where the user wants to plug together different parts of a system; in the case of Java, the components may be downloaded from the Internet. Although useful, this technique can mean the compiler cannot make certain assumptions that it otherwise could if all information about an application were available to it. For example, if a method of an object were being called and there was only one possible class for that object, it would be preferable to avoid doing dynamic method dispatch and directly call the method. There are specific optimizations to deal with this situation in optimizing runtime environments, and we describe them later in "On-stack replacement."

Garbage Collection Is Slower Than Explicit Memory Management

Automatic garbage collection is an advanced area of computer science research. Memory is requested and reclaimed when no longer needed. Explicit memory management requests and then reclaims memory using explicit commands to the runtime environment. Although explicit memory management is error-prone, it is often argued that it is needed for performance. However, this is overlooking the many complications caused by explicit memory management. With explicit memory management, it is necessary to track what blocks of memory are in use and maintain lists for those that are not. When this is combined with many threads concurrently requesting memory, the problem of memory fragmentation, and the fact that merging smaller regions can yield larger regions of memory to be allocated, the job of an explicit memory manager can become complex. The explicit memory manager also cannot move things around in memory—for example, to reduce fragmentation.

The requirements of a memory manager are application-specific, and in the context of a metacircular runtime, a simple JIT compiler doesn't need to perform much memory allocation, so either explicit or automatic garbage collection schemes would work well. For a more sophisticated optimizing compiler, the picture isn't as clear, other than the fact that garbage collection reduces the potential for bugs. For other parts of the runtime system there are further complications, which we describe later in "Magic, Annotations, and Making Things Go Smoothly." Although we haven't been able to refute the claim that garbage collection may be slower than explicit memory management, one thing it definitely improves is the "hackability" of the system.

Summary

As development tools are themselves applications, our original question of how best to develop an application becomes self-referential. Managed languages design out faults and improve the productivity of the developer. Simplifying the development model, exposing more opportunities to optimize the application and the runtime, and doing this in a metacircular way allow the developer to gain from the features they introduce without encountering barriers between application, runtime, and compiler views of the system. In the next sections we look at Jikes RVM, a runtime that puts these principles together.

A Brief History of Jikes RVM

Jikes RVM stems from an IBM project called Jalapeño. The Jalapeño project was started in November of 1997 with the goal of developing a flexible research infrastructure to explore ideas in high-performance virtual machine design. By early 1998, an initial functional prototype was bootstrapped and capable of running small Java programs. In the spring of 1998, work was begun on the optimizing compiler, and the project rapidly grew in size. By early 2000, project members had published several academic papers describing aspects of Jalapeño, and university researchers began to express interest in getting access to the system to use as the basis for their own research efforts.

By the time the system went open source in October of 2001, there were already 16 universities using Jikes RVM under license from IBM. This community rapidly expanded and now includes hundreds of researchers at well over 100 institutions. Jikes RVM has been the basis for over 188 papers that have appeared in peer-reviewed publications, and it has formed a foundation for at least 36 university dissertations.

Version 2 was the original open source Jikes RVM and had support for both the Intel and PowerPC architectures. A range of different garbage collection algorithms were available, including reference counting, mark-sweep, and semi-space. A year later, version 2.2 of Jikes RVM was released. One of the main enhancements was a completely new implementation of the memory management subsystem, called the *Memory Management Toolkit* (MMTk). MMTk has become a very widely used framework in the garbage collection research community, and has been ported to other runtimes besides Jikes RVM. We discuss MMTk and garbage collection techniques further in "Garbage Collection." The optimizing compiler and adaptive optimization system also had significant improvements, and the development of the runtime was simplified by a switch to the open source GNU Classpath standard class libraries. In April 2003, Jikes RVM 2.2.1 was one of the first open source Java runtimes capable of running significant portions of the Eclipse IDE.

In the almost four years between versions 2.2 and 2.4.6, a number of significant improvements in both functionality and performance were made, but with a source structure and architecture that were mostly unchanged.

Jikes RVM 3.0 was released in August 2008 and represents almost two years of concerted community effort to modernize and improve the system. Java 5.0 language features were adopted across the codebase, the build system switched to using Apache Ant, and a greatly improved testing infrastructure was developed to increase system stability and performance. In addition, a number of functional and performance improvements were made, resulting in performance for many programs that is competitive with that achieved by modern production JVMs (implemented in traditional runtime system languages such as C/C++).

There are too many contributions to Jikes RVM to mention, but we thank the Jikes RVM development community for all their work. Just under 100 people have contributed code to Jikes RVM, and 19 people have served as members of the Jikes RVM core team. For a list of full credits, it is worth going to the Jikes RVM website. More details about the early history of Jikes RVM and the growth of its open source community can be found in a 2005 *IBM System Journal* paper (Alpern et al. 2005).

Bootstrapping a Self-Hosting Runtime

Compared to the bootstrap of a traditional compiler (Figure 10-1), the bootstrap of a metacircular runtime involves a few more tricks. Figure 10-2 shows a T-diagram depicting the process.

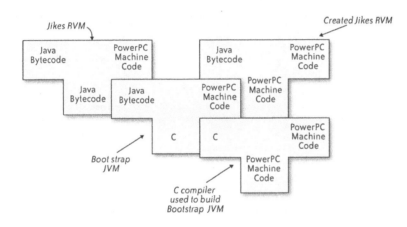

FIGURE 10-2. A T-Diagram showing the bootstrapping of Jikes RVM on an existing JVM written in C

The *boot image* contains several files that represent memory when the system is hosting itself (the rightmost T in Figure 10-2). The contents of the boot image are code and data, similar to what is found in a regular compiler's object file. An extra section in Jikes RVM's boot image contains the root map, which is created by the garbage collector. We describe the root map later in "Garbage Collection." The *boot image writer* is a program that uses Jikes RVM's compilers to create the boot image files, executing on a bootstrap JVM. A loader is responsible for loading the boot image into the correct area of memory, and in Jikes RVM, the loader is known as the *boot image runner*.

Object Layout

The boot image writer must lay out the objects on disk as they will be used in the running Jikes RVM. The object model in Jikes RVM is configurable, allowing different design alternatives to be evaluated while keeping the rest of the system fixed. An example of a design alternative is whether to provide more bits per object for the hashing of objects or more bits per object for implementing fast locking for synchronization. An overview of the Jikes RVM object layout is shown in Figure 10-3.

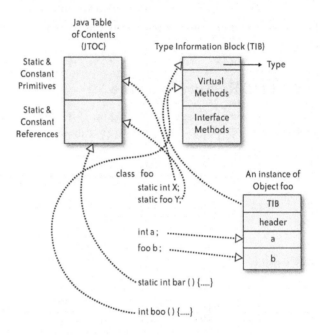

FIGURE 10-3. The layout of objects in Jikes RVM

The default 32-bit object model in Jikes RVM currently uses two words for the object header: the first references the *Type Information Block* (TIB), and the second holds status information for the object on locking, hashing, and garbage collection. After the object header comes the fields of the object. For an array, the first field is the array length and the remainder are the array elements. To avoid a displacement for accesses to an array, the size of which would be the size of the object header and the array length field, all references to objects actually reference a location three words into the object. This allows element zero of an array to be at offset zero within the object, but it also means the object header is always three words behind an object's reference and that the first field of an object is always at a negative offset from the object's reference.

The TIB is responsible for holding the data that is common to every object of a particular type. This data is primarily used for virtual and interface *method dispatch*. Method dispatch is the process of identifying the method that is associated with and should be called for a particular object. Methods within classes are allocated locations within the TIB, in order to enable fast and efficient method dispatch. The TIB also holds values that allow fast runtime type information to be determined, which speeds Java's `instanceof` and `checkcast` operations. It also holds special methods to process an object during garbage collection.

As well as objects, static data belonging to classes must be tracked. Static data is held in a location called the *Java Table Of Contents* (JTOC). The JTOC is organized in two directions: positive offsets in the JTOC hold the contents of static fields that contain references and reference literals. A literal value in Java is something like a string literal, something that the bytecode can directly reference but that doesn't have a field location. Negative addresses in the JTOC are responsible for holding primitive values. Splitting the values in this way allows the garbage collector to easily consider which references from static fields should prevent an object from being considered garbage.

Runtime Memory Layout

The boot image writer is responsible for laying out the memory of the virtual machine when it first boots. All of the objects needed in the boot image are there because they are referenced by code that will start the Java application. Figure 10-4 gives an overview of the regions of memory that are in use when Jikes RVM executes.

A number of key items can be seen in Figure 10-4:

Boot image runner
  The boot image runner and its stack comprise the loader responsible for loading the boot image.

Native libraries
  Memory is required for native libraries that are used by the class library. This is described further in the later section "Native Interface."

MMTk spaces
  These are the different garbage-collected heaps in use by MMTk to support the running application.

Root map
  Information for the garbage collector on fields that may be reachable from the boot image. More information is given later in "Garbage Collection."

Code image
  The executable code for static and virtual methods that are reachable directly from the JTOC or from TIBs, respectively. Code is written to a separate region of the boot image to provide support for memory protection.

Data image
  The data image is created by first writing the boot record and the JTOC and then performing a traversal of the objects reachable from the JTOC. The objects are traversed in the bootstrap JVM using Java's reflection API. The order in which objects are traversed can impact locality, and thereby performance, so the traversal mechanism is configurable (Rogers et al. 2008).

JTOC

As described earlier in "Object Layout," the JTOC is responsible for holding literal and static field values. Traversing the JTOC produces the boot image.

Boot record

A table at the beginning of the data image that contains data shared between the boot image runner and Jikes RVM. These values typically cannot be determined during the bootstrap.

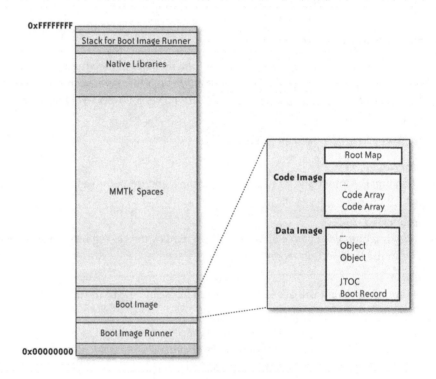

FIGURE 10-4. The runtime memory layout of Jikes RVM

Compiling the Primordials and Filling in the JTOC

The *primordials* are a collection of classes that must be built into the boot image for it to run. The most important primordial is `org.jikesrvm.VM`, which is responsible for starting the virtual machine. If something isn't part of the boot image, and therefore isn't a primordial, then it is referenced. When a referenced object is accessed at runtime, it causes the class loader to load and link the referenced class.

The list of primordial classes is produced during the bootstrap by searching directories and from reading an explicit list of classes to compile. The explicit list is particularly important for array types. It would be possible to produce the list of primordial classes by repeated compilation and growing the set of classes included into the boot image, but this would significantly increase the time it takes to build Jikes RVM. A proposed alternative is to use Java annotations to mark which classes are primordials.

▶ 但要注意，倘若使用了annotation来标识原生类，又会因为增加了反射的成本而增加了构建Jikes RVM的时间。

Before traversing the object graph and writing the boot image, the boot image writer compiles the primordials. Compiling a primordial involves loading its class with Jikes RVM's class loader, which will automatically allocate space in the JTOC and the TIB as necessary, and then iterating over all the methods and compiling them with one of Jikes RVM's compilers. As this is all pure Java code, the boot image writer takes advantage of Java's concurrency API to perform this task in parallel if possible.

Once compilation of the core set of the primordal classes is complete, the object graph in the host JVM's heap represents sufficient functionality for Jikes RVM to bootstrap itself, allocate additional objects, and start loading and executing user classes. To complete the bootstrap process, this core object graph is traversed and written out to disk in Jikes RVM's object model using the capabilities of the Java reflection API provided by the host JVM.

The Boot Image Runner and VM.boot

As mentioned in "Bootstrapping a Self-Hosting Runtime," the boot image runner is responsible for loading the compiled images into memory. The exact details of this vary depending on the operating system, but the images are set up to be demand-paged into memory. Demand paging means that pages from the boot image remain on disk until they are required.

Once in memory, the boot image runner initializes the boot record and then loads the machine registers to transfer execution over to the Jikes RVM method org.jikesrvm.VM.boot (or VM.boot for short). Jikes RVM is responsible for all memory layout, enabling efficient garbage-collection techniques and a stack organization that is efficient at dealing with Java exceptions (see the later section "Exception Model"). Once the VM.boot method is entered, special wrappers are needed to transfer between native code in the boot image runner and C libraries (these are described further in the upcoming section "Native Interface").

The job of VM.boot is to ensure that the VM is in a ready state to execute a program. It does this by initializing the components of the RVM that couldn't be initialized when the boot image was written. Some components must be started explicitly—for example, the garbage collector. The remaining components are a small subset of the primordial classes that were not fully written into the boot image because of inconsistencies in the bootstrap and Jikes RVM class files. To initialize these classes, the static initializer of the class must be run.

Initializing the threading system is an important part of the `VM.boot` method. It creates the necessary garbage-collection threads, a thread for running object finalizer methods, and threads responsible for the adaptive optimization system. A debugger thread is also created, but is scheduled for execution only if a signal is sent to Jikes RVM from the operating system. The final thread to be created and to start execution is the main thread, which is responsible for running the Java application.

Runtime Components

The previous section described getting Jikes RVM to a point where it is ready for execution. In this section, we look at the main runtime components of Jikes RVM, beginning with those directly responsible for executing Java bytecode, and then look at some of the other virtual machine subsystems that support this execution.

Basic Execution Model

Jikes RVM does not include an interpreter; all bytecodes must first be translated by one of Jikes RVM's compilers into native machine code. The unit of compilation is the method, and methods are compiled lazily when they are first invoked by the program. This initial compilation is done by Jikes RVM's *baseline compiler*, a simple nonoptimzing compiler that generates low-quality code very quickly. As execution continues, Jikes RVM's *adaptive system* monitors program execution to detect program hot spots and selectively recompiles them with Jikes RVM's optimizing compiler. This is a significantly more sophisticated compiler that generates higher-quality code, but at a significantly larger cost in compile time and compiler memory footprint than the baseline compiler.

This selective optimization model is not unique to Jikes RVM. All modern production JVMs rely on some variant of selective optimization to target optimizing compilation resources to the subset of the program's methods where they will yield the most benefit. As discussed earlier, selective optimization is the key to enabling the deployment of sophisticated optimizing compilers as dynamic compilers.

Adaptive Optimization System

Architecturally, the Adaptive Optimization System is implemented as a collection of loosely synchronized entities. Because it is implemented in Java, we are able to utilize built-in language features such as threads and monitors to structure the code.

As the program executes, timer-based samples are accumulated by the running Java threads into sampling buffers. Two types of profile data are collected: samples of the currently executing method (to guide identification of candidate methods for optimizing compilation) and call-stack samples (to identify important call graph edges for profile-directed inlining). When a sampling buffer is full, the low-level profiling agent signals a higher-level *Organizer* (implemented as separate Java threads) to summarize and record the raw profile data. Periodically, a *Controller* thread analyzes the current profile data and uses an analytic model to determine which, if any, methods should be scheduled for optimizing compilation. These decisions are made using a standard 2-competitive solution to the "ski rental" problem[*] from online algorithms. A method is not selected for optimization until the expected benefit (speedup in future invocations) from optimizing it exceeds the expected cost (compile time). These cost-benefit calculations are made by combining the online profile data (how often has the candidate method been sampled in the current execution?) with (offline) empirically derived constants that describe the expected relative speedup and compilation costs of each optimization level of the optimizing compiler (known as the compiler's DNA).

Optimizing Compilation

As a metacircular runtime, Jikes RVM compiles itself instead of relying on another compiler to ensure good performance. Metacircularity creates a virtuous cycle: our strong desire to write clean, elegant, and efficient Java code in the virtual machine implementation has driven us to develop innovative compiler optimizations and runtime implementation techniques. In this section, we present the optimizing compiler, which is composed of many phases that are organized into three main stages:

1. High-level Intermediate Representation (HIR)
2. Low-level Intermediate Representation (LIR)
3. Machine-level Intermediate Representation (MIR)

All of the stages operate on a control-flow graph composed of basic blocks, which are composed of a list of instructions, as shown in Figure 10-5. An instruction is composed of operands and an operator. As the operator gradually becomes more machine-specific, the operator can be changed (mutated) during the lifetime of the instruction. The operands are organized into those that are defined and those that are used. The main kinds of operands are constant operands and operands that encode a register. A basic block is a list of instructions where branch instructions may only occur at the end of the list. Special instructions mark the beginning and end of the basic block. The control-flow graph connects the different regions of code by edges. Exceptions are treated specially and are described later in "Factored control flow graph." Because each of the three main stages of compilation use the same basic intermediate representation, a number of optimizations are applied in more than one of the stages. We describe the main tasks of the three compiler stages next.

---

[*] See *http://en.wikipedia.org/wiki/Ski_rental_problem* for more information.

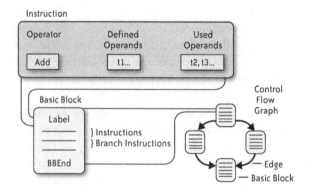

FIGURE 10-5. Overview of the optimizing compiler's intermediate representation

HIR

The High-level Intermediate Representation (HIR) is generated by a compiler phase, called BC2IR, that takes bytecode as input. This initial phase performs propagation based on the bytecode operand stack. Rather than generate separate operations, the stack is simulated and bytecodes are folded together to produce HIR instructions. The operators at the HIR level are equivalent to operations performed in bytecode but using an unbounded number of symbolic registers instead of an expression stack.

Once instructions are formed, they are reduced to simpler operations (if simpler operations exist). Following this, remaining call instructions are considered for inlining. The main part of the BC2IR phase is written recursively, so inlining performs a recursive use of the BC2IR phase.

Local optimizations are performed in the HIR phase. A local optimization is one that considers reducing the complexity of a basic block. As just a basic block is considered, the dependencies within the basic block are considered. This simplifies the compiler phase from having to consider the effect of branches and loops, which can introduce different kinds of data dependencies. Local optimizations include:

Constant propagation
    Propagating constant values avoids their placement in registers.

Copy propagation
    Replacing copies of registers with uses of the original register.

Simplification
    Reducing operations to less complex operations based on their operands.

Expression folding
    Folding trees of operations together. For example, "x=y+1; z=x+1" can have the expression "y+1" folded into "z=x+1", producing "x=y+1; z = y+2".

▶ Expression Folding可以理解为是一个预编译过程，通过 Substitution Model 对表达式中的变量进行了替换。但是，在这个过程中，需要考虑表达式和变量的范围。

Common subexpression elimination
> Finding instructions that perform the same operation on the same operands and removing the latter with a copy of the former's result.

Dead code elimination
> If a register is defined and then redefined without an intervening use, then the instruction performing the original definition is dead code and can be removed.

Branch optimizations consider improving the control-flow graph so that the most likely paths are laid out in a manner that is optimal for the target of the compiler. Other branch optimizations look to remove redundant tests and unroll loops.

Escape analysis looks at whether objects can be accessed within just the context of the code being compiled and also whether an object that isn't just accessed in the local context can be shared among threads. If an object is just accessed locally, the requirement to allocate memory for it can be removed and the object's fields moved into registers. If an object is accessed in just one thread, then synchronization operations upon it can be removed.

A number of optimizations that use extra information added to the intermediate form are described later in "Scalar and extended array SSA forms."

LIR

The Low-level Intermediate Representation (LIR) converts the previous bytecode-like operations into more machine-like operations. Operations upon fields are changed into load and store operations; operators such as new and checkcast are expanded into calls into runtime services that provide those operations and are possibly inlined. Most of the optimizations applicable to HIR are also applicable to LIR, but optimizations such as escape analysis are not. Due to the kind of operations supported, there are some small differences between LIR for different architectures. As with HIR instructions, an unbounded number of symbolic registers can be used in LIR instructions.

MIR

Creating the final Machine-level Intermediate Representation (MIR) involves the three interdependent yet competing transformations of a compiler backend. The transformations are said to compete because their order can determine the performance of the generated machine code. In Jikes RVM, the first transformation is instruction selection. Instruction selection is the process whereby the RISC-like LIR instructions are converted to instructions that exist on the actual machine. Sometimes more than one instruction is required to perform an LIR instruction; for example, an LIR long addition requires two instructions on 32-bit architectures. At other times, the tree pattern-matching instruction selector, known as a *Bottom Up Rewrite System* (BURS), merges several instructions into one instruction. Figure 10-6 shows an example of pattern matching. Two patterns are found for Intel's IA32 architecture: one that creates three instructions with a cost of 41, and another that encodes the load and store into a memory operand of a single instruction and costs just 17. The cheapest pattern is selected.

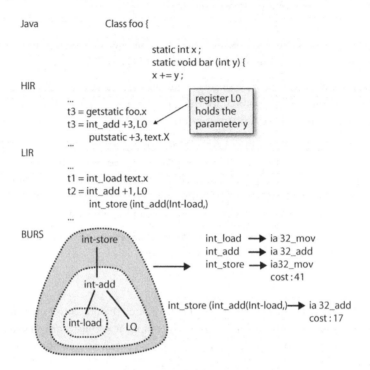

FIGURE 10-6. BURS instruction selection

After instruction selection, the unlimited number of registers need to be mapped onto the finite number of registers provided by a real machine. This process is known as register allocation. When there are not enough registers, values can be held in memory on the stack; swapping an actual register for a value on the stack is known as spilling and filling. A register allocator needs to minimize spills and fills, as well as taking into account any architectural requirements—for example, ones that multiply and divide must occur using certain registers.

Jikes RVM has a linear scan register allocator that performs quick register allocation, but with the possibility of generating extra copy and memory operations. This is a greater problem when the number of registers available on the machine is small. With increases in the number of registers and improvements to the linear scan algorithm, for certain code it is less clear that it would be beneficial to perform a more expensive register allocation.

The final part of MIR instruction creation is instruction scheduling. Scheduling separates instructions in order to allow the processor to exploit instruction-level parallelism.

At the level of MIR, few optimizations are performed. Because of the large number of side effects associated with machine instructions, it is difficult to reason and therefore optimize the instruction's behavior. The other compiler phases at the MIR level are concerned with ensuring that calling, exception, and other conventions are adhered to.

Factored control flow graph

Java programs create runtime exceptions if a null pointer is used to access memory or if an array index is out-of-bounds. These runtime exceptions control the flow of code and so should end basic blocks. This results in small basic blocks with less scope for local optimizations (described earlier in "HIR"). To increase the size of basic blocks, the control-flow dependence of runtime exceptions is turned into an artificial data dependence at the HIR and LIR levels. This dependence ensures that operations are ordered correctly for exception semantics, and therefore the basic block may be left during mid-execution to handle a runtime exception. When the intermediate form has exceptional exit points in the middle of blocks, the control flow graph is said to be factored (Choi et al. 1999).

Instructions are created that explicitly test the runtime exceptions and generate synthetic guard results. Instructions that then require ordering make use of the guard. An instruction that can generate an exception is known as a *Potentially Exceptioning Instruction* (PEI). All PEIs generate a synthetic guard value, as do instructions that can be used to remove PEIs if they are redundant. For example, a branch that tests for null makes null pointer tests on the same value redundant. The guard from the branch is used in place of the guard from the null pointer tests to ensure that instructions cannot be reordered to before the branch.

Figure 10-7 shows a single array assignment having its constituent runtime exceptions turned into PEIs and the guard dependencies that ensure the code is executed in the correct order.

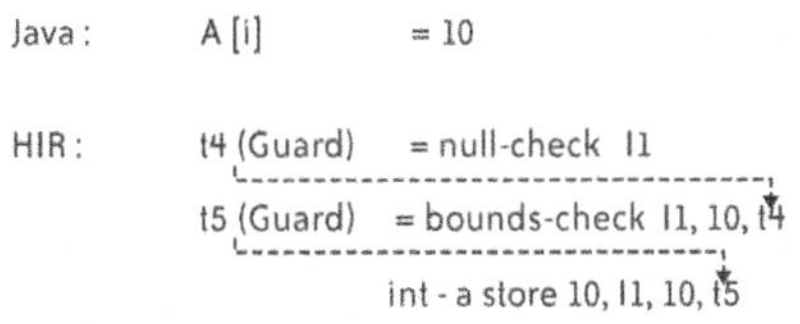

```
Java :     A [i]        = 10

HIR :      t4 (Guard)   = null-check  I1
           t5 (Guard)   = bounds-check I1, 10, t4
                        int - a store 10, I1, 10, t5
```

FIGURE 10-7. An example of instructions in a factored control flow graph

Scalar and extended array SSA forms

Static Single Assignment (SSA) form reduces the dependencies that compiler optimizations need to be concerned about. The form ensures that any registers (more commonly known as variables; in the HIR and LIR phases of Jikes RVM, all variables will be held in registers) are written to at most once. A compiler transformation must handle three kinds of dependence:

True dependence
    Where a register is written and then read.

Output-dependence
> Where a register is written and then written to again (the second write must occur after the first following any transformation).

Anti-dependence
> Where a register is read and then written to (the write must occur after the read following any transformation).

Knowing a register is written once means output-dependence and anti-dependence don't occur. This property means that our previous local optimizations can be applied globally. To handle loops and other branches, special phi instructions encode the merge of particular values from a different place on the control-flow graph.

Array SSA form is an extension to SSA form in which loads and stores are considered to be defining a special variable called a heap. Modeling memory accesses in this way allows the compiler to reason that if two reads occur from the same array location with a heap with the same name, the second read can be replaced with a copy of the first. Redundant stores are handled in a similar way. It also allows accesses to nonrelated heaps to be reorganized—for example, with floating-point and integer operations. Array SSA form was originally devised for FORTRAN; extended array SSA form adds to the form factors, such as Java's fields being unable to alias with one another (Fink et al. 2000).

Also in the SSA form, Jikes RVM constructs pi instructions, placed after a branch uses an operand. The pi instruction uses the same operand as the branch and gives it a new name to be used in place of the operand in subsequent instructions. Using pi instructions, the compiler can reason that a branch performs a test, such as a null test, and that any null tests using the register result of the pi instruction are redundant. Similarly, array bound checks can be removed (Bodik et al. 2000).

The loop versioning optimization in the HIR SSA form also removes possible exceptions. Loop versioning moves exception-checking code out of loops and explicitly tests whether the exceptions can occur before the loop is entered. If an exception can occur, then a version of the loop with exception-generating code is executed. If an exception cannot occur, then a loop version without exceptions is executed.

Partial evaluation

The HIR optimizations, including the SSA optimizations, are able to reduce the complexity of a region of code, but this can be limited by what constant values are available. Often a value cannot be determined to be constant when it comes from an array, which in Java can always have their values altered. The most common occurrence of this is with strings. Jikes RVM introduces the pure annotation as an extension to Java that enables a method with constant arguments to be evaluated, using reflection, at compile time. Leveraging annotations and reflection for this purpose is straightforward in a metacircular runtime.

Partial evaluation has the ability to totally remove some overheads, such as a runtime security check. A security check typically walks the stack to determine what called a restricted method and then checks to see whether that method had permissions to call the restricted method. As the method on the stack at the point of the call can be determined within the compiler, if it is on the stack as a result of inlining, the precise method is provided by simplification. By evaluating the code of the security check or, if the check is pure, performing a reflective call, the result of the security check can be ascertained and removed if it will always pass.

On-stack replacement

On-Stack Replacement (OSR) is the process of swapping executing code while it is executing on the stack. For example, if a long running loop were being executed by code created by the JIT baseline compiler, it would be beneficial to swap that code for optimized code when the optimizing compiler produced it. Another example of its use is in invalidating an executing method if it had unsafely assumed properties of the class hierarchy, such as a class having no subclasses, which can allow improvements in inlining.

OSR works by introducing new bytecodes (in the case of the JIT baseline compiler) or new instructions (in the case of the optimizing compiler) that save the live state of execution of a method. Once the execution state is saved, the code can be swapped for newly compiled code, and the execution can continue by loading in the saved state.

Summary

This section has given brief descriptions of many of the advanced components of Jikes RVM's optimizing compiler. Writing these components in Java has provided a number of inherent benefits to these systems. Threading is available, allowing all of the components of the system to be threaded. The systems are designed to be thread-safe, allowing, for example, multiple compiler threads to run concurrently. Another advantage has come from Java's generic collections, which provide easy-to-understand libraries that have simplified development and allow the components of the system to concentrate on their role, rather than underlying data structure management.

Exception Model

Are exceptions things that are exceptional? Although many programmers program with the belief that they are, virtual machines have had to optimize for a use-case that is common in benchmarks. The use-case in question is to read from an input stream and then throw an exception when a sought pattern occurs. This pattern occurs in both the SPECjvm98[†] jack benchmark and the DaCapo 2006[‡] lusearch benchmark.

[†] *http://www.spec.org/jvm98/*

[‡] *http://dacapobench.org/*

Let's first consider what's needed for an exception. Every exception needs to create a list of methods that are on the stack at the point of the exception (known as the stack trace) and to pass control to a catch-block that handles the exception. Null pointer exceptions are efficiently handled by allowing memory protection failures to occur—for example, by reading or writing to a page of memory that doesn't exist. When a fault is generated, the address of the faulting instruction is provided to a handler. From this, the stack trace and handler information must be determined.

A VM that is written in C may try to use the C calling conventions for the Java JIT compiled code. A problem here is that the C calling conventions don't record what code is running. To solve this problem, a C VM can try to use the return address to compute what method is running. Every method within the VM must be searched to see whether its corresponding compiled code contains a given return address. This must be repeated for every method that is on the stack.

As Jikes RVM controls its memory layout, the stack is laid out to contain extra information to handle exceptions quickly. In particular, Jikes RVM places an identifier on the stack that is just a simple lookup away from providing method information. To determine the bytecode where the exception occurred and the handler locations, a computation is made using the return address or faulting instruction address. This means handling an exception's speed is proportional to the depth of the stack, and not dependent on the number of methods inside the VM.

In common with other VMs, Jikes RVM will optimize away exceptions when possible, in the best case reducing them to branches. It will also inline many methods into one method, so placing an identifier on the stack occurs infrequently, only when going between noninlined methods.

Magic, Annotations, and Making Things Go Smoothly

> ▶ VM Magic库有些像是C#中提供的unsafe操作。

The code generated by the compiler has instructions that directly access memory. On other occasions memory needs to be accessed directly—for example, to implement garbage collection routines or access the object header for locking. To allow strongly typed memory accesses, Jikes RVM uses a library called *VM Magic*, which has been taken as a standard interface and used in other JVM projects. VM Magic defines classes that are either compiler pragma annotations, which extend the Java language, or unboxed types, which represent access to the underlying memory system. The unboxed type Address is used to directly access memory.

The VM magic unboxed types are handled specially within Jikes RVM's compilers. All the method calls upon an unboxed type are treated as manipulating a word-sized value directly. For example, the plus method is treated as an addition of the underlying word size of the machine, and directly acts upon what would normally be the this pointer of the object. As the value being manipulated isn't an object reference, the compiler records that the places in which unboxed types are held are of the same interest to the garbage collector as a memory location that holds a primitive field, such as an int.

Having strongly typed access to memory isn't quite enough to allow Jikes RVM to run without issue. Some special methods are required that indicate to the compilers that either something intrinsic is occurring or they should shortcut some parts of Java's strong type semantics. These special methods exist in org.jikesrvm.runtime.Magic. An example of an intrinsic operation is the square root method, for which Java provides no bytecode but many computer architectures have an instruction, or the routines that directly access fields for reflection. An example of working around Java's strong semantics is to avoid casts during thread scheduling, as a runtime exception wouldn't be permissible at certain key points.

All magic operations are compiled differently than regular methods, and they can't be compiled as standalone methods. Methods exist in their place, but if these were ever executed, they'd fail with an exception. Instead, the compiler determines from the method which magic operation is being performed, and then it looks up what operations the compiler must provide for that. During boot image creation, replacements for some of the unboxed magic types and methods are provided. For example, addresses that are in the boot image are unknown until objects are laid out. The boot image unboxed types keep track of identifiers and which objects they map to. During boot image creation, these identifiers are linked to the object they reference.

Although magic operations and unboxed types allow Java code in Jikes RVM the access to memory that a pointer would have in C or C++, the pointers are strongly typed and cannot be used in place of references. Strong typing allows programmer errors to be detected at compile time, good IDE integration, and support from bug finding tools, all of which have aided the development of Jikes RVM.

Thread Model

Java has integral threading support. In 1998, operating system support for threading was not well-suited for many Java applications. In particular, typical operating-system threading implementations did not scale to support hundreds of threads in a single process, and locking operations (required to implement Java synchronized methods) were prohibitively expensive. To avoid this overhead, some JVMs implemented threading themselves, using a mode of operation called *green threading*. In green threading, threads are implemented by having the JVM itself manage the process of multiplexing multiple Java threads onto a smaller number of operating system threads (often one, or one per CPU in a multiprocessor system). The primary advantage of green threading is that it allows the implementations of locking and thread scheduling that are specialized to Java semantics and that can be made highly scalable. The primary disadvantage is that the operating system is not aware of what the JVM is doing, which can lead to a number of performance pathologies, especially if the Java application is interacting heavily with native code (which may itself be making assumptions about the threading model). However, green-threaded JVMs have provided the starting point for a number of Java operating system projects that could be able to avoid some of the performance pathologies of green threading on top of a standard operating system.

As processor performance and operating system implementations have improved, the advantage of managing threads in the JVM has diminished—although it is still advantageous to manage uncontended locking operations within the JVM.

Using Java, a clean threading API has been created that allows Jikes RVM to have different underlying threading models, such as those provided by different operating systems. In the future, having a flexible interface between the language and operating system threads can allow Jikes RVM to adapt to new programmer behaviors (for example, by supporting thousands of threads).

Native Interface

Unfortunately, staying in Java code isn't always possible for Jikes RVM. Accessing operating system routines is necessary to allocate pages of memory. The class library interfaces Java code with existing libraries, such as those for windowing. Jikes RVM provides two means of accessing native code:

Java Native Interface (JNI)
> JNI is a standard that allows Java applications to integrate with native applications typically written in C. A feature of JNI is that Java objects can be handled, and Java methods can be called. To do the bookkeeping, a list of objects accessed in native code is required to prevent these objects from being garbage collected. This introduces an overhead when using JNI for native methods.

SysCalls
> These are similar to native methods in their declaration, except they have an extra annotation. They allow a more efficient transition in and out of native code, with the restriction that the native code is not able to call back into the VM via the JNI interfaces. Jikes RVM implements a SysCall as a simple procedure call to a native method using the default operating-system calling conventions.

Class Loaders and Reflection

The class loader is responsible for loading classes into Jikes RVM and interfacing this process with the object model. Reflection allows an application to query the types of objects and even perform method calls on objects whose types were not known at static compile time. Reflection occurs through applications using objects such as `java.lang.Class` or those in the package `java.lang.reflect`, which are API wrappers to routines within Jikes RVM's runtime.

As some optimizations rely on the class hierarchy, important runtime hooks exist in the class loader that can trigger recompilation if previously held assumptions are now incorrect. We described this in more detail in the earlier section "On-stack replacement."

Garbage Collection

The Memory Management Toolkit (MMTk) is a framework for building high-performance Memory Management implementations (Blackburn et al., May 2004). Over the last five years, MMTk has become a heavily used piece of core infrastructure for the garbage collection research community. Its highly portable and configurable architecture has been the key to its success. In addition to serving as the memory management system for Jikes RVM, it has also been ported to other language runtimes, such as Rotor. In this section, we will just touch on some of the key design concepts of MMTk. For general information on garbage collection algorithms and concepts, the best reference is the book *Garbage Collection: Algorithms for Automatic Dynamic Memory Management* (Jones and Lins 1996).

An underlying meta-principle of MMTk is shared with Jikes RVM as a whole: by relying on an aggressive optimizing compiler, high-performance code can be written in a high-level, extensible, and object-oriented style. This principle is most developed in MMTk, where even the most performance-critical operations, such as fast path object allocation and write barriers, are cleanly expressed as well-structured, object-oriented Java code. On the surface, this style appears to be completely incompatible with high-performance (to allocate a single object, there are dozens of source-level virtual calls and quite a bit of complex numeric computation). However, when the allocation sequence is recursively inlined and optimized by the optimizing compiler, it yields tight inline code sequences (on the order of 10 machine instructions to allocate an object and initialize its header) that are identical to those that can be achieved by handcoding the fast path sequence in assembly code or hard-wiring it into the compilers.

MMTk organizes memory into *Spaces*, which are (possibly discontiguous) regions of virtual memory. There are a number of different implementations of Space provided by MMTk, each embodying a specific primitive mechanism for allocating and reclaiming chunks of the virtual memory being managed by the Space. For example, the CopySpace utilizes bump pointer allocation and is collected by copying out all live objects to another Space; the MarkSweepSpace organizes memory by using free lists to chain together free chunks of memory and supports collection by "sweeping" unmarked (dead) objects and relinking them to the appropriate free list.

A *Plan* is MMTk's terminology for what one typically thinks of as a garbage-collection algorithm. Plans are defined by composing one or more Space instances together in different ways. For example, MMTk's GenMS Plan implements the fairly standard generational mark-sweep algorithm, composing a CopySpace to implement the nursery and a MarkSweep space to implement the mature space. In addition to the Spaces used to manage the user-level Java heap, all MMTk Plans also include several spaces used for virtual-machine-level memory. In

▶ 这种提供多种内存空间管理策略的方式更为灵活，可根据需要选择不同方式组织内存。正如后文所说，这种Space Instance还可以组合，例如CopySpace 与 MarkSweepSpace 的组合，而Plan的引入又进一步丰富了内存管理机制。我们可以认为Space是MMTK中内存管理的原子元素，通过组合又可以支持不同的垃圾回收算法，并以Plan的形式提供调用。

particular, special Spaces are used to allocate JIT-generated code, the Jikes RVM Bootimage, and low-level VM implementation objects such as TIBs. There are 15 Plans (i.e., 15 different GC algorithms) predefined and distributed in MMTk version 3.0. Furthermore, a number of other Plans have been developed and described in the academic literature by the MMTk research community.

MMTk and its hosting runtime system interact through two narrowly defined interfaces that specify what API MMTk exposes to the host virtual machine and what virtual machine services MMTk expects the host to provide. To maintain MMTk's portability to multiple host virtual machines, the build process strictly enforces these interfaces by separately compiling MMTk against a stub implementation of the VM interface. Perhaps the most complex bit of these interfaces is the portion that is used to allow MMTk and Jikes RVM to collaboratively identify the *roots* to be used for a garbage collection. Most of MMTk's Plans represent tracing collectors. In a tracing collector, the garbage collector starts from a set of root objects (typically the program's global variables and references on thread stack frames) and does a transitive closure of their reference fields to find all reachable objects. Reachable objects are considered live and as such are not collected. Any objects that are not reached in the transitive closure are dead; these may be safely reclaimed by the collector and used to satisfy future memory allocation requests. In Jikes RVM, the roots for garbage collection come from registers, the stack, the JTOC, and the boot image. The references in the boot image are determined from the root map, and the root map compresses all of the offsets in the boot image that contain references. To determine the references on the stack and in registers, the method and the location within it are determined in the same manner as is used for the exception model (see the earlier section "Exception Model"). From the method the compiler is determined, and it can give an iterator to MMTk that processes the registers and stack returning the references.

Jikes RVM integration

Jikes RVM integrates with MMTk during the initial creation of object representations, providing iterators to process references, object allocation, and barrier implementations. Object allocation and barriers can influence performance; as MMTk is written in Java, the associated code can be directly linked into the code being compiled for efficiency. Barriers are necessary in garbage collection schemes for a variety of reasons. For example, read barriers are necessary to catch the use of objects that are possibly being copied by a concurrent garbage collector, and write barriers are used in generational collectors where a write to an old object means the old object needs to be considered as a root for collection of the young generation.

Summary

MMTk provides a powerful and popular set of precise garbage collectors. The ease of being able to link together different modules of Jikes RVM and MMTk eases development and reduces overhead, with Jikes RVM inlining parts of MMTk for performance reasons. Writing garbage-collection algorithms in Java allows the garbage collector implementor to ignore what occurs inside the compilers. Java's inherent threading has meant all garbage collectors are parallel and integrate with the runtime model. Java's libraries provide inspiration for the interfaces of the collector, which means that writing new garbage collectors is significantly simplified in MMTk.

Lessons Learned

Jikes RVM is a successful research virtual machine, providing performance close to the state-of-the-art in a flexible and easy-to-extend manner. Being written in the language supported by the runtime allows close integration and reuse of components. The use of Java allows a simple-to-understand codebase, good modularity, and the use of high-quality tools, such as IDEs.

The Java language and the implementations of JVMs are developing, and Jikes RVM has and will continue to develop along with these changes. One interesting development is the X10 programming language, which tackles the issue of how programmers will develop applications for many core systems by providing thread-safety guarantees, just as garbage collection provides memory-safety guarantees in Java. Jikes RVM's codebase already provides an excellent test bed for development ideas with X10. As JVM implementations create new optimizations, such as new garbage-collection techniques or object inlining, the ability to slot these optimizations into a framework such as Jikes RVM means the runtime, compilers, and codebase also improve and demonstrate the strength of metacircularity.

The extensibility of a metacircular environment make Jikes RVM an excellent platform for multilanguage virtual machine research. This extension would also allow aspects of Jikes RVM to be written in different programming languages.

There are many other exciting extensions to Jikes RVM and related projects, including support for languages such as C and C++ through binary translation, extensions to provide aspect-oriented programming within the virtual machine, and making the entire virtual machine into an operating system to remove barriers to runtime optimization. Although initially controversial, garbage collection and adaptive and link-time optimization are now desired by developers. Jikes RVM shows how this can be achieved through one common metacircular environment that will help develop, and adapt to, these features. Jikes RVM supports a large research community and achieves high-performance. By having a beautiful architecture, Jikes RVM can continue to provide a platform for future runtime environments.

References

Abelson, Harold, Gerald Jay Sussman, and Julie Sussman. 1985. *Structure and Interpretation of Computer Programs*. Cambridge, MA: MIT Press.

Aho, Alfred, Ravi Sethi, and Jeffrey Ullman. 1986. *Compilers, Principles, Techniques, and Tools*. Boston, MA: Addison-Wesley.

Alpern, Bowen, et al. 2005. "The Jikes Research Virtual Machine project: Building an open-source research community." *IBM Systems Journal*, vol. 44, issue 2.

Blackburn, Steve, Perry Cheng, and Kathryn McKinley. 2004. *Oil and water? High performance garbage collection in Java with MMTk* (pp. 137–146). International Conference on Software Engineering, Edinburgh, Scotland. ACM, May '04.

Bodik, Rastislav, Rajiv Gupta, and Vivek Sarkar. 2000. *ABCD: eliminating array-bounds checks on demand*. ACM SIGPLAN Conference on Programming Language Design and Implementation (PLDI 2000), Vancouver, British Columbia, Canada. ACM '00.

Choi, Jong-Deok, et al. 1999. *Efficient and Precise Modeling of Exceptions for the Analysis of Java Programs*. ACM SIGPLAN-SIGSOFT Workshop on Program Analysis for Software Tools and Engineering (PASTE '99), Toulouse, France: ACM, Sept. '99.

Fink, Stephen, Kathleen Knobe, and Vivek Sarkar. 2000. *Unified Analysis of Array and Object References in Strongly Typed Languages*. Static Analysis Symposium (SAS 2000), Santa Barbara, CA: Springer Verlag.

Ingalls, Daniel, et al. 1997. "Back to the future: the story of Squeak, a practical Smalltalk written in itself." *ACM SIGPLAN Notices*, vol. 13, issue 10: 318–326.

Jones, Richard and Rafael Lins. 1996. *Garbage Collection: Algorithms for Automatic Dynamic Memory Management*. Hoboken, NJ: John Wiley and Sons.

McCarthy, John, et al. 1962. *LISP 1.5 Programmer's Manual*. Cambridge, MA: MIT Press.

Piumarta, Ian, and Fabio Riccardi. 1998. "Optimizing direct threaded code by selective inlining." *ACM SIGPLAN Notices*, vol. 33, issue 5: 291–300.

Rogers, Ian, Jisheng Zhao, and Ian Watson. 2008. *Boot Image Layout For Jikes RVM*. Implementation, Compilation, Optimization of Object-Oriented Languages, Programs and Systems (ICOOOLPS '08), Paphos, Cyprus. July '08.

PART IV

第 4 部分

End-User Application Architectures
终端用户应用架构

Chapter 11   GNU Emacs: Creeping Featurism Is a Strength
第 11 章   GNU Emacs：滋生的特性为其优势

Chapter 12   When the Bazaar Sets Out to Build Cathedrals
第 12 章   当集市开始构建教堂

Principles and properties		Structures	
✓	Versatility	✓	Module
	Conceptual integrity		Dependency
✓	Independently changeable		Process
	Automatic propagation		Data access
	Buildability		
✓	Growth accommodation		
✓	Entropy resistance		

CHAPTER ELEVEN

GNU Emacs: Creeping Featurism Is a Strength

Jim Blandy

> I use Emacs, which might be thought of as a thermonuclear word processor. It was created by Richard Stallman; enough said. It is written in Lisp, which is the only computer language that is beautiful. It is colossal, and yet it only edits straight ASCII text files, which is to say, no fonts, no boldface, no underlining…. If you are a professional writer—i.e., if someone else is getting paid to worry about how your words are formatted and printed —Emacs outshines all other editing software in approximately the same way that the noonday sun does the stars. It is not just bigger and brighter; it simply makes everything else vanish.
>
> —Neal Stephenson

THE GNU EMACS TEXT EDITOR IS UNMATCHED IN ITS NOTORIETY. Its proponents swear nothing else comes close, and are oddly resistant to the charms of more modern alternatives. Its detractors call it obscure, complex, and outdated compared to more widely used development environments, such as Microsoft's Visual Studio. Even its fans blame their wrist injuries on its contorted keyboard command set.

Emacs provokes such strong reactions partly because there's so much of Emacs to react to. The current Emacs sources include 1.1 million lines of code written in Emacs's own programming language, Emacs Lisp. This corpus includes code to help you edit programs in C, Python, and other languages, as you might expect from a programmer's text editor. But it also includes code to help you debug running programs, collaborate with other programmers, read electronic mail and news, browse and search directories, and solve symbolic algebra problems.

Looking under the hood, the story gets stranger. Emacs Lisp has no object system, its module system is just a naming convention, all the fundamental text editing operations use implicit global arguments, and even local variables aren't quite local. Almost every software engineering principle that has become generally accepted as useful and valuable, Emacs flouts. The code is 24 years old, huge, and written by hundreds of different people. By rights, the whole thing should blow up.

But it works—and works rather well. Its bazaar of features grows; the user interface acquires addictive new behavior; and the project survives broad changes to its fundamental architecture, infrequent releases, leadership conflicts, and forks. What is the source of this vigor? What are its limitations?

And finally, what can other software learn from Emacs? When we encounter a new architecture that shares some goal with Emacs, what question can we ask to gauge its success? Over the course of this chapter, I'll pose three questions that I think capture the most valuable characteristics of Emacs's architecture.

Emacs in Use

Before discussing its architecture, let's look briefly at what Emacs is. It's a text editor that you can use much like any other. When you invoke Emacs on a file, a window appears, displaying the file's contents. You can make your changes, save the revised contents, and exit. However, Emacs is not very effective when used this way: it is slower to start than other popular text editors, and its strengths don't come to bear. When I need to work in this fashion, I don't use Emacs.

Emacs is meant to be started once, when you begin work, and then left running. You can edit as many files as you like within one Emacs session, saving changes as you go. Emacs can keep files in memory without displaying them, so what you see reflects what you're working on at present, but your other tasks are ready to be called up just as you left them. The experienced Emacs user closes files only if the computer seems to be running low on memory, so a long-running Emacs session may have hundreds of files open. The screenshot in Figure 11-1 shows an Emacs session with two frames open. The left frame is split into three windows, showing the Emacs splash screen, a browsable directory listing, and a Lisp interaction buffer. The right frame has a single window, displaying a buffer of source code.

FIGURE 11-1. Emacs in use

There are three essential kinds of objects involved here: frames, windows, and buffers.

Frames are what Emacs calls the windows of your computer's graphical user interface. The screenshot shows two frames side by side. If you use Emacs in a text terminal, perhaps via a *telnet* or *ssh* connection, that terminal is also an Emacs frame. Emacs can manage any number of graphical frames and terminal frames simultaneously.

Windows are subdivisions of frames.* New windows are created only by dividing existing windows in two, and deleting a window returns its space to the adjacent windows; as a consequence, a frame's windows (or window) always fill its space completely. There is always a currently selected window, to which keyboard commands apply. Windows are lightweight; in typical use, they tend to come and go frequently.

Finally, *buffers* hold editable text. Emacs holds each open file's text in a buffer, but a buffer need not be associated with a file: it might contain the results of a search, online documentation, or simply text entered by hand and not yet saved to any file. Each window displays the contents of some buffer, and a buffer may appear in zero, one, or more windows.

It's important to understand that, aside from the mode line at the bottom of each window and other similar decorations, the only way Emacs ever displays text to users is by placing it in a buffer and then displaying that buffer in some window. Help messages, search results, directory

---

* Note that what most graphical user interfaces call a window, Emacs calls a frame, since Emacs uses the term "window" as described earlier. This is unfortunate, but Emacs's terminology was established well before the widespread use of graphical user interfaces, and Emacs's maintainers seem uninclined to change it.

listings, and the like all go into buffers with appropriately chosen names. This may seem like a cheap implementation trick—it does simplify Emacs internally—but it's actually quite valuable because it means that these different kinds of content are all ordinary editable text: you can use the same commands to navigate, search, organize, trim, and sort this data that are available to you in any other text buffer. Any command's output can serve as any another command's input. This is in contrast with environments such as Microsoft Visual Studio, where the results of, say, a search can only be used in the ways the implementors anticipated would be useful. But Visual Studio is not alone in this; most programs with graphical user interfaces have the same shortcoming.

▶ 一种抽象，正如 UNIX 中的 Pipeline 将传输的数据统一抽象为 Stream 一样。

For example, in the screenshot, the middle window of the frame on the left shows a directory listing. Like most directory browsers, this window provides terse keyboard commands for copying, deleting, renaming, and comparing files, selecting groups of files with globbing patterns or regular expressions, and (of course) visiting the files in Emacs. But unlike most directory browsers, the listing itself is plain text, held in a buffer. All the usual Emacs search facilities (including the excellent *incremental search* commands) apply. I can readily cut and paste the listing into a temporary buffer, delete the metadata on the left to get a plain list of names, use a regular expression search to winnow out the files I'm not interested in, and end up with a list of filenames to pass to some new operation. Once you've gotten used to working this way, using ordinary directory browsers becomes annoying: the information displayed feels out of reach, and the commands feel constrained. In some cases, even composing commands in the shell can feel like working in the dark because it's not as easy to see the intermediate results of your commands as you go.

▶ 评注见 P267

This suggests the first question of the three I promised at the beginning of this chapter, a question we can ask of any user interface we encounter: *how easy is it to use the results of one command as input to another?* Do the interface's commands compose with each other? Or have the results reached a dead end once they've been displayed? I would argue that one of the reasons many programmers have been so loyal to the Unix shell environment, despite its gratuitous inconsistencies, anorexic data model, and other weaknesses, is that nearly everything can be coaxed into being part of some larger script. It's almost easier to write a program that is readily scriptable than not, so dead ends are rare. Emacs achieves this same goal, although it takes a radically different route.

▶ 虽然MVC模式有许多变种，例如MVVM、MVP等，但其核心思想都是所谓的"职责分离"。在软件系统架构设计中，我认为最终的两个词，就是"分离"与"抽象"。上文提及的buffer的设计，则是一种抽象。

Emacs's Architecture

Emacs's architecture follows the widely used *Model-View-Controller* pattern for interactive applications, as shown in Figure 11-2. In this pattern, the *Model* is the underlying representation of the data being manipulated; the *View* presents that data to the user; and the *Controller* takes the user's interactions with the View (keystrokes, mouse gestures, menu selections, and so on) and manipulates the Model accordingly. Throughout this chapter, I'll capitalize Model, View, and Controller when I'm referring to elements of this pattern. In Emacs, the Controller is almost entirely Emacs Lisp code. Lisp primitives manipulate buffer contents (the Model) and the window layout. Redisplay code (the View) updates the display without explicit guidance from the Lisp code. Neither the buffer's implementation nor the redisplay code can be customized by Lisp code.

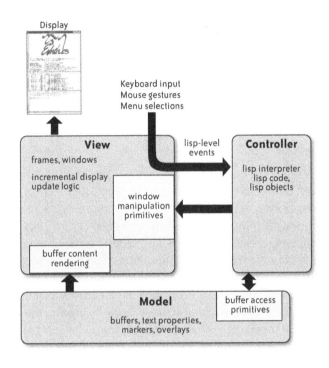

FIGURE 11-2. The Model-View-Controller pattern in Emacs

The Model: Buffers

Emacs edits text files, so the heart of Emacs's Model is the *buffer* type, which holds text. A buffer is simply a flat string, where newline characters mark line endings; it is not a list of lines, nor is it a tree of nodes, like the document object model that web browsers use to represent HTML documents. Emacs Lisp has primitive operations on buffers that insert and delete text, extract portions of buffer text as strings, and search for matches of exact strings or regular expressions. A buffer can hold characters from a wide variety of character sets, including those needed to write most Asian and European scripts.

Each buffer has a *mode*, which specializes the buffer's behavior for editing a given kind of text. Emacs includes modes for editing C code, XML text, and hundreds of other kinds of content. At the Lisp level, modes use *buffer-local key bindings* to make mode-specific commands available to the user, and use *buffer-local variables* to maintain state specific to that buffer.

▶ Emac的架构设计思想就是将模型与展现分离，这有助于实现的简洁。与UI有关的所有框架设计都遵循了这样的思想。我倾向于将模型描述为某种数据结构，控制器则负责侦听该数据结构的变化。视图以Composite模式来render该数据结构。一旦数据结构的值或者节点（例如数据结构为Tree）发生了变化，就会重新render以反映模型的变化。如此一来，复杂的UI操作逻辑就简化为对数据结构的操作，实现时，也无须考虑数据结构中各个节点之间的依赖关系了。若以immutable方式进行设计，则可以对数据结构执行函数的transform（或者reduce）操作，获得一个新的数据结构。每次变更实则都是创建了一个不变的对象。这些变更历史可以放到如前所说的"Undo Log"中，以Memento模式轻松地支持Redo、Undo。React和Redux的结合正是这样的设计思路。

Used together, these features make a buffer's mode closely analogous to an object's class: the mode determines which commands are available and provides the variables upon which those commands' Lisp implementations depend.

For each buffer, the text-editing primitives maintain an "undo" log, which holds enough information to revert their effects. The undo log remembers the boundaries between user commands, so a user command that carries out many primitive operations can be undone as a unit.

When Emacs Lisp code operates on a buffer, it can use integers to specify character positions within a buffer's text. This is a simple approach, but insertions and deletions before a given piece of text cause its numeric position to change from one moment to the next. To track positions within a buffer as its contents are modified, Lisp code can create *marker* objects, which float with the text they're next to. Any primitive operation that accepts integers as buffer positions also accepts markers.

Lisp code can attach *text properties* to the characters in a buffer. A text property has a name and a value, and both can be arbitrary Lisp objects. Logically, each character's properties are independent, but Emacs's representation of text properties stores a run of consecutive characters with identical text properties efficiently, and Emacs Lisp includes primitives to quickly find positions where text properties change, so in effect, text properties mark ranges of text. Text properties can specify how Emacs should display the text, and also how Emacs should respond to mouse gestures applied to it. Text properties can even specify special keyboard commands available to the user only when the cursor is on that text. A buffer's undo log records changes to its text properties, in addition to the changes to the text itself. Emacs Lisp strings can have text properties, too: extracting text from a buffer as a string and inserting that string into a buffer elsewhere carries the text properties along.

An *overlay* represents a contiguous stretch of text in a buffer. The start and end of an overlay float with the text around them, as markers would. Like text properties, overlays can affect the display and mouse sensitivity of the text they cover, but unlike text properties, overlays are not considered part of the text: strings cannot have overlays, and the undo log doesn't record changes to overlay endpoints.

The View: Emacs's Redisplay Engine

As the user edits text and manipulates windows, Emacs's redisplay engine takes care of keeping the display up to date. Emacs redisplay has two important characteristics:

Emacs updates the display automatically
> Lisp code need not specify how to update the display. Rather, it is free to manipulate buffer text, properties, overlays, and window configurations as needed, without regard for which portions of the display will become out of date. When the time comes, Emacs's redisplay code looks over the accumulated changes and finds an efficient set of drawing operations that will bring the display into line with the new state of the Model. By relieving Lisp code

of responsibility for managing the display, Emacs greatly simplifies the task of writing correct extensions.

Emacs updates the display only when waiting for user input

A single command may apply any number of buffer manipulation primitives, affecting arbitrary portions of the buffer—or perhaps of many different buffers. Similarly, the user may invoke a keyboard macro that carries out a long series of prerecorded commands. In cases like these, rather than showing the intermediate states of the buffer flickering by as it works, Emacs puts off updating the display until it needs to pause for user input, and refreshes the display in a single step at that point. This relieves Lisp programmers from the temptation to optimize or arrange their uses of buffer and window primitives to get smoother display behavior. The simplest code will display well.

By updating the display to reflect the effects of editing primitives automatically and efficiently, Emacs drastically simplifies the work of Lisp authors. Some other systems have picked up on this behavior: most notably, JavaScript programs embedded in web pages can edit the page's contents, and the browser will update the display accordingly. However, a surprising number of systems that invite extensions require careful cooperation between the extension code and the display update code, laying this burden on the extension author's back.

▶ Lisp代码相当于MVC中的controller,将视图呈现的逻辑与控制逻辑分开,乃题中应有之义,现在看来,已经不足为奇了。然而,正是所谓"职责分离"的基本设计原则,很好地保障了Emacs的可扩展性。这就值得我们深思了。

When Emacs was first written, its audience used computers via standalone terminals connected to the computer over serial lines or modems. These connections often were not terribly fast, so a text editor's ability to find the optimal series of drawing operations sufficient to update the screen could have a big impact on the editor's usability. Emacs went to a great deal of trouble to do well here, having a hierarchy of update strategies ranging from the quick-but-limited to the more-work-but-exhaustive. The latter used the same dynamic programming techniques employed by file comparison programs such as *diff* to find the smallest set of operations sufficient to transform the old screen into the new screen. In modern times, although Emacs still uses these algorithms to minimize update work, most of this effort is wasted, as faster processors and faster links between the computer and the display allow simpler algorithms to perform perfectly well.

▶ 无论提供的策略是否因为时间推移而逐渐失效,这种抽象为策略的设计思想值得借鉴,它能够更好地应对架构演化。通常,我们应从应用场景来驱动这种设计。

The Controller: Emacs Lisp

The heart of Emacs is its implementation of its own dialect of Lisp. In Emacs's realization of the Model-View-Controller pattern, Lisp code dominates the Controller: almost every command you invoke, whether from the keyboard, a menu, or by name, is a Lisp function. Emacs Lisp is the key to Emacs's ability to successfully accommodate the wide range of functionality that Emacs has grown to offer.

THE FIVE-MINUTE LISP TUTORIAL

People new to Lisp often find the language hard to read. This is mostly because Lisp has fewer syntactic constructs than most languages, but works those constructs harder to get an even richer vocabulary of features. To give you a reader's intuition, consider the following rules for converting code from Python to Lisp:

1. Write control constructs as if they were function calls. That is, while x*y < z: q() becomes while (x*y < z, q()). This is just a change in notation; it's still a while loop.

2. Write uses of infix operators as if they were also function calls, where the functions have odd-looking names. The expression x*y < z would become <(*(x, y), z). Now any parentheses present only for grouping (as opposed to those that surround function arguments) are redundant; remove them.

3. Now everything looks like a function call, including control constructs and primitive operations. Move the opening parenthesis of every "call" from after the function name to before the function name, and drop the commas. For example, f(x, y, z) becomes (f x y z). To continue the previous example, <(*(x, y), z) becomes (< (* x y) z).

Taking all three rules together, the Python code while x*y < z: q() becomes the Lisp code (while (< (* x y) z) (q)). This is a proper Emacs Lisp expression, with the same meaning as the original Python statement.

There's more, obviously, but this is the essence of Lisp's syntax. The bulk of Lisp is simply a vocabulary of functions (such as < and *) and control structures (such as while), all used as shown here.

Here is the definition of an interactive Emacs command to count words in the currently selected region of text. If I explain that "\\<" is an Emacs regular expression matching the beginning of a word, you can probably read through it:

```
(defun count-region-words (start end)
  "Count the words in the selected region of text."
  (interactive "r")
  (save-excursion
    (let ((count 0))
      (goto-char start)
      (while (re-search-forward "\\<" end t)
        (setq count (+ count 1))
        (forward-char 1))
      (message "Region has %d words." count))))
```

There's an argument to be made that Lisp gives its human readers too few visual cues as to what's really going on in the code, and that it should abandon this bland syntax in favor of one that makes more apparent distinctions between primitive operations, function calls, control structures, and the like. However, some of Lisp's most important features (which I can't go into here) depend fundamentally on this syntactic uniformity; the many attempts that have been made to leave it behind generally haven't fared well. With experience, many Lisp programmers come to feel that the syntax is a reasonable price to pay for the language's power and flexibility.

In Emacs, a command is simply an Emacs Lisp function that the programmer has marked (with a brief annotation at the top of the definition) as suitable for interactive use. The name of the command that the user types after Meta+x is the name of the Lisp function. *Keymaps* bind key sequences, mouse clicks, and menu selections to commands. Core Emacs code takes care of looking up the user's keys and mouse gestures in the appropriate keymaps and dispatching control to the indicated command. Along with the automatic display management provided by Emacs's View, the keymap dispatching process means that Lisp code is almost never responsible for handling events from an event loop, a welcome relief for many user interface programmers.

Emacs and its Lisp have some critical characteristics:

- Emacs Lisp is light on bureaucracy. Small customizations and extensions are easy: one-line expressions placed in the *.emacs* file in your home directory, which Emacs loads automatically when it starts, can load existing packages of Lisp code, set variables that affect Emacs's behavior, redefine key sequences, and so on. You can write a useful command in under a dozen lines, including its online documentation. (See the "The Five-Minute Lisp Tutorial" sidebar for the complete definition of a useful Emacs command.)

- Emacs Lisp is interactive. You can enter definitions and expressions into an Emacs buffer and evaluate them immediately. Revising a definition and reevaluating it puts your changes in place; there is no need to recompile or restart Emacs. Emacs is, in effect, an integrated development environment for Emacs Lisp programs.

- The Emacs Lisp programs you write yourself are first-class citizens. Every nontrivial editing feature of Emacs itself is written in Lisp, so all the primitive functions and libraries Emacs itself needs are equally available for use in your own Lisp code. Buffers, windows, and other editing-related objects appear as ordinary values in Lisp code. You can pass them to and return them from functions, store them in data structures, and so on. Although Emacs's Model and View components are hardcoded, Emacs arrogates no special privileges to itself within the Controller.

- Emacs Lisp is a full programming language. It is comfortable to write reasonably large programs (hundreds of thousands of lines) in Emacs Lisp.

- Emacs Lisp is safe. Buggy Lisp code may cause an error to be reported, but it can't crash Emacs. The damage a bug can do is contained in other ways as well: for example, buffers' built-in undo logging lets you revert many unintended effects. This makes Lisp development more pleasant and encourages experimentation.

▶ 从某种角度讲，LISP是决定Emacs生命长青的关键。

- Emacs Lisp code is easy to document. A function's definition can include a *docstring* (for "documentation string"), text explaining the function's purpose and use. Almost every function provided with Emacs has a docstring, meaning that help on a given facility is never more than a command away. And when Emacs displays a function's docstring, it includes a hyperlink to the function's source code, making Lisp code easy to browse as well. (Naturally, docstrings aren't adequate for large Lisp packages, so those usually include a more traditionally structured manual as well.)

- Emacs Lisp has no module system. Instead, established naming conventions help unrelated packages avoid interfering with each other when loaded into the same Emacs session, so users can share packages of Emacs Lisp code with each other without the coordination or approval of Emacs's developers. It also means that every function in a package is visible to other packages. If a function is valuable enough, many other packages may come to use it and depend on the details of its behavior.

Oddly, this is not nearly as much of a problem as one might expect. One might hypothesize that Emacs Lisp packages are mostly independent, but of the roughly 1,100 Lisp files included in the standard Emacs distribution, 500 of them have functions used by some other package. My guess is that the maintainers of the packages that are distributed along with Emacs only take care to remain compatible with other such packages, and that those developers form a sufficiently tightly knit group that it's practical to negotiate incompatible changes as you make them. Packages not included with Emacs probably just break—creating an incentive for developers to get their packages included and join the cabal.

Creeping Featurism

Emacs's creeping featurism is a direct consequence of its architecture. Here's the life cycle of a typical feature:

1. When you first notice a feature that would be nice to have, it's easy to try implementing it; Emacs's lack of bureaucracy makes the barrier to entry extremely low. Emacs provides a pleasant, interactive environment for Lisp development. The simple buffer Model and automatic display update let you focus on the task at hand.

2. Once you have a command definition that works, you can put it in your *.emacs* file to make it available permanently. If you use it frequently, you can include code there to bind it to a key.

3. Eventually, what began life as a single command may grow into a suite of commands that work together, at which point you can gather them up in a package to share with your friends.

4. Finally, as the most popular packages come to be included in the stock Emacs distribution, its standard feature set expands.

Similar processes are at work within the established codebase. Experience writing your own commands makes the code of Emacs itself that much more comprehensible, so when you notice a potential improvement to an existing command, you might bring up the command's source code (linked to from its help text, as mentioned earlier) and try to implement your improvement. You can redefine Emacs Lisp functions on the fly, making it easy to experiment. If your idea works out, you can put your redefinition in your *.emacs* file for your own use, and post a patch for inclusion in the official sources for everyone else.

Of course, truly original ideas are rare. For long-time users it's a common experience to think of an improvement to Emacs, look through the documentation, and find that someone else has already implemented it. Since the Lisp-friendly cohort of Emacs's user base has been adjusting and adapting Emacs for almost 20 years now, it's usually a given that many people have already been wherever you are now, and someone may have done something about it.

But whether Emacs grows by acquiring new packages or by incorporating patches its users contribute, the growth is a grass-roots process, reflecting its users' interests: the features exist, from the obvious to the strange, simply because someone wrote them and others found them useful. The role of Emacs's maintainers, beyond fixing bugs, accepting patches, and adding useful new primitives, is essentially to select the most popular, well-developed packages that the community is already using for inclusion in the official sources.

If this were the whole story, creeping featurism wouldn't be much of a problem. However, it usually has two unpleasant side effects: the program's user interface becomes too complex to understand, and the program itself becomes difficult to maintain. Emacs manages to ameliorate the former with mixed success, but it escapes the latter rather effectively.

Creeping Featurism and User Interface Complexity

There are two dimensions along which one can assess the complexity of an application's user interface: the complexity of the Model being manipulated, and the complexity of the command set that operates on that Model.

How complex is the Model?

How much does the user need to learn before he can be confident that he's put the application's Model in the state he wants? Is there hidden or obscure state that affects the Model's meaning?

Microsoft Word documents have a complex model. For example, Word has the ability to automatically number the sections and subsections of a document, keep the numbering current as pieces come and go, and update references to particular sections in the text. However, making this feature work as expected requires a solid understanding of Word style sheets. It is easy to make a mistake that has no visible effect on the document's contents, but that prevents renumbering from working properly. (For another example, ask a help desk staffer about "automatically updating styles" in the 2003 edition of Word.)

Emacs avoids this kind of problem by taking the easy way out: it doesn't support style sheets, automatically numbered sections, headers or footers, tables, or any number of other features expected from modern word processors. It's purely a text editor, and an Emacs buffer is just a string of characters. In almost every circumstance, all the buffer state that matters is readily apparent. As a result, one is rarely confused over what Emacs has done to the contents of one's files.

How complex is the command set?

How easy is it to discover the actions that are relevant and useful at any given point? How easy is it to discover features one hasn't used yet? In this sense, Emacs's user interface is quite complex. A freshly started Emacs session without customization recognizes around 2,400 commands and 700 key bindings, and will usually load many more as the session goes on.

Fortunately, a new user need not confront this horde all at once, any more than a Unix user needs to learn every shell command. It's perfectly possible for a new user to treat Emacs like any other editor with a graphical user interface, selecting text with the mouse, moving the cursor with the arrow keys, and loading and saving files with menu commands. Commands left unused don't impinge on the visibility of essential functionality.

However, using Emacs in this fashion isn't much better than using any other editor. Becoming a proficient Emacs user entails reading the manual and the online documentation, and learning to search through these resources effectively. Features such as *grep* and compilation buffers, interactive debugging, and source code indexing are what distinguish Emacs from its peers, but they don't reveal themselves unless you explicitly request them—which you wouldn't do unless you already knew they existed.

To make this kind of exploration a little easier, Emacs also includes the *apropos* family of commands, which prompt for a string or regular expression, and then list commands and customization variables with matching names or documentation strings. Although they're no substitute for reading the manual, the *apropos* commands are effective when you have a general idea what you're looking for.

In terms of this kind of complexity, the Emacs user interface has many of the characteristics common to command-line interfaces: it's possible to have many commands available, the user need not know all of them (or even many of them), and it takes deliberate effort to discover new functionality.

Fortunately, the Emacs community has been effective at establishing conventions that commands should follow, so there's a good deal of consistency from one package to the next. For example, almost all Emacs commands are modeless: the standard commands for moving, searching, switching buffers, rearranging windows, and so on are always available, so you needn't worry about how to "exit" a command. As another example, most commands use standard Emacs facilities to prompt the user for parameters, making prompting behavior consistent from one package to the next.

Creeping Featurism and Maintainability

Obviously, the more code you have, the more effort it takes to maintain it. When a developer's Lisp package is selected for inclusion in the standard Emacs distribution, the lead maintainers invite that package's author to continue maintaining it, so as the number of packages expands, the number of maintainers expands to match. If someone relinquishes responsibility for a package (perhaps because she's too busy or no longer interested), then the lead maintainers must either find a new volunteer or remove the package.

The key to this solution is that Emacs is a collection of packages, not a unified whole. In a sense, the dynamics of Emacs maintenance more closely resemble those of a platform (such as an operating system) than a single application. Instead of a single design team choosing priorities and allocating effort, there is a community of self-directed developers, each pursuing his own ends, and then a process of selection and consolidation that brings their efforts into a single release. In the end, no single person bears the weight of maintaining the entire system.

In this process, the Lisp language acts as an important abstraction boundary. Like most popular interpreted languages, Emacs Lisp largely insulates code from the details of the Lisp interpreter and the underlying processor architecture. Likewise, the editing primitives available to Lisp conceal the implementations of buffers, text properties, and other editing objects; the characteristics visible to Lisp are, for the most part, restricted to those the developers are willing to commit to supporting in the long term. This gives Emacs's core of C code a good deal of freedom to improve and expand, without the risk of breaking compatibility with the existing body of Lisp packages. For example, Emacs buffers have acquired support for text properties, overlays, and multiple character sets while remaining largely compatible with code written before those features existed.

Two Other Architectures

Many applications allow user extensions. Extension interfaces appear in everything from collaborative software development website systems (such as Trac plugins) to word processing software (Open Office's Universal Network Objects) to version control software (Mercurial's extensions). Here I compare Emacs with two architectures that support user extensions.

Eclipse

Although most people know Eclipse as a popular open source integrated development environment for Java and C++, Eclipse proper includes almost no functionality. Rather, it is a framework for plug-ins, allowing components that support specific aspects of development—writing Java code, debugging a running program, or using version control software—to communicate easily with each other. Eclipse's architecture allows programmers with a solid solution to one part of the problem to join their work with others' in order to produce a unified and full-featured development environment.

As a development environment, Eclipse provides valuable features that Emacs lacks. For example, the Java Development Tools plug-ins provide extensive support for refactoring and code analysis. In comparison, Emacs has only a limited understanding of the semantic structure of the programs it edits, and can't offer comparable support.

Eclipse's architecture is nothing if not open-ended, as plug-ins provide nearly all significant functionality. There are no second-class citizens: the popular plug-ins are built on the same interfaces available to others. And because Eclipse gives each plug-in relatively low-level control over its input and display, the plug-in is free to choose whatever Model, View, and Controller best suits its purpose.

However, this approach has a number of drawbacks:

- Eclipse plug-in development is not safe, in the sense attributed to Emacs Lisp code. A buggy plug-in can easily cause Eclipse to crash or lock up. In Emacs, the Lisp interpreter ensures that the user can interrupt runaway Lisp code, and the strong boundary between Lisp code and the Model implementation protects the user's data from the more insidious forms of corruption.

- Because the interfaces that Eclipse plug-ins offer each other are relatively complex, writing an Eclipse plug-in is more like adding a module to a sophisticated application than writing a script. Certainly, these interfaces are what have made Eclipse's features possible, but plug-in authors must pay that price for both ambitious and simple-minded projects.

- A plug-in requires enough boilerplate code that Eclipse includes a plug-in to help people write plug-ins. The Eclipse Plug-in Development Environment can generate code skeletons, write default XML configuration files and manifest files, and create and tear down test environments. Providing a "wizard" to generate boilerplate code automatically can help get a plug-in developer started, but it doesn't reduce the complexity of the underlying interfaces.

The overall effect is to leave Eclipse plug-ins as a rather coarse-grained extension facility. Plug-ins are not suited for quick-and-dirty automation tasks, and they're not especially friendly to casual user-interface experimentation.

This suggests the second question of the three I promised at the beginning of the chapter, one we can ask of any plug-in facility: *what sort of interfaces are available for plug-ins to use?* Are they simple enough to allow the kind of rapid development associated with scripting languages? Can a plug-in developer work at a high level of abstraction, close to the problem domain? And how is the application's data protected from buggy plug-in code?

Firefox

The current generation of sophisticated web applications (Google Mail, Facebook, and so on) makes heavy use of techniques such as *dynamic HTML* and *AJAX* to provide a smooth user experience. These applications' web pages contain JavaScript code that responds to the user's input locally, and then communicates with the underlying servers as needed. The JavaScript code uses a standard interface, the *Document Object Model*, to navigate and modify the web page's contents, and further standards dictate the page's visual appearance. All modern web browsers implement these standards to some degree.

Although a web browser is not a text editor, there are some striking resemblances between Emacs's architecture and that of a browser:

- Although Emacs Lisp and JavaScript don't resemble each other much at the syntactic level, their semantics have many essential traits in common: like Emacs Lisp, JavaScript is interpreted, highly dynamic, and safe. Both are garbage-collected languages.
- As with Emacs Lisp, it's very practical to begin with a small fragment of JavaScript on a page to improve some minor aspect of its behavior, and then grow that incrementally into something more sophisticated. The barrier to entry is low, but the language scales up to larger problems.
- As in Emacs, display management is automatic. JavaScript code simply edits the tree of nodes representing the web page, and the browser takes care of bringing the display up to date as needed.
- As in Emacs, the process of dispatching input events to JavaScript code is managed by the browser. Firefox takes care of deciding which element of the web page an event was directed at, finds an appropriate handler, and invokes it.

However, Firefox takes the ideas behind these modern web applications a bit further: Firefox's own user interface is implemented using the same underlying code that displays web pages and handles their interactions. A set of packages known as *chrome* describe the interface's structure and style, and include JavaScript code to bring it to life.[†] This architecture allows third-party developers to write *add-ons* that extend Firefox's user interface with new chrome packages. Taking the same techniques even further, developers can replace the standard Firefox chrome altogether and radically reshape the entire user interface—to adapt it for use on mobile devices, for example.

Like Eclipse plug-ins, Firefox chrome packages include a significant amount of metadata. And, resembling Eclipse's Plug-in Development Environment plug-in, there is a Firefox extension to help people write Firefox extensions. So there is still a significant amount of work required up front before one can extend or fix Firefox. However, Firefox's automatic display management and simplified event handling mean that the effort required is still not as high as that needed to write an Eclipse plug-in.

---

† Naturally, JavaScript code used in chrome can read and write preference files, bookmark tables, and ordinary user files—privileges that would never be granted to code downloaded from a web page.

Firefox's developers are working to improve the performance of its JavaScript implementation. Although this obviously helps users visiting JavaScript-heavy websites, it also allows the Firefox developers to migrate more and more of the browser itself from C++ to JavaScript, a much more comfortable and flexible language for the problem. In this sense, Firefox's architecture is evolving to look more like that of Emacs, with its all-Lisp Controller.

This suggests the third and final question, one we can ask of any extension language we encounter: *is the extension language the preferred way to implement most new features for the application?* If not, what restrictions discourage its use? Is the language itself weak? Is its interface to the Model cumbersome? Whatever the case, these same weaknesses probably affect extension developers in similar ways, leaving extensions as second-class citizens. (One could ask the analogous question of plug-in interfaces as well.) Like Emacs, Firefox places its extension language at the heart of its architecture, a strong argument that the language's relationship with the application has been designed properly.

As an avid Emacs user, but one concerned about its future, I'm especially interested in Firefox because it seems so close to Emacs in many ways: a View that provides automatic display management, a Controller based on an interpreted, dynamic language, and a Model that does everything Emacs's does, and much more. If one were willing to leave behind the accumulated corpus of Emacs Lisp code, a few days' worth of chrome programming could produce a text editor with an architecture very similar to Emacs's, but with a much richer model and a much stronger connection to the current frontiers of technology. The most valuable lessons Emacs's architecture has to teach need not be forgotten.

Principles and properties		Structures	
	Versatility	✓	Module
✓	Conceptual integrity	✓	Dependency
✓	Independently changeable	✓	Process
	Automatic propagation		Data access
✓	Buildability		
✓	Growth accommodation		
✓	Entropy resistance		

CHAPTER TWELVE

When the Bazaar Sets Out to Build Cathedrals

How ThreadWeaver and Akonadi were shaped by the KDE community and how they shape it in turn

Till Adam
Mirko Boehm

Introduction

THE KDE PROJECT IS ONE OF THE BIGGEST FREE SOFTWARE[*] **EFFORTS IN THE WORLD.** Over the course of 10 years, its very diverse community of contributors—students, seasoned professionals, hobbyists, companies, government agencies, and others—has produced a vast amount of software addressing a wide variety of issues and tasks, ranging from a complete desktop environment with a web browser, groupware suite, file manager, word processor, spreadsheet, and presentation tools, to highly specialized applications, such as a planetarium tool. The basis of these applications is provided by a collection of shared libraries that are maintained by the project collectively. Beyond their primary intended use by the members of

---

[*] Also referred to as open source software.

the KDE developer community itself, they are also used by many third-party developers, both commercial and noncommercial, to produce thousands of additional applications and components.

Although the initial focus of the KDE project was to provide an integrated desktop environment for free Unix operating systems, notably GNU/Linux, the scope of KDE has broadened considerably, and much of its software is now available on not just various flavors of Unix, but also on Microsoft Windows and Mac OS X, as well as embedded platforms. This implies that the code written for the KDE libraries has to work with many different tool chains, cope with the various platform peculiarities, integrate with system services flexibly and in extensible ways, and make judicious and careful use of hardware resources. The broad target audience of the libraries also means that they have to provide an API that is understandable, usable, and adaptable by programmers with diverse backgrounds. Someone accustomed to dealing with Microsoft technologies on Windows will have different preconceptions, biases, habits, and tools from an embedded programmer with a Java background or an experienced Mac developer. The goal is to make all programmers able to work comfortably and productively, to allow them to solve the problems at hand, but also (and some say more importantly) to benefit from their contributions should they choose to give back their suggestions, improvements, and extensions.

This is a very agile, very diverse, and very competitive ecosystem, one in which most active contributors are interested in collaborating to improve their software and their skills by constantly reviewing each other's work. Opinions are freely given, and debates can become quite heated. There are always better ways to do one thing or another, and the code is under constant scrutiny by impressively smart people. Computer science students analyze implementations in college classes. Companies hunt down bugs and publish security advisories. New contributors try to demonstrate their skills by improving existing pieces of code. Hardware manufacturers take the desktop and squeeze it onto a mobile phone. People feel passionately about what they do and how they think things should be done, and one is, in many ways, reminded of the proverbial bazaar.[†] Yet this wild and unruly bunch is faced with many of the same challenges that those in more traditional organizations who are tasked with maintaining a large number of libraries and applications must overcome.

Some of these challenges are technical. Software has to deal with ever-increasing amounts of data, and that data becomes more complex, as do the workflows individuals and organizations require. The necessity to interoperate with other software (Free and proprietary) used in corporations or government administrations means that industry paradigm shifts, such as the move towards service-oriented architecture (SOA), have to be accommodated. Government and corporate mission-critical uses pose stringent security requirements, and large deployments need good automation. Novice users, children, or the elderly have different needs

---

† *http://en.wikipedia.org/wiki/The_Cathedral_and_the_Bazaar*

entirely, yet all of those concerns are valid and of equal importance.‡ Much like the rest of the industry, Free Software in general and KDE in particular are coming to terms with pervasive concurrency and increasingly distributed processing. Users rightfully demand easier-to-use, cleaner, and more beautiful interfaces; responsive, well-thought-out, and not overly complex interactions with software; and high reliability, stability, and data safety. Developers expect language-neutral extension points, well-maintained APIs, binary and source compatibility, clean migration paths, and functionality on par with the commercial offerings they are frequently accustomed to from their day jobs.

But even more daunting are the social and coordinative aspects of a larger Free Software project. Communication is key, and it is hampered by time zones, cultural barriers, more or less reflected preferences and prejudices, and also simply by physical distance. Discussions that can be finished in 15 minutes in a stand-up office meeting may take days of arguing on mailing lists to allow the opinions of team members from the other hemisphere to be heard. The process of reaching consensus often is at least as important as the resulting decision itself, to keep the cohesion of the group intact. Naturally, more patience, understanding, and rhetoric wit is necessary to gain a favorable consensus. This is much to ask from people who aced math and physics classes without flinching, but avoid getting a haircut because they do not like the talking to nonprogrammers that such an activity entails. With this in mind, the amicable spirit of the annual Akademy conferences is a wonderful experience (see the next section). We have never seen such a massively diverse group of old and young people from all over the globe, from rivaling countries, countless nations, seemingly different worlds even, and with all shades of skin set aside differences to argue—heatedly, but rationally—about arcane C++ subtleties.

And then, aside from technical and personal aspects, there is a third major influence on how a Free Software project fares: structure. In some ways, structure is inevitable for Free Software groups. Larger and better-known software projects especially need to hold trademarks, receive donations, and organize conferences. The time when this is possible solely through the good will of spouses and weekend trips is usually already over when a project is mentioned publicly for the first time. In some respects, structure (or the lack thereof) determines the fate of the community. An important decision has to be made—whether to go corporately dull or geekily chaotic. The answer is not trivial, because multiple trade-offs are involved: funding versus freedom from external influence; stability of the development process versus attracting the most brilliant cave dwellers;§ visibility and mind-share versus focus on impressive technical results. Beyond the point where structure is a bare necessity, there are options for the levels of bureaucracy. Some projects do not even have a board of directors, whereas some have

---

‡ This is fundamentally different from software that is entirely produced for a market. Free Software can cater to the needs of the blind, for example, without having to justify the considerable effort needed for that with corresponding sales figures and expected returns on the investment. That is one of the reasons why KDE is available in so many more languages than proprietary competitors, for example.

§ We use that term with a lot of affection. It cannot be denied, though, that many of our dearest friends could do with a tad more exposure to sunlight and healthy food, overall.

dictators who claim to be benevolent. There are examples for successful and unsuccessful communities on both ends of this continuum.

Before we look in detail at two concrete examples of how the technical, social, and structural issues were approached and dealt with by groups within KDE, it seems useful to provide some background on the history of the KDE project in these three areas. The following section will thus describe what KDE is today and how the community arrived at that point.

History and Structure of the KDE Project

KDE, or the K Desktop Environment, was originally conceived out of despair. At a time when FVWM was considered a desktop, Xeyes was a stock inventory item on the screen, and Motif was the toolkit competing with XForms for the higher body count of developer's brain cells and for lack of sexiness, KDE was founded to achieve a revolutionary goal: to marry the raw power of Unix with a slick, eye-candied user experience. This goal was now considered achievable because the Norwegian startup Trolltech was about to release the first version of its groundbreaking object-oriented GUI toolkit for C++, Qt. Qt set out to allow GUI programming the way it was meant to be: systematic, object-oriented, elegant, easy to learn, well-documented, and efficient. In 1996 Matthias Ettrich, at the time a student at Tuebingen University, first emphasized the potential offered by using Qt to develop a complete desktop environment. The idea quickly attracted a team of about 30 developers, and that group has been growing steadily ever since.

Version 1.0 of KDE was released in 1998. Although nimble in its functionality from today's point of view, it needs to be measured in relation to the competition: Windows 3.1 did not have memory protection at the time, Apple was struggling to find a new kernel, and Sun swept the sorry remnants of CDE into the gutter. Also, this was before the first Linux hype, and the momentum of Free Software was not yet understood by all in the software industry.

▶ 开源项目如果缺乏一个持之以恒的推动者，就很难持续演进；但若推动者以领袖身份参与太多管理，又可能因为理念不同给团队带来诸多限制，从而影响项目的开发。一个成功的开源项目团队，一定是一个优秀的自组织（Self-Organized）团队。

Even in the process of finishing KDE 1, the developers had already redesigned it for 2.0. Some major elements where missing: the component model, the network abstraction layer, the desktop communication protocol, the UI style API, and more. KDE 2 was the first release that was architected, designed, and implemented in a reasonably rational process. Corba was considered for the component model and rejected. This was also the first time the contributor group fluctuated. Interestingly, although some early core developers left, the size of the active KDE development team slowly but steadily grew. KDE e.V., the organization of the KDE contributors, was founded in 1996 and grew to represent the vast majority of the KDE committers by 1998. From the beginning, the organization was meant to support the contributors, but not to influence the technical direction of KDE. The course of development is supposed to be the result of the work of the active developers, not a management team. This idea proved to be one of the most important factors, if not the most important, toward making KDE one of the very few major Free Software projects that are not massively influenced by funding corporate bodies. External funding is not regarded as problematic in KDE, but many other projects that relied too much on it ceased to exist when the major funding sponsor keeled over or lost interest. Because of that, KDE survived many trends and hypes in the Free Software world and continuously maintained its development momentum.

In April 2002, KDE 3 was ready. Since KDE 2 was considered well-designed, version 3 was more evolutionary and matured closer and closer to perfection over five major releases and six years. Important applications that became standard on free desktops have been developed based on the KDE 3 technologies: K3B, the disk burning program; Amarok, one of the slickest music players in general; Kontact, a full personal communication suite. Most interestingly, for the first time these applications use KDE not only as one target desktop, but also as the platform on top of which end-user applications are built. With version 3, KDE started to separate into two things: the desktop and the environment, usually called the platform. But since KDE was still confined to X11, this split was not easily recognized by users. That was the next step.

In 2004, one of the toughest calls in its history had to be made by the KDE team. Trolltech was about to release version 4.0 of Qt, and it was very advanced and very different compared to both previous releases and any other toolkit on the market. Because of the massive changes in the toolkit, going from Qt 3 to Qt 4 was not an adaptation, but a port. The question was whether KDE 4 was going to be a straight port of KDE 3 from Qt 3 to Qt 4 or a major redesign in the process of porting. Both options had many supporters, and it was clear to the vast majority of those involved that, either way, an immense amount of work had to be done. The decision was made in favor of a complete redesign of KDE. Even if it is now accepted that this was the right choice, it was a very risky one, because it meant providing KDE 3 as the main line for an extended period of time in parallel until completing the huge porting effort.

One major new feature of Qt 4 needs particular emphasis. The GPL and Commercial dual licensing scheme Trolltech was using already for the X11 version was now extended to all target platforms Qt supports, most notably to Windows, Mac OS X, and embedded platforms. KDE 4 thus had the potential to become something relevant beyond the Unix world. Although the Unix desktop remains its home turf, applications developed for KDE can now run on Windows and Mac OS X computers. This possibility was received controversially. One argument against it was that the Free Software community would provide neat applications for those proprietary desktops, thus reducing the incentive to switch away from them to free alternatives. Another one was, "What do we care?" or more politely, "Why should we invest scarce development time in supporting nonfree target systems?" Proponents argued that providing the same applications everywhere would ease the transition for users from proprietary to free operating systems and allow gradual replacements of key applications. In the end the trend was set according to KDE's long-term mantra of "those who do the work decide." There was enough interest in the new target platforms to gain the attention of sufficient contributors, so in the end, there was no reason to deprive the KDE users of a capability many obviously longed for.

To become platform-independent, KDE 4 was rearchitected and separated into an application development platform, the desktop built on top of it, and the surrounding applications. Of course the dividing lines are blurry at times.

▶ 架构师在进行技术选型时，会经常遇到版本升级带来的迁移问题。坦白说，决策者往往很难正确地预见未来，但我们应该将这一问题列入"风险列表"中，随时关注。

Given this history, let us look at some of the ways in which a project like KDE tries to deal with issues that go beyond what a single developer, however talented, can solve. How can knowledge and experience be retained and used to maximum effect? Incorporating strategic thinking, how can a large, diverse group make difficult decisions and agree on steering toward the overall destination without jeopardizing the sense of fun? How can we incorporate the idea that decisions are made on equal footing among peers and the other aspects that have made KDE and other Free Software projects so successful? In other words, how to build a cathedral when everyone involved has a tendency to consider themselves, rightfully or not, an architect, and how to fill the role of architect without wearing a funny-looking hat. We start with looking at the contributors because we are convinced that Free Software communities are first and foremost social structures as opposed to technical ones. People are the most scarce and valuable resource we have.

There are different roles to fill for every project, and there is no authority that decides in the recruiting process. Every Free Software project needs a number of poster girls and boys and a large bunch of motivated, skilled, down-to-earth hackers, artists, administrators, writers, translators, and more. The one thing that seems to be a common denominator among all of these trades is that they all require a great deal of self-motivation, skill, and self-guidance. To be able to contribute seems to attract many extraordinary individuals of all ages, from high school students to retirees. What makes them join the project is the fascination of being part of a group that creates tomorrow's technologies, to meet other people interested in and driving these technologies, and often to do something with a reason, instead of writing throw-away college papers.

A well-functioning Free Software community is about the most competitive environment to be in. Most commercial engineering teams we have encountered are way more regulated, and policies protect the investment employees have made in their position. Not so for Free Software developers. The only criteria for a certain code contribution's inclusion in the software is its quality and if it is being maintained. For a short while, a mediocre piece of code made by a respected developer may stay in the source code, but in the long term, it will get replaced. The code is out in the open, constantly scrutinized. Even if a certain implementation is of good quality but not close enough to perfection in a few people's perception, a group of coders will take it on and improve it. Since creation is what motivates most, Free Software developers constantly need to be aware that their creation is made obsolete by the next one. This explains why code written in Free Software projects is often of higher quality than what can be found in commercial projects—there is no reason to hope the embarrassing bits will not be discovered.

Surprisingly, few coders want exterior motivation in the form of recognition from users or the media. Often, they turn away from some piece of work shortly before it is finally released. As with marathon runners, it seems their satisfaction is internal—the knowledge of having reached the finishing line. Sometimes, this is misunderstood as shying away from the public, but on the contrary, it only underlines the disinterest in praise. Because of this set of personal values, in software projects that are really free, it is almost impossible to assign priorities to development goals. Code artists are intrinsically motivated and choose and decide their own individual priorities. Software still gets finished because the most challenging elements attract the highest attention. This coordination from within favors the journey instead of the goal and still leads to results. The chosen path is often different from the one a well-managed commercial software project would have taken, and usually it also incurs fewer compromises of quality versus deadlines. If such compromises are accepted on a regular basis, it is generally a good sign that a Free Software project is in the process of turning into a commercial, or "less Free," one. Typically, this happens when either a major group of developers spin a company off the project, a funding partner asserts control over major contributors, or the project itself stops innovating and goes into maintenance mode.

Most kinds of structure and organization imposed on a Free Software project are met with resistance. Contributors join KDE for many reasons, but enjoying the bureaucracy of the project is not one of them. People who join a Free Software project usually are very aware of using and protecting their own freedom of choice. At the same time, the necessity of a minimum amount of formal organization is recognized. Although almost all technical challenges are no match for a good team, formal structure is where centrifugal forces are most effective. Many Free Software projects have dissolved because decisions of political scope were forced through by influential team members. Finding a formal structure that solves the problems the project faces once it becomes significant but at the same time does not hinder the further technical development is one of the most (or the single most) important steps KDE had to take in its history. The fact that a very stable and accepted structure was found surely contributed significantly to the long-term stability of the KDE community.

In 1996, KDE founded KDE e.V. as the representation of the KDE contributors. An e.V., or "eingetragener Verein," is the classical not-for-profit organization in Germany, where most of the KDE contributors were based at the time and where the contributors met. The main force leading to its creation was that a legally capable body representing the project was needed to enter into the Free Qt Foundation. The Free Qt Foundation was an agreement between Trolltech, the makers of Qt, and KDE with the purpose of ensuring that KDE always had access to Qt as a free toolkit. It guaranteed (and still does) that, should Trolltech stop to develop and publish the free version of Qt, the latest released free version can be continued without Trolltech. This agreement was even more important when the free version of Qt was not licensed under the GPL. Today it is, and the foundation is now dealing mostly with resolving copyright subtleties, developer rights, and other similar issues. It was important because many contributors would have hesitated to spend their time on a project that ran the risk of being commercialized, but it also served as an example that KDE as a project must be able to carry legal rights. Later on, the KDE trademark was registered and is held by KDE e.V., which assumed many other responsibilities such as serving as the host of the large annual KDE conference.

Regarding the architecture and design of KDE, the organization is still rather influential, even if it is not supposed to manage the development. Since it is assumed that most of the "accomplished" KDE contributors are members, the KDE e.V.'s opinion weights rather heavy. Also, it is the host of the annual Akademy conference, which for most contributors is the only chance to meet and discuss things in person. It also raises funds, mostly donations and membership fees of sponsors, and thus has the ability to fund (or not fund, in very rare cases) developer activities such as sprints or targeted meetings. Still, the budget the organization spends annually is surprisingly small compared to the effects created by it. The developer meetings are where most of the coordination takes place, which does give the membership a lever. Still, the initiative of groups of contributors gets most things done, not a statement at the general assembly.

Akademy has become an institution that is well-known even outside the KDE community. It is where most of the coordination that really requires in-person meetings takes place. Since neither human resources nor funds can be directly assigned to any development activity, many discussions of broader scope are saved for the conference. It has become a routine to use this annual gathering for decision-making and to only loosely coordinate in the meantime. The conference takes place quite reliably around summer, and contributors to other Free Software projects use that opportunity to coordinate with KDE. One of the decisions made during the 2007 conference was to switch to six-month release cycles, which was suggested and championed by Marc Shuttleworth of Ubuntu.

Akademy is the only global conference KDE organizes. In addition to this large meeting, many small gatherings of subgroups and sprints take place. These meetings are usually more frequent, more local, and more focused, so whereas architectural issues are debated at Akademy, design issues for certain modules or applications are discussed here. Some subgroups, such as the KOffice or Akonadi developers, usually meet at three-month intervals.

This reiterative process of coordinated high- and medium-level reviews has proven to be quite effective and also provides a good understanding of the goals and next actions among the developers. Most attendees express that the annual conference gives them a boost in motivation and in the effectiveness of their development work.

The organization and structure KDE shows today is not the brain child of a group of executives who asked themselves how a Free Software project should be organized. It is the result of an iterative process of trying to find a suitable structure for the main nontechnical goals—to remain free and to ensure the longevity of the project and sustainable growth of the community. Freedom is used here not only in the sense of being able to provide the software for free and as Free Software, but also to be free of dominating influences from third parties. External parties such as companies or governmental groups are regularly present at the conferences, and the project is interested in their findings, experiences, and contributions. However, these stakeholders must be prevented from investing enough resources to be able to determine the outcome of votes in the community. This may seem paranoid, but it actually happened to other projects, and the KDE community is aware of that. So staying active and healthy as a Free Software project is directly related to protecting the freedom of the project itself. The main technical goal, to support the developers and other contributors with funds, materials, organizations, and other resources, are comparatively simple to achieve.

The result is a living, active, vibrant community with an established development process, a stable and healthy fluctuation of contributors, and a lot of fun on the way. And this in turn helps to attract and secure the most important resource for such an environment: new contributors. KDE has succeeded in preventing the typical tendency to create larger and larger formal structures and positions and also does not have any dictators, as benevolent as they might be. KDE is an archetypal Free Software community.

In order to understand how this community functions in practice, how architectural decisions are arrived at, and how the peculiar process influences the outcomes technologically, we will look at two examples in some detail: Akonadi, the personal information management infrastructure layer for KDE 4, and ThreadWeaver, a small library for high-level concurrency management.

Akonadi

The KDE 4 platform, both as a development platform and as a runtime environment for the execution of integrated applications, rests on a number of so-called "pillars." These key pieces of infrastructure provide the central services that applications expect to have easy and ubiquitous access to on a modern desktop. There is the *Solid* hardware interaction layer, which is responsible for providing the desktop with information about and notification from the hardware, such as a USB stick becoming available or the network going down. *Phonon* provides an easy-to-program multimedia layer for playback of various media types and user interface elements for their control. *Plasma* is a library for rich, dynamic, scalable (SVG-based) user interfaces that go beyond the standard office look.

The personal information of the user—her email, appointments, tasks, journal entries, blog posts and feeds, bookmarks, chat transcripts, address book entries, etc.—contains not only a large amount of primary information (the content of the data). It also weaves a rich contextual fabric from which much about the user's preferences, social interactions, and work contexts can be learned, and it can be used to make the interaction experience of many applications on the desktop more valuable and interesting, provided that it is readily, pervasively, and reliably accessible. The *Akonadi* framework aims to provide access to the user's personal information, the associated metadata, and the relationships between all that data, as well as services that operate on them. It aggregates information from a variety of sources, such as email and groupware servers, web and grid services, or local applications that feed into it, caches that information, and provides access to it. Akonadi thus serves as another pillar of the KDE 4 desktop but, as we shall see, aims to go beyond that.

In the following sections, we will explore the history of this large and powerful framework, the social, technical, and organizational struggles that were and are involved in making it happen, and the vision the authors have for its future. On the way, we will provide some detail on the technical solutions that were found and the reasons why they were chosen.

Background

From the earliest conversations of the initial group of KDE developers about which applications would need to be included in order for a desktop offering to be considered complete by users, email handling, calendering, task list management, and an address book were always considered obvious and important user needs. Consequently, KMail, KAddressbook, and KOrganizer (applications to handle email, contacts, and events and tasks, respectively) were among the first projects to be taken on, and the first usable versions emerged quite quickly. User needs were comparatively simple then, and the standard modes of receiving email were through local mail spools or from POP-3 servers. Mail volume was low, and mail folders were generally stored in the mbox format (one continuous plain text file per folder). HTML content in emails was overwhelmingly frowned upon by the user community that KDE was targeting, multimedia content of any kind was very rare, and encryption and digital signatures were equally exotic. Along similar lines, the custom formats used for address and calendaring data were text-based, and the overall volume of the information to be stored was easily manageable. It was thus relatively straightforward to write basic applications that were already powerful enough to be quickly adopted by other KDE developers and, after the first releases of KDE, also by the user community.

▶ 这实际上是一个架构设计的反例，证明了我们在进行架构设计之前去识别风险的重要性。此外，所谓 "the overall complexity of the code increased"，除了由软件本身随着规模的扩大带来的复杂性之外，还有技术债（Technical Debt）的缘故。技术债是不可避免的，而在项目演进过程中，我们要用积极的态度去管理技术债，并采用重构、自动化测试、持续集成、代码评审等诸多手段驱使团队及时清理债务。

The early and continuous success of the personal information management (PIM) applications would prove to be a double-edged sword in the years to follow. As the Internet and computer use in general skyrocketed, the PIM problem space started to become a lot more complex. New forms of access to email, such as IMAP, and storage of email, such as the maildir format, had to be integrated. Workgroups were starting to share calendars and address books via so-called *groupware* servers, or store them locally in new standard formats such as vcal/ical or vcard. Company and university directories hosted on LDAP servers grew to tens of thousands of entries. Yet users still expected to use the KDE applications they had come to appreciate under those changed circumstances and get access to new features quickly. As a result, and given the fact that only a few people were actively contributing to the PIM applications, their architectural foundations could not be rethought and regularly cleaned up and updated as new features added, and the overall complexity of the code increased. Fundamental assumptions—that access to the email storage layer would be synchronous and never concurrent, that reading the whole address book into memory would be reasonably possible, that the user would not go back and forth between timezones—had to be upheld and worked around at times, because the cost of changing them would have been prohibitive given the tight time and resource constraints. This is especially true for the email application, KMail, whose codebase subsequently evolved into something rather unpleasant, hard-to-understand, hard-to-extend and maintain, large, and ever more featureful. Additionally, it was a stylistically diverse collection of work by a series of authors, none of whom dared to change things too

> much internally, for fear of bringing the wrath of their fellow developers upon them should they break their ability to read and write email.

As PIM applications became more and more widely and diversely used, working on these applications became something that was mainly continued by those who had been involved with it for a while out of a sense of dedication and loyalty, rather than an attractor for new contributors. Simply put, there were other places in KDE and Free Software where those wishing to enter the community could do so with a more shallow learning curve and with less likelihood of bloodying their nose by, for example, inadvertently breaking some enterprise encryption feature they would have no way of testing themselves, even if they knew about it. In addition to (and maybe at least partially because of) the technical unattractiveness, the social atmosphere (especially around KMail) was seen as unsavory. Discussions were often conducted rudely, ad hominem attacks were not infrequent, and those who attempted to join the team seldom felt welcome. Individuals worked on the various applications mostly in isolation. In the case of KMail there was even a rather unpleasant dispute over maintainership. All of this was highly unusual within the KDE community, and the internal reputation of this group was thus less than stellar.

As the need for more integration between the individual applications grew, though, and as users were increasingly expecting to be able to use the email and calendaring components from a common shell application, the individuals working on these components had to make some changes. They had to start interacting more, agree on interfaces, communicate their plans and schedules, and think about aspects such as branding of their offerings under the umbrella of the *Kontact* groupware suite and consistent naming in their areas of responsibility. At the same time, because of commercial interest in the KDEPIM applications and Kontact, outside stakeholders were pushing for a more holistic approach and for the KDEPIM community to speak with one voice, so they could interact with it reliably and professionally. These developments catalyzed a process that would lead the PIM group toward becoming one of the tightest knit, friendliest, and most productive teams within KDE. Their regular meetings have become a model for other groups, and many contributors have become friends and colleagues. Personal differences and past grudges were put aside, and the attitude toward newcomers drastically changed. Today most communication and development talk happens on a combined mailing list (*kde-pim@kde.org*), the developers use a common IRC channel (#kontact), and when asked what they are working on will answer "KDEPIM," not "KMail" or "KAddressBook."

Personal information management is pivotal in an enterprise context. Widespread adoption of the KDE applications in large organizations has led to a steady demand for professional services, code fixes, extensions, packaging (as part of distributions and specifically for individual use cases and deployments), and the like, which in turn has resulted in an ecosystem of companies offering these services. They have quite naturally hired those people most qualified to do such work—namely the KDEPIM developers. For this reason, the majority of the active code contributors in KDEPIM are now doing it as part of their day jobs. And among those who are not directly paid for their KDE work, there are many who work as C++ and Qt developers full-time. There are still volunteers, especially among the newer contributors, but the core group consists of professionals.[||]

Because of the universal importance of PIM infrastructure for most computer users, be it in a personal or business context, the guiding principle of the technical decision-making process in KDEPIM has become practicability. If an idea cannot be made to work reliably and in a reasonable timeframe, it is dropped. Changes are always scrutinized for their potential impact on the core functionality, which must be maintained in a working state at all times. There is very little room for experimentation or risky decisions. In this, the project is very similar to most commercial and proprietary product teams and somewhat dissimilar from other parts of KDE and Free Software in general. As noted in the introduction, this is not always purely positive, as it has the potential to stifle innovation and creativity.

The Evolution of Akonadi

The KDEPIM community meets quite regularly in person, at conferences and other larger developer gatherings, as well as for small hacking sprints where groups of 5 to 10 developers get together to focus on a particular issue for a few days of intensive discussion and programming. These meetings provide an excellent opportunity to discuss any major issues that are on people's minds in person, to make big decisions, and to agree on roadmaps and priorities. It is during such meetings that the architecture of a big concerted effort such as Akonadi first emerges and later solidifies. The remainder of this section will trace some of the important decision points of this project, starting from the meeting that brought the fundamental ideas forward for the first time.

When the group met for its traditional winter meeting in January 2005, parts of the underlying infrastructure of KDEPIM were already showing signs of strain. The abstraction used to support multiple backend implementations for contact and calendaring information, KResources, and the storage layer of the email application KMail were built around a few basic assumptions that were starting to no longer hold. Specifically, it was assumed that:

- There would be only a very limited number of applications interested in loading the address book or the calendar: namely the primary applications serving that purpose, KAddressbook and KOrganizer. Similarly, there was the assumption that only KMail would need to access the email message store. Consequently there was no, or very limited, support for change notification or concurrent access, and thus no proper locking.

[||] This might be surprising, since it goes against the intuitive idea of Free Software being produced largely by students with too much time on their hands, but lately there have been substantiated claims that by now the majority of successful Free Software is in fact written and maintained by professional software engineers. See for example Karim Lakhani, Bob Wolf, Jeff Bates, and Chris DiBona's Hacker Survey v0.73 at http://freesoftware.mit.edu/papers/lakhaniwolf.pdf (24.6.2002, Boston Consulting Group).

- There would be only a very limited amount of data. After all, how many contacts would users typically have to manage, and how many appointments or tasks? The assumption was that this would be in the order of magnitude of "a few hundred."
- That access would be required only by C++ and Qt libraries and KDE applications.
- That the various backend implementations would work "online," accessing the data stored on a server without duplicating a lot of that data locally.
- That read and write access to the data would be synchronous and fast enough not to block the caller for a noticeable length of time in the user interface.

There was little disagreement among those present at the 2005 meeting that the requirements imposed by real-world usage scenarios of the current user base, and even more so the probable requirements of future use cases, could not be met by the current design of the three major subsystems. The ever-increasing amounts of data, the need for concurrent access of multiple clients, and more complicated error scenarios leading to a more pressing need for robust, reliable, transactional storage layers cleanly separated from the user interface were clearly apparent. The use of the KDEPIM libraries on mobile devices, transferring data over low-bandwidth, high-latency, and often unreliable links, and the ability to access the user's data not just from within the applications traditionally dealing with such data, but pervasively throughout the desktop were identified as desirable. This would include the need to provide access via other mechanisms than C++ and Qt, such as scripting languages, interprocess communication, and possibly using web and grid service technologies.

Although those high-level issues and goals were largely undisputed, the pain of the individual teams was much more concrete and had a different intensity for everyone, which led to disagreement about how to solve the immediate challenges. One such issue was the fact that in order to retrieve information about a contact in the address book, the whole address book needed to be loaded into memory, which could be slow and take up a lot of memory if the address book is big and contains many photos and other attachments. Since access was by way of a library, with a singleton address book instance per initialization of the library and thus per process, a normal KDE desktop running the email, address book, and calendaring applications along with helpers such as the appointment reminder daemon could easily have the address book in memory four or more times.

To remedy the immediate problem of multiple in-memory instances of the address book, the maintainer of that application proposed a client/server-based approach. In a nutshell, there would be only one process holding the actual data in memory. After loading it from disk once, all access to the data would be via IPC mechanisms, notably DCOP, KDE's remote procedure call infrastructure at the time. This would also isolate the communication with contact data backends (such as groupware servers) in one place, another concern with the old architecture. Lengthy discussion revealed, though, that while the memory footprint issue might be solved by this approach, several major problems would remain. Most notably, locking, conflict resolution, and change notification would still need to be implemented on top of the server somehow. It was also felt that the heavily polymorphic (thus pointer-based) APIs in use for calendaring in particular would make it very hard to serialize the data, which would be needed for the transfer over DCOP. Transferring the data through an IPC interface all the time might also end up being slow, and this was raised as a concern as well, especially if the server was also going to be used for access to email, which had been suggested as an option.

The conclusion of the meeting was that the memory footprint problem might be better solved through a more clever sharing mechanism for both the on-disk cache and the in-memory representation, possibly through the use of memory mapped files. The complexity of changing to a client/server-based architecture was deemed too challenging, and it was felt that the benefits that would come from it were not enough to justify the risk of breaking an essentially working (albeit not satisfactorily) system. As a possible alternative, the Evolution Data Server (EDS) as used by the Evolution team, a competing PIM suite associated with the GNOME project and written in C using the glib and GTK library stack, was agreed to be worth investigating.

The proponents of the data server idea left the meeting somewhat disappointed, and in the following months not much progress was made one way or the other. A short foray into EDS's codebase with the aim of adding support for the libraries used by KDE for the address book ended quickly and abruptly when those involved realized that bridging the C and C++ worlds, and especially mindsets and API styles, to arrive at a working solution would be inelegant at best and unreliable and incomplete at worst. EDS's use of CORBA, a technology initially adopted by KDE and then replaced with DCOP for a variety of reasons, was not very appealing either. It should be noted, in all fairness, that the rejection of EDS as the basis for KDE's new PIM data infrastructure was based on technical judgement as well as personal bias against C, a dislike of the implementation and its perceived maintainability, along with a certain amount of "not invented here" syndrome.

Toward the end of 2005, the problems discussed at the beginning of the year had only become more pressing. Email was not easily accessibly to applications such as desktop search agents or to semantic tagging and linking frameworks. The mail handling application, including its user interface, had to be started in order to access an attachment in a message found in a search index and to open it for editing. These and similar usability and performance issues were adding to the unhappiness with the existing infrastructure.

At a conference in Bangalore, India, where a significant part of the team of developers working on Evolution was located at the time, we had the opportunity to discuss some of these issues with them and ask them about their experiences with EDS. In these meetings it became quickly apparent that they were facing many of the same issues the KDEPIM team had identified, that they were considering similar solutions, and that they did not feel that extending EDS to support mail or porting it away from CORBA were feasible options. The general message from the Evolution team was that if the KDEPIM developers were to build a new infrastructure for PIM data access, they would be interested in sharing it at least in concept, and possibly also in implementation, provided that it wasn't available solely as a C++ and KDE library.

The possibility to share such a crucial piece of infrastructure across the whole Free desktop in the future, plus the fact that if done right this would be useful to and appreciated by the wider development community beyond KDEPIM, gave new weight to the idea of a client/server design. It would provide an integration point in the form of a protocol or IPC mechanism rather than a library that would need to be linked against, thus opening the door to other toolkits, languages, and paradigms. This scenario of a shared PIM server between KDE and GNOME also seemed more and more realistic in light of the emergence of the DBUS stack as a cross-desktop and cross-platform IPC and desktop bus system, which was at the time actively being adopted by both desktop projects and shared via the Freedesktop.org coordination site.

In an additional interesting instance of cross-project pollination, some developers from the PostgreSQL database project were also present at the conference that year and happened to participate in some of the discussions around this topic. They were interested in it from the perspective of the efficient management of, access to, and ability to query into a large amount of data with a lot of structure, such as the user's email, events, and contacts. They felt that their expertise and the software they were building could become an important part of such a system. They brought up many interesting aspects related to handling searches efficiently and designing the system with type extensibility in mind, but also related to more operational concerns, such as how to back up such a system and how to ensure its data integrity and robustness.

A few months later, at the annual meeting in Osnabrueck, Germany, the concept of a PIM data server was brought forward again, this time with some additional backing by those working on email. This group had been among those who were more skeptical of the idea the year before and most conscious of the possible performance impact and added complexity.[#] The added perspective beyond the KDE project, the fact that the alternative solutions suggested the year before had not been realised or even tried, and the ever-increasing pressure on the team to address the skeletons in their closet eventually prompted the opponents to rethink their position and seriously consider this disruptive change.

This was helped considerably by a nontechnical aspect that had crystallized through many conversations among the developers over the course of the previous year. It had become clear to the group that their biggest problem was the lack of new developers coming into the project, and that that was largely due to the unwieldy and not-very-pleasant-to-work-on libraries and applications that made up the project. There was quite some fear that the current developers would not be able to contribute forever, and that they would not be able to pass on what they had learned to the next generation and keep the project alive. The solution, it was felt, was to focus as much attention and energy as possible on encoding the combined experience and knowledge of those currently working on KDEPIM into something that would be the basis for the next generation of contributors and upon which others in the KDE project and beyond could build better PIM tools. The goal would be to produce something in the process that would

---

[#] http://pim.kde.org/development/meetings/osnabrueck4/overview.php

be fun to work on, better documented, less arcane, and more modern. It was hoped that this would attract new contributors and allow creative work in the PIM space again by limiting the unpleasantness and complexity involved to those who wished to make use of the infrastructure form. The client/server approach seemed to facilitate this.

It should be noted that this fundamental decision to rework the whole data storage infrastructure for KDEPIM in a very disruptive way essentially entailed completely rewriting large parts of the system. We quite consciously accepted the fact that this would mean a lot less resources would be available to focus on maintaining the current codebase, keeping it stable and working, and also making it available as part of the KDE 4.0 release somehow, since that would almost certainly happen before such a major refactoring could possibly be finished. As it would turn out, this meant that KDE 4.0 was in fact released without KDEPIM, a somewhat harsh but probably necessary sacrifice in retrospect.

Once the group had agreed on the general direction, they produced the following mission statement:

> We intend to design an extensible cross-desktop storage service for PIM data and meta data providing concurrent read, write, and query access. It will provide unique desktop-wide object identification and retrieval.

▶ 在架构之初，需要明确确定设计目标，特别是要定义好系统的质量属性，明确哪些质量属性是架构的主要目标，以便于确定设计的优先级，更有利于在做设计决策时作为 trade off 的参考。

The Akonadi Architecture

Some key aspects that would remain in the later iterations of the architecture were already present in the first draft of the design produced in the meeting. Chief among those was the decision to not use DBUS, the obvious choice for the IPC mechanism in Akonadi, for the transport of the actual payload data. Instead, a separate transport channel and protocol would be used to handle bulk transfers, namely IMAP. Some reasons for this were that it could control traffic out of band with respect to the data, thus allowing lengthy data transfers to be canceled, for example, since the data pipe would never block the control pipe. Data transfers would have a lot less overhead with IMAP compared to pushing them through an IPC mechanism, since the protocol is designed for fast, streaming delivery of large amounts of data. This was a reaction to concerns about the performance characteristics of DBUS in particular, which explicitly mentioned in its documentation that it was not designed for such usage patterns. It would allow existing IMAP library code to be reused, saving effort when implementing the protocol, both for the KDEPIM team itself and for any future third-party adopters wishing to integrate with Akonadi. It would retain the ability to access the contents of the mail store with generic, non-Akonadi-specific tools, such as the command-line email applications, pine or mutt. This would counter the subjective fear users have of entrusting their data to a system that would lock them in by preventing access to their data by other means. Since IMAP only knows about email, the protocol would need to be extended to support other mime types, but that seemed doable while retaining basic protocol compatibility. An alternative protocol option that was discussed was to use http/webdav for the transport, possibly reusing an existing http server implementation such as Apache, but this approach did not get much support.

A central notion in Akonadi is that it constitutes the one central cache of all PIM data on the system and the associated metadata. Whereas the old framework assumed online access to the storage backends to be the normal case, Akonadi embraces the local copy that has to be made as soon as the data needs to be displayed to the user, for example, and tries to retain as much of the information it has already retrieved as possible, in order to avoid unnecessary redownloads. Applications are expected to retrieve only those things into memory that are actually needed for display (in the case of an email application, the header information of those few email messages that are currently visible in the currently shown folder, for example) and to not keep any on disk caches of their own. This allows the caches to be shared, keeps them consistent, reduces the memory footprint of the applications, and allows data not immediately visible to the user to be lazy-loaded, among other optimizations. Since the cache is always present, albeit potentially incomplete, offline usage is possible, at least in many cases. This greatly increases the robustness against problems with unreliable, low-bandwidth, and high-latency links.

> ▶ 很多系统，特别是互联网系统，面临不同的使用场景需要制定不同的解决方案。最典型的就是offline与online的分离。在大数据处理场景下，则可以借鉴Lambda Architecture的思想，在Batch Layer中通过batch processing对海量数据进行预运算，以获得batch view，并作为real time需要操作的数据。此外，还可以采用Streaming架构，将时间窗缩小，以流式方式处理数据。

To allow concurrent access that does not block, the server is designed to keep an execution context (thread) for each connection, and all layers take a potentially large number of concurring contexts into account. This implies transactional semantics of the operations on the state, proper locking, the ability to detect interleavings of operations that would lead to inconsistent state, and much more. It also puts a key constraint on the choice of technology for managing the on-disk persistence of the contents of the system, namely that it supports heavily concurrent access for both reads and writes. Since state might be changed at any time by other ongoing sessions (connections), the notification mechanism that informs all connected endpoints of such changes needs to be reliable, complete, and fast. This is yet another reason to separate the low-latency, low-bandwidth, but high-relevance control information from the higher-bandwidth, higher-latency (due to potential server roundtrips), and less time-critical bulk data transfer in order to prevent notifications from being stuck behind a large email attachment being pushed to the application, for example. This becomes a lot more relevant if the application is concurrently able to process out-of-band notifications while doing data transfer at the same time. Although this might not be the case for the majority of applications that currently exist for potential users of Akonadi, it can be reasonably assumed that future applications will be a lot more concurrency-aware, not least because of the availability of tools such as ThreadWeaver, which is discussed in the next section of this chapter. The high-level convenience classes that are part of the KDE-specific Akonadi access library already make use of this facility.

Another fundamental aspect of Akonadi already present in this first iteration of the design is the fact that the components providing access to a certain kind of storage backend, such as a groupware server, run as separate processes. This has several benefits. The potentially error-prone, slow, or unreliable communication with the server cannot jeopardize the stability of

> ▶ 这种将服务运行在单独进程（separate process）中的方式颇类似于微服务的架构思想。

the overall system. Agents* can crash without taking down the whole server. If synchronous interaction with the server is more convenient or maybe even the only possible way to get data from the other end, it can block without blocking any other interaction with Akonadi. They can link against third-party libraries without imposing dependencies on the core system, and they can be separately licensed, which is important for those cases where access libraries do not exist as Free Software. They are also isolated from the address space of the Akonadi server proper and thus less of a potential security problem. They can more easily be written by third parties and deployed and tested easily against a running system, using whatever programming language seems suitable, as long as support for DBUS and IMAP is available. The downside, of course, is the impact of having to serialize the data on the way into the Akonadi store, which is crossing process boundaries. This is less of a concern in practice, though, for two reasons. First, it happens in the background, from the user's perspective, without interrupting anything at the UI level. Second, the data will either be already available locally, in the cache, when the user asks for it, or it will be coming from a network socket that can pass the data onto the Akonadi server socket unparsed, in many cases, and potentially even without copying it. The case where the user asks to see data is one of the few that need to be as fast as possible, to avoid noticeable waiting times. In most cases the interaction between agents and the store is not performance-critical.

From a concurrency point of view, Akonadi has two layers. At the multiprocessing level, each user of Akonadi, generally an application or agent, has a separate address space and resource acquisition context (files, network sockets, etc.). Each of these processes can open one or more connections to the server, and each is represented internally by a thread. The trade-off between the benefits of using threads versus processes is thus side-stepped by using both: processes where the robustness, resource, and security isolation is important, and threads where the shared address space is needed for performance reasons and where the code is controlled by the Akonadi server implementation itself (and thus assumed to be less likely to cause stability problems).

The ability to add support for new types of data to the system with reasonable effort was among one of the first desired properties identified. Ideally, the data management layer should be completely type agnostic, and knowledge about the contents and makeup of the data should be centralized in one place for each kind of data (emails, events, and contacts initially, but later notes, RSS feeds, IM conversations, bookmarks, and possibly many more). Although the desire to achieve this was clear from the start, how to get there was not. Several iterations of the design of the core system itself and the access library API were needed until the goal was reached. We describe the end result a bit further on in this section.

During the initial architecture discussions, the group approached the big picture in a pretty traditional way, separating concerns as layers from the bottom up. As the diagram of the white

---

* Entities interacting with the Akonadi server and reading and writing its data are referred to as *agents*. A general example would be a data mining agent. Agents that deal specifically with synchronizing data between the local cache and a remote server are called *resources*.

board drawing from that discussion in Figure 12-1 shows, there would be a storage (persistence) layer at the bottom, with logic to access it on top, a transport and access protocol layer above that, on up to application space.

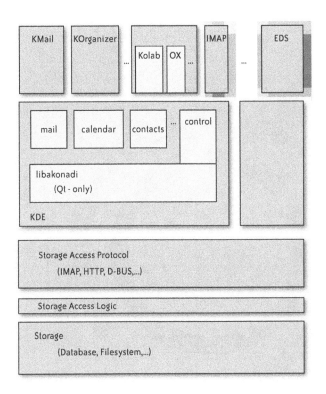

FIGURE 12-1. Initial draft of the layers in the Akonadi framework

While debating what the API for the applications' access to the store would look like (as opposed to that used by agent or resources delivering into the system), some were suspicious that there would need to be only one access API, used by all entities interacting with the store, whether their primary focus was providing data or working with the data from a user's point of view. It quickly emerged that that option is indeed preferable, as it makes things simpler and more symmetric. Any operation by an agent on the data triggers notifications of the changes, which are picked up by all other agents monitoring that part of the system. Whatever the resources need in addition to the needs of the applications—for example, the ability to deliver a lot of data into the store without generating a storm of notifications—is generic enough and useful enough to be worthwhile in one unified API. Keeping the API simple for both application access and resource needs is required anyhow, in order to make it realistically possible for third parties to provide additional groupware server backends and to get application developers to embrace Akonadi. Any necessary special cases for performance or error recovery reasons should be handled under the hood as much as possible. The example of avoiding notification storms is taken care of by a configurable notification compression and update monitoring system, which users of the system can subscribe to with some granularity.

> The next version of the high-level architecture diagram that was drawn, shown in Figure 12-2, thus reflects this notion by portraying the layers of the system as concentric rings or parts of rings.

▶ 这是我认同的架构模式。分层架构模式观察的视角始终太单一了。图12-2启发我们从内外这个角度去思考，类似于Cockburn提出的六边形架构模式，又或者是所谓的内核模式。从内外的角度去思考，有利于我们获得更加内聚的可重用的内核，尤其是六边形架构还可以驱动出模块（Bounded Context）之间的集成关系与通信方式，识别出属于infrastructure的模块。

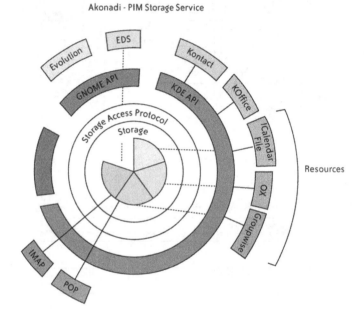

FIGURE 12-2. Inside out, not bottom up

Given the requirements outlined so far, it was fairly obvious that a relational database would make implementation of the lowest (or innermost) layer much easier. At least for the metadata around the actual PIM items, such as retrieval time, local tags, and per folder policies, just to name a few, which are typed, structured, and benefit from fast, indexed random access and efficient querying, no other solution was seriously considered. For the payload data itself, the email messages, contacts, and so on, and their on-disk storage, the decision was less clear cut. Since the store is supposed to be type-independent, a database table holding the data would not be able to make any assumptions about the structure of the data, thus basically forcing it to be stored as BLOB fields. When dealing with unstructured (from the point of view of the database) data, only some of the benefits of using a database can be leveraged. Efficient indexing is basically impossible, as that would require the contents of the data fields to be parsed. Consequently querying into such fields would not perform well. The expected access patterns also do not favor a database; a mechanism that handles continuous streaming of data with a lot of locality of reference might well be preferable. Not using a database for the data items would mean that transactional semantics on the operations on the store would have to be implemented manually. A proposed solution for this was the extension of the principles employed by the maildir standard, which essentially allows lock-free ACID access by relying on atomic rename operations on the filesystem.† For the first pass implementation, it was decided that the database would store both data and metadata, with the intent of optimizing this at a later point, when the requirements of searching in particular would be more well defined.

Given that the Akonadi store serves as the authoritative cache for the user's personal information, it can effectively offer and enforce advanced cache lifetime management. Through the concept of cache policies, very fine-grained control over which items are retained and for how long can be exposed to users to allow a wide variety of usage patterns. On one end of the spectrum—for example, on an embedded device with a bad link and very limited storage—it might make sense to develop a policy to never store anything but header information, which is pretty lightweight, and only download a full item when it needs to be displayed, but to keep already downloaded items in RAM until the connection is severed or power to the memory is shut down. A laptop, which often has no or only unreliable connectivity but a lot more disk space available, might proactively cache as much as possible on disk to enable working offline productively and only purge some folders at regular intervals of days, weeks, or even months. On the other hand, on a desktop workstation, with an always-on, broadband Internet connection or local area network access to the groupware server, fast online access to the data on the server can be assumed. This means that unless the user wants to keep a local copy for reasons of backup or to save network bandwidth, the caching can be more passive, perhaps only retaining already downloaded attachments for local indexing and referencing. These cache policies can be set on a per-folder, per-account, or backend basis and are enforced by a component running in its own thread in the server with lower priority, regularly inspecting the database for data that can be purged according to all policies applicable to it.

Among the major missing puzzle pieces that were identified in the architecture at the 2007 meeting was how to approach searching and semantic linking. The KDE 4 platform was gaining powerful solutions for pervasive indexing, rich metadata handling, and semantic webs with the *Strigi* and *Nepomuk* projects, which could yield very interesting possibilities when integrated with Akonadi. It was unclear whether a component feeding data into Strigi for full indexing could be implemented as an agent, a separate process operating on the notifications from the core, or would need to be integrated into the server application itself for performance reasons. Since at least the full text index information would be stored outside of Akonadi, a related question was how search queries would be split up, how and where results from Strigi and Akonadi itself would be aggregated, and how queries could be passed through to backend server systems capable of online searching, such as an LDAP server. Similarly, the strategy for how to divide responsibilities with Nepomuk—for example, whether tagging should be entirely

† http://pim.kde.org/development/meetings/osnabrueck4/icaldir.php

delegated to it—was to be discussed. To some extent, these discussions are still going on, as the technologies involved are evolving along with Akonadi, and the current approaches have yet to be validated in production use. At the time of this writing, there are agents feeding into Nepomuk and Strigi; these are separate processes that use the same API to the store as all other clients and resources. Incoming search queries are expressed in either XESAM or SPARQL, the respective query languages of Strigi and Nepomuk, which are also implemented by other search engines (such as Beagle, for example) and forwarded to them via DBUS. This forwarding happens inside the Akonadi server process. The results come in via DBUS in the form of lists of identifiers, which Akonadi can then use to present the results as actual items from the store to the user. The store does not do any searching or content indexing itself at the moment.

The API for the KDE-specific C++ access library took a while to mature, mostly because it was not clear from the start how type-independent the server would end up being and how much type information would be exposed in the library. By April 2007, it was clear that the way to extend the access library to support new types would be to provide so-called serializer plug-ins. These are runtime-loadable libraries capable of converting data in a certain format, identified by a mime type, into a binary representation for storage as a blob in the server and conversely, capable of restoring the in-memory representation from the serialized data. This is orthogonal to adding support for a new storage backend, for example, and the data formats used by it, which happens by implementing a new resource process (an agent). The responsibility of the resource lies in converting what the server sends down the wire into a typed, in-memory representation that it knows how to deal with, and then using a serializer plug-in to convert it into a binary datastream that can be pushed into the Akonadi store and converted back on retrieval by the access library. The plug-in can also split the data into multiple parts to allow partial access (to only the message body or only the attachments, for example). The central class of that library is called `Akonadi::Item` and represents a single item in the store. It has a unique ID, which allows it to be identified globally on the user's desktop and associated with other entities as part of semantic linking (for example, a remote identifier). This maps it to a source storage location, attributes, a data payload, and some other useful infrastructure, such as flags or a revision counter. Attributes and payload are strongly typed, and the methods to set and get them are templatized. `Akonadi::Item` instances themselves are values, and they are easily copiable and lightweight. `Item` is parameterized with the type of the payload and attributes without having to be a template class itself. The template magic to enable that is somewhat involved, but the resulting API is very simple to use. The payload is assumed to be a value type, to avoid unclear ownership semantics. In cases where the payload needs to be polymorphic and thus a pointer, or when there is already a pointer-based library to deal with a certain type of data (as is the case for `libkcal`, the library used for events and tasks management in KDE), shared pointers such as `boost::shared_ptr` can be used to provide value semantics. An attempt to set a raw pointer payload is detected with the help of template specialization and leads to runtime assertions.

The following example shows how easy it is to add support for a new type to Akonadi, provided there is already a library to deal with data in that format, as is frequently the case. It shows the complete source code of the serializer plug-in for contacts, or KABC::Addressee objects as the KDE library calls them:

```
bool SerializerPluginAddressee::deserialize( Item& item,
                                             const QByteArray& label,
                                             QIODevice& data,
                                             int version )
{
    if ( label != Item::FullPayload || version != 1 )
        return false;

    KABC::Addressee a = m_converter.parseVCard( data.readAll() );
    if ( !a.isEmpty() ) {
        item.setPayload<KABC::Addressee>( a );
    } else {
        kWarning() << "Empty addressee object!";
    }
    return true;
}

void SerializerPluginAddressee::serialize( const Item& item,
                                           const QByteArray& label,
                                           QIODevice& data,
                                           int &version )
{

    if ( label != Item::FullPayload
       || !item.hasPayload<KABC::Addressee>() )
        return;
    const KABC::Addressee a = item.payload<KABC::Addressee>();
    data.write( m_converter.createVCard( a ) );
    version = 1;
}
```

The typed payload, setPayload, and hasPayload methods of the Item class allow developers to use the native types of their data type libraries directly and easily. Interactions with the store are generally expressed as jobs, an application of the command pattern. These jobs track the lifetime of an operation, provide a cancelation point and access to error contexts, and allow progress to be tracked. The Monitor class allows a client to watch for changes to the store in the scope it is interested in, such as per mime type or per collection, or even only for certain parts of particular items. The following example from an email notification applet illustrates these concepts. In this case the payload type is a polymorphic one, encapsulated in a shared pointer:

```
Monitor *monitor = new Monitor( this );
monitor->setMimeTypeMonitored( "message/rfc822" );
monitor->itemFetchScope().fetchPayloadPart( MessagePart::Envelope );
connect( monitor, SIGNAL(itemAdded(Akonadi::Item,Akonadi::Collection)),
         SLOT(itemAdded(Akonadi::Item)) );
connect( monitor, SIGNAL(itemChanged(Akonadi::Item,QSet<QByteArray>)),
         SLOT(itemChanged(Akonadi::Item)) );

// start an initial message download for the first message to show
ItemFetchJob *fetch = new ItemFetchJob( Collection( myCollection ), this );
```

```
fetch->fetchScope().fetchPayloadPart( MessagePart::Envelope );
connect( fetch, SIGNAL(result(KJob*)), SLOT(fetchDone(KJob*)) );

...

typedef boost::shared_ptr<KMime::Message> MessagePtr;

void MyMessageModel::itemAdded(const Akonadi::Item & item)
{
    if ( !item.hasPayload<MessagePtr>() )
        return;
    MessagePtr msg = item.payload<MessagePtr>();
    doSomethingWith( msg->subject() );
    ...
}
```

The First Release and Beyond

When the group congregated in reliably cold and rainy Osnabrueck once more, in January 2008, the first application uses of Akonadi could be presented by their developers, who had been invited to attend. The authors of Mailody, a competitor to the default email application in KDE, had decided some time before that Akonadi could help them build a better application, and they had become the first to try out its facilities and APIs. Their feedback proved very valuable in finding out what was still too complicated, where additional detail was needed, and where concepts were not yet well documented or not well implemented. Another early adopter of Akonadi present at the meeting was Kevin Krammer, who had taken up the interesting task of trying to allow users of the legacy libraries for PIM data in KDE to access Akonadi (as well as the other way around, to access the data stored with the old infrastructure through Akonadi) by providing compatibility agents and resources for both frameworks. The issues he encountered while doing that exposed some holes in the API and validated that at least all of the existing functionality would be possible with the new tools.

A notable outcome of this meeting was the decision to drop backward compatibility with IMAP in the protocol. It had evolved so far away from the original email-only standard that maintaining the ability of the Akonadi server to act as a standard conforming IMAP server for email access was a burden that outweighed the benefits of that feature. The IMAP protocol had served as a great starting point, and many of its concepts remain in the Akonadi access protocol, but it can no longer justifiably be called IMAP. It is possible that this mechanism will return in later versions of the server, probably implemented as a compatibility proxy server.

With the KDE 4.1 release rapidly approaching, the team met again in March 2008 to do a full review of the API before its first public release, which would commit them to keep it stable and binary compatible for the foreseeable future. Over the course of two days, an amazing number of small and large inconsistencies, missing pieces of documentation, implementational quirks, and unhappy namings were identified, and they were rectified in the weeks that followed.

At the time of this writing, the KDE 4.1 release is imminent, and the Akonadi team is excitedly looking forward to the reaction of the many application and library developers in the wider KDE community who comprise its target audience. Interest in writing access resources for various storage backends is increasing, and people have started work on support for Facebook address books, Delicious bookmarks, the MS Exchange email and groupware server via the OpenChange library, RSS blog feeds, and others. It will be fascinating to see what the community will be able to create when data from all of these sources and many others will be available easily, pervasively, and reliably; when it will be efficiently queryable; when it will be possible to annotate the data, link items to each other, and create meaning and context among them; and when they can exploit that richness to make users do more with their software and enjoy using it more.

Two related ideas for optimization remain unimplemented so far. The first is to avoid storing the payload data in blobs in the database by keeping only a filesystem URL in the table and storing the data itself directly on the filesystem, as mentioned earlier. Building on that, it should be possible to avoid copying the data from the filesystem into memory, transferring it through a socket for delivery to the client (which is another process), thus creating a second in-memory copy of it only to release the first copy. This could be achieved by passing a file handle to the application, allowing it to memory map the file itself for access. Whether that can be done without violating the robustness, consistency, security, and API constraints that make the architecture work remains to be seen. An alternative for the first part is to make use of an extension to MySQL for blob streaming, which promises to retain most of the benefits of going through the relation database API while maintaining most of the performance of raw file system access.

Although the server and KDE client libraries will be released for the first time with KDE 4.1, the intent is still to share it with as much of the Free Software world as possible. To this end, a project on Freedesktop.org has been created, and the server will be moved there as soon as that process is finished. The DBUS interfaces have all been named in a desktop-neutral fashion; the only dependency of the server is the Qt library in version 4, which is part of the Linux Standard Base specification and available under the GNU GPL for Linux, Windows, OS X, and embedded systems, including Windows CE. A next major step would be to implement a second access library—in Python, for example, which comes with a lot of infrastructure that should make that possible with reasonable effort, or maybe using Java, where the same is true.

ThreadWeaver

ThreadWeaver is now one of the KDE 4 core libraries. It is discussed here because its genesis contrasts in many ways with that of the Akonadi project, and thus serves as an interesting comparison. ThreadWeaver schedules parallel operations. It was conceived at a time when it was technically pretty much impossible to implement it with the libraries used by KDE, namely Qt. The need for it was seen by a number of developers, but it took until the release of Qt 4 for it to mature and become mainstream. Today, it is used in major applications such as KOffice and KDevelop. It is typically applied in larger-scale, more complex software systems, where the need for concurrency and out-of-band processing becomes more pressing.

ThreadWeaver is a job scheduler for concurrency. Its purpose is to manage and arbitrate resource usage in multi-threaded software systems. Its second goal is to provide application developers with a tool to implement parallelism that is similar in its approach to the way they develop their GUI applications. These goals are high-level, and there are secondary ones at a smaller scale: to avoid brute-force synchronization and offer means for cooperative serialization of access to data; to make use of the features of modern C++ libraries, such as thread-safe, implicit sharing, and signal-slot-connections across threads; to integrate with the application's graphical user interface by separating the processing elements from the delegates that represent them in the UI; to allow it to dynamically throttle the work queue at runtime to adapt to the current system load; to be simplistic; and many more.

The ThreadWeaver library was developed to satisfy the needs of developers of event-driven GUI programs, but it turned out to be more generic. Because GUI programs are driven by a central event loop, they cannot process time-consuming operations in their main thread. Doing so would freeze the user interface until the operation is finished. In some windowing environments, the user interface can be drawn only from the main thread, or the windowing system itself is single-threaded. So a natural way of implementing responsive cross-platform GUI applications is to perform all processing in worker threads and update the user interface from the main thread when necessary. Surprisingly, the need for concurrency in user interfaces is rarely ever as obvious as it should be, although it has been emphasized for OS/2, Windows NT, and Solaris eons ago. Multithreaded programming is more complicated and requires a better understanding of how the written code actually functions. Multithreadings also seems to be a topic well understood by software architects and designers, and badly disseminated to software maintainers and less-experienced programmers. Also, some developers seem to think that most operations are fast enough to be executed synchronously, even reading from mounted filesystems, which is a couple of orders of magnitude slower than anything processed in the CPU. Such mistakes surface only under extraordinary circumstances, such as when the system is under heavy I/O load or, more commonly, a mounted filesystem has been put to sleep to save power or—heavens!—when, all of a sudden, the filesystem happens to be on the network.

The following section will describe the architecture of the library along with its underlying concepts. At the end of the chapter, we will explore how it found its way into KDE 4.

Introduction to ThreadWeaver: Or, How Complicated Can It Be to Load a File?

To convince programmers to make use of concurrency, it needs to be conveniently available. Here is a typical example for an operation performed in a GUI program—loading a file into a memory buffer to process it and display the results. In an imperative program, where all individual operations are blocking, it is of little complexity:

1. Check whether the file exists and is readable.
2. Open the file for reading.
3. Read the file contents into memory.
4. Process them.
5. Display the results.

To be user-friendly, it is sufficient to print a progress message to the command line after every step (if the user requested verbose mode).

In a GUI program, things appear in a different light because during all of these steps, it is necessary to be able to update the screen, and users expect a way to cancel the operation. Although it sounds unbelievable, even recent documentation for GUI toolkits mentions the "check for events occasionally" approach. The idea is to periodically check for events while processing the aforementioned steps in chunks and update or abort if necessary. A lot of care needs to be applied in this situation because the application state can change in unexpected ways. For example, the user might decide to close the program, unaware that the program checks for an event in the midst of a call stack of an operation. To put it shortly, this approach of polling has never worked very well and is generally out of fashion.

A better approach is to use a thread (of course). But without a framework to help, this often leads to weird implementations as well. Since GUI programs are event-based, every step just listed starts with an event, and an event notifies its completion. In the C++ world, signals are often used for the notification. Some programs look like this:

1. The user requests to load the file, which triggers a handler method by a signal or event.
2. The operation to open and load the file is started and connected to a second method for notification about its completion.
3. In this method, processing the data is started, connected to a third method.
4. The last method finally displays the results.

This cascade of handler methods does not track the state of the operation very well and is usually error-prone. It also shows a lack of separation of operations and the view. Nevertheless, it is found in many GUI applications.

This is exactly where ThreadWeaver is there to help. Using jobs, the implementation will look like this:

1. The user requests to load the file, which triggers a handler method by a signal or event.
2. In the handler method, the user creates a sequence of jobs (a sequence is a job container that executes its jobs in the order they were added). He adds a job to load the file and one to process its contents. The sequence object is a job itself and sends a signal when all its

contained jobs are completed. Up to this point, no processing has taken place; the programmer only declared what needs to be done in what order. Once the whole sequence is set up, the user queues it into the application-global job queue (a lazy initialized singleton). The sequence is automatically executed by worker threads.

3. When the done() signal of the sequence is received, the data is ready to be displayed.

Two aspects are apparent. First, the individual steps are all declared in one go and then executed. This alone is a major relief to GUI programmers because it is a nonexpensive operation and can easily be performed in an event handler. Second, the usual issues of synchronization can largely be avoided by the simple convention that the queuing thread only touches the job data after it has been prepared. Since no worker thread will access the job data anymore, access to the data is serialized, but in a cooperative fashion. If the programmer wants to display progress to the user, the sequence emits signals after the processing of every individual job (signals in Qt can be sent across threads). The GUI remains responsive and is able to dequeue the jobs or request cancellation of processing.

Since it is much easier to implement I/O operations this way, ThreadWeaver was quickly adopted by programmers. It solved a problem in a nice, convenient way.

Core Concepts and Features

In the previous example, job sequences have been mentioned. Let us look at what other constructs are provided in the library.

▶ 相当于GoF设计模式中的composite模式的体现。

Sequences are a specialized form of job collections. Job collections are containers that queue a set of jobs in an atomic operation and notify the program about the whole set. Job collections are composites, in the way that they are implemented as job classes themselves. There is only one queuing operation in ThreadWeaver: it takes a Job pointer. Composite jobs help keep the queue API minimal.

▶ 现代支持future的语言例如Java、Scala、Python、JavaScript等，都不需要程序员显式去控制并行多任务之间的依赖，程序会帮我们去控制，若不存在前后依赖，则可以并行执行。

Job sequences use dependencies to make sure the contained jobs are executed in the correct order. If a dependency is declared between two jobs, it means that the depending job can be executed only after its dependency has finished processing. Since dependencies can be declared in an m:n fashion, pretty much all imaginable control flows of depending operations (which are all directed graphs, since repetition of jobs is not allowed) can be modeled in the same declarative fashion. As long as the execution graph remains directed, jobs may even queue other jobs while being processed. A typical example is that of rendering a web page, where the anchored elements are discovered only once the text of the HTML document itself is processed. Jobs can then be added to retrieve and prepare all linked elements, and a final job that depends on all these preparatory jobs renders the page for display. Still, no mutex necessary. Dependencies are what distinguish a scheduling system like ThreadWeaver from mere tools for parallel processing. They relieve the programmer of thinking how to best distribute the individual suboperations to threads. Even with modern concepts such as futures, usually the programmer still needs to decide on the order of operations. With ThreadWeaver, the worker threads eagerly execute all possible jobs that have no unresolved dependencies. Since the execution of concurrent flow graphs is inherently undeterministic, it is very unlikely that a manually defined order is flexible enough to be the most efficient. A scheduler can adapt much better here. Computer scientists tend to disagree with this thesis, whereas economists, who are more used to analyzing stochastic systems, often support it.

Priorities can be used to influence the order of execution as well. The priority system used is quite simple: of all integer priorities assigned to jobs in the queue, the highest ones are first handed to an available worker thread. Since the job base class is implemented in a way that allows writing decorators, changing a job's priority externally can be done by writing a decorator that bumps the priority without touching the job's implementation. The combination of priorities and dependencies can lead to interesting results, as will be shown later.

Instead of relying on direct implementations of queueing behavior, ThreadWeaver uses queue policies. Queue policies do not immediately affect when and how a particular job is executed. Instead, they influence the order in which jobs are taken from the queue by the worker threads. Two standard implementations come with ThreadWeaver. One is the dependencies discussed earlier. The other is resource restrictions. Using resource restrictions, it can be declared that of a certain subset of all created jobs (for example, local filesystem I/O-expensive ones), only a certain amount can be executed at the same time. Without such a tool, it regularly happens that some subsystems get overloaded. Resource restrictions act much like semaphores in traditional threading, except that they do not block a calling thread, and instead simply mark a job as not yet executable. The thread that checked whether the job can be executed is then able to try to get another job to execute.

Queue policies are assigned to jobs, and the same policy object can be assigned to many. As such, they are composed, and every job can be managed by any combination of the available policies. Inheriting specialized policy-driven job base classes would not have provided such flexibility. Also, this way, job objects that do not need any extra policies are in no way affected by a possible performance hit of evaluating the policies.

Declarative Concurrency: A Thumbnail Viewer Example

Another example explains how these different ThreadWeaver concepts play together. It uses jobs, job composites, resource restrictions, priorities, and dependencies, all to render wee little thumbnail images in a GUI program. Let us first look at what operations are required to implement this function, how they depend, and how the user expects to be presented with the results. The example is part of the ThreadWeaver source code.

In this example, it is assumed that loading the thumbnail preview for a digital photo involves three operations: to load the raw file data from disk, to convert the raw data into an image representation without changing its size, and then to scale to the required size of the thumbnail. It is possible to argue that the second and third steps could be merged into one, but that is (a) not the point of the exercise (just like streamed loading of the image data) and (b) would impose the restriction that only image formats can be used where the drivers support scaling during load. It is also assumed that the files are present on the hard disk. Since the processing of each file does not influence or depend on the processing of any other, all files can be processed in parallel. The individual three steps to process one file need to be performed in sequence.

But that is not all. Since this is an example with a graphical user interface, the expectations of the user have to be kept in mind. It is assumed that the user is interested in visual feedback, which also gives him the impression of progress. Image previews should be shown as soon as they are available, and progress information in the form of a reliable progress bar would be nice. The user also expects that the program will not grind his computer to a halt, for example by excessive I/O operations.

Different ThreadWeaver tools can be applied to this problem. First of all, processing an individual file is a sequence of three job implementations. The jobs are quite generic and can be part of a toolbox of premade job classes available to the application. The jobs are (the class names match the ones in the example source code):

- A `FileLoaderJob`, which loads a file on the file system into an in memory byte array
- A `QImageLoaderJob` to convert the image's raw data into the typical representation of an in-memory image in Qt applications (providing the application access to all available image decoders in the framework or registered by the application)
- A `ComputeThumbNailJob`, which simply scales the image to the wanted size of the preview

All of those are added to a `JobSequence`, and each of these sequences is added to a `JobCollection`. The composite implementation of the collection classes allow for implementations that represent the original problem very closely and therefore feel somewhat natural and canonical to the programmer.

This solves part one of the problem, the parallel processing of the different images. It could easily lead to other problems, though. With the given declaration of the problem to the ThreadWeaver queue, there is nothing that prevents it from loading all files at once and only then starting to process images. Although this is unlikely, we haven't told the system otherwise yet. To make sure that only so many file loaders are started at the same time, a resource restriction is used. The code for it looks like this:

```
#include "ResourceRestrictionPolicy.h"

...

static QueuePolicy* resourceRestriction()
{
    static ResourceRestrictionPolicy policy( 4 );
    return &policy;
}
```

File loaders simply apply the policy in their constructor or, if generic classes are used, when they are created:

```
fileloader->assignQueuePolicy( resourceRestriction() );
```

But that still does not completely arrange the order of the execution of the jobs exactly as wanted. The queue might now start only four file loaders at once, but it still might load all the files and then calculate the previews (again, this is a very unlikely behavior). It needs one more tool, and a bit of thinking against the grain, to solve the problem, and this is where priorities come into play. The problem, translated into ThreadWeaver lingo, is that file loader jobs have lowest priority but need to be executed first; image loader jobs have precedence over file loaders, but a file loader must have finished first before an image loader can be started; and finally, thumbnail computer jobs have highest priority, even if they depend on the other two phases of processing. Since the three jobs are already in a sequence, which will make sure they are executed in the right order for every image, assigning priority one to file loaders, two to image loaders, and three to thumbnail computers finally solves the problem. Basically, the queue will now complete one thumbnail as soon as possible, but will not stop to load the images if slots for file loading become available. Since the problem is mostly I/O bound, this means that the total time until the thumbnails for all images are shown is a little more than the time it takes to load them from the hard disk (other factors aside, such as extremely high-resolution RAW images). In any sequential solution, the behavior would likely be much worse.

The description of the solution might have felt complex, so lightening it up with a bit of code is probably in order. This is how the jobs are generated after the user has selected a couple of hundred images for processing:

```
m_weaver->suspend();
for (int index = 0; index < files.size(); ++index)
{
    SMIVItem *item = new SMIVItem ( m_weaver, files.at(index ), this );
    connect ( item, SIGNAL( thumbReady(SMIVItem* ) ),
              SLOT ( slotThumbReady( SMIVItem* ) ) );
}
m_startTime.start();
m_weaver->resume();
```

To give correct progress feedback, processing is suspended before the jobs are added. Whenever a sequence is completed, the item object emits a signal to update the view. For every selected file, a specialized item is created, which in turn creates the job objects for processing one file:

```
m_fileloader = new FileLoaderJob ( fi.absoluteFilePath(), this );
m_fileloader->assignQueuePolicy( resourceRestriction() );
m_imageloader = new QImageLoaderJob ( m_fileloader, this );
m_thumb = new ComputeThumbNailJob ( m_imageloader, this );
m_sequence->addJob ( m_fileloader );
m_sequence->addJob ( m_imageloader );
m_sequence->addJob ( m_thumb );
weaver->enqueue ( m_sequence );
```

The priorities are virtual properties of the job objects and are set there. It is important to keep in mind that all these objects are set to not process until they are queued, and in this case, until the processing is explicitly resumed. So the whole operation to create all these sequences and jobs really takes only a very short time, and the program returns to the user immediately, for all practical matters. The view updates as soon as a preview image is available.

From Concurrency to Scheduling: How to Implement Expected Behavior Systematically

The previous examples have shown how analyzing the problem completely really helps to solve it (I hope this does not come as a surprise). To make sure concurrency is used to write better programs, it is not enough to provide a tool to move stuff to threads. The difference is scheduling: to be able to tell the program what operations have to be performed and in what order. The approach is remotely reminiscent of PROLOG programming lessons, and sometimes requires a similar way of thinking. Once the minds involved are sufficiently assimilated, the results can be very rewarding.

One design decision of the central `Weaver` class has not been discussed yet. There are two very disjunct groups of users of the `Weaver` classes API. The internal `Thread` objects access it to retrieve their jobs to process, whereas programmers use it to manage their parallel operations. To make sure the public API is minimal, a combination of decorator and facade has been applied that limits the publicly exposed API to the functions that are intended to be used by application programmers. Further decoupling of the internal implementation and the API has been achieved by using the PIMPL idiom, which is generally applied to all KDE APIs.

A Crazy Idea

It has been mentioned earlier that at the time ThreadWeaver started to be developed, it was not really possible to implement all its ideas. One major obstacle was, in fact, a prohibitive one: the use of advanced implicit sharing features, which included reference counting, in the Qt library. Since this implicit sharing was not thread-safe, the passing of every simple Plain Old Data object (POD) was a synchronization point. The author assumed this to be impractical for users and therefore recommended against using the prototype developed with Qt 3 for any production environments. The developers of the KDEPIM suite (the same people who now develop Akonadi) thought they really knew better and immediately imported a preliminary ThreadWeaver version into KMail, where it is used to this day. Having run into many of the problems ThreadWeaver promised to solve, the KMail developers eagerly embraced it, willing to live with the shortcomings pointed out by its author, even against his express wishes.

The fact that an imperfect version of the library was in active use in KDE served as a motivating factor for quickly porting it to Qt4 when that became usable in beta versions. Thus it was available rather early in the KDE 4 development cycle, if only in a secondary module and not yet as part of KDELibs. Over the course of two years, the author gave a number of presentations on the library, presenting an ever-easier and more complete API as he kept improving it. It was, one could say, a solution looking for a problem. The majority of developers working on KDE needed time to realize that this library was not only academic, but could improve their software significantly given that they make the investment in taking a step back and rethinking some of their architectural structures. There was no concrete need by a group of developers driving the library's progress; it was progressed by an individual because of his belief in the growing relevance of the problem and the importance of making available a good solution for the KDE 4 platform. Especially following the 2005 Akademy conference in Malaga, Spain, more programs started to use ThreadWeaver, including KOffice and KDevelop, which created enough momentum for it to be integrated into the main KDE 4 set of libraries.

ThreadWeaver represents the case of an alternative solution to a problem that once it had matured to critical point and once the author and the prospective user community agreed that the time had come for it to be adopted by developers in their projects, it was quickly promoted to a cornerstone of KDE 4. After that, the attitudes of community members changed from mild amusement to appreciation and recognition of the effort that had gone into it. This is an example of how efficient this community can be at making technical decisions and adapting its stance when an approach proves itself in practice. There can be no doubt that ThreadWeaver is a much better library now than it would have been if it not taken three to four years of rubbing up against the KDE project until its inclusion. And this includes the rogue premature adoption by the KMail developers. There is also little doubt that applications written for KDE 4 can deal with concurrency a lot better and thus provide a better experience to their users, because it succeeded in the end.

ThreadWeaver will be extended mostly by adding GUI components to visually represent queue activity, and by including more predefined job classes. Another idea is the integration with operating system IPC mechanisms (to allow for host-global resource restrictions, for example), but those are hindered by the requirement to be cross-platform. The approaches taken by the different operating systems are very diverse. With the public availability of the KDE 4 line, it became visible to a large audience. Since ThreadWeaver is not really KDE-specific, the question of where to go next (Freedesktop.org?) is in the air. For now, the focus remains to provide developers of applications and the desktop with a reliable scheduler for concurrency.

PART V
第 5 部分

Languages and Architecture
语言与架构

Chapter 13   Software Architecture: Object-Oriented Versus Functional
第 13 章   软件架构：面向对象 vs. 面向函数

Chapter 14   Rereading the Classics
第 14 章   重读经典

CHAPTER THIRTEEN

Software Architecture: Object-Oriented Versus Functional

Bertrand Meyer

ONE OF THE ARGUMENTS FOR FUNCTIONAL PROGRAMMING IS BETTER MODULAR DESIGN. By analyzing publications advocating this approach, in particular through the example of a framework for financial contracts, we access its strengths and weaknesses, and compare it with object-oriented design. The overall conclusion is that object-oriented design, especially in a modern form supporting high-level routine objects or "agents," *subsumes* the functional approach, retaining its benefits while providing higher-level abstractions more supportive of extension and reuse.

Overview

"Beauty," as a slogan for a software architecture, is not strictly for the beholder to judge. Clear objective criteria exist (Meyer 1997):

Reliability
    Does the architecture help establish the correctness and robustness of the software?

Extendibility
    How easy is it to accommodate changes?

Reusability
> Is the solution general, or better yet, can we turn it into a *component* to be plugged in directly, off-the-shelf, into a new application?

The success of object technology has largely followed from the marked improvements it brings—if applied properly as a method, not just through the use of an object-oriented programming language—to the reliability, extendibility, and reusability of the resulting programs.

The *functional programming* approach predates object-oriented thinking, going back to the Lisp language available for almost 50 years. To those fortunate enough to have learned it early, functional programming will always remain like the memory of a first kiss: sweet, and the foretaste of even better experiences. Functional programming has made a comeback in recent years, with the introduction of new languages such as Scheme, Haskell, OCaml and F#, sophisticated type systems, and advanced language mechanisms such as monads. Functional programming even seems at times to be presented as an improvement over object-oriented techniques. The present discussion compares the two approaches, using the cited software architecture criteria. It finds that the relationship is the other way around: object-oriented architecture, particularly if enriched with recent developments such as *agents* in Eiffel terminology ("closures" or "delegates" in other languages), subsumes functional programming, retaining its architectural advantages while correcting its limitations.

To qualify this finding, it is important to note both the study's limitations and arguments to mitigate some of them. The limitations include:

Few data points
> The analysis is primarily based on two examples of functional design. This could cast doubts on the generality of the lessons drawn.

Lack of detail
> The source of the examples consists of an article (Peyton Jones et al. 2000) and a PowerPoint presentation (Eber et al. 2001)—referred to from now on as "the article" and "the presentation"—complemented in the section "Assessing the Modularity of Functional Solutions," later in this chapter, by ideas from a classic functional programming paper (Hughes 1989). Uses of the presentation may miss some details and nuances that would be present in a more discursive document.

Specific focus
> We only consider the issue of modularity. The case for functional programming also relies on other criteria, such as the elegance of a declarative approach.

Experimenter bias
> The author of the present chapter is a long-time contributor to and exponent of object technology.

The following observations counterbalance some of this possible criticism:

- The functional examples come from industrial practice; specifically, a company whose business appears to rest on the application of functional programming techniques. The principal example—specifying sophisticated financial instruments—addresses complex problems faced by the financial industry, which current tools do not address well according to the presentation's author, an expert in that industry. This suggests that it is representative of the state of the art. (The first example—specifying puddings—is academic, intended only as a pedagogical stepping stone.)
- One of the authors of the article (S. Peyton Jones), also acknowledged in the presentation as coauthor of the underlying theoretical work, is the lead designer of the Haskell language and one of the most notable figures in functional programming, bringing considerable credibility. The paper used as a subsidiary example in the later section "Assessing the Modularity of Functional Solutions" has been extremely influential and was written by another leading member of the functional programming community (J. Hughes).
- In spite of the reservations expressed below, the solutions described in these documents are elegant and clearly the result of considerable reflection.
- The examples do not exercise the notion of changeable *state*, which would favor an imperative object-oriented programming style.

We must also note that mechanisms such as agents, which provide essential ingredients of the full object-oriented solution, were openly inspired by functional programming ideas. So the conclusion will not be a dismissal of the functional school's contribution, simply the observation that the object-oriented (OO) style is more suited for defining the overall architecture of reliable, extendible, and reusable software, while the building blocks may involve a combination of OO and functional techniques.

Further observations about the following discussion:

- Object technology as used here takes the form of Eiffel. We have not attempted to analyze what remains if one *removes* mechanisms such as multiple inheritance (absent in Java and C#), genericity (absent in earlier versions of these languages), contracts (absent outside of Eiffel except in JML and Spec#), or agent-style facilities (absent in Java), or if one *adds* mechanisms such as overloading and static functions, which threaten the solidity of the OO edifice.
- The discussion is about architecture and design. In spite of its name, functional programming is (like object technology) relevant to these tasks and not just to "programming" in the restricted sense of implementation. The Eiffel approach explicitly introduces a continuum from specification to design and implementation through the concept of seamless development. Implementation-oriented properties of either approach, while important in practice, will not be considered in any detail.

- Also relevant in practice are issues of expressiveness and notation. They are taken into account to the extent that they affect the key criteria of architecture and design. For the most part, however, the discussion considers semantics rather than syntax.

Two more preliminary notes. First, terminology: by default, the term "contract" refers to financial contracts, relevant to the application domain of the article and presentation, and not to be confused with the software notion of Design by Contract* (the idea [Meyer 1997] of including elements of specification such as preconditions, postconditions, or invariants). In case of possible ambiguity, the terms used here will be *financial contracts* and *software contracts*.

Second, a semi-apology: when the discussion moves to OO territory in its second half, it includes more references to and repetitions from the author's previous publications than discretion would command. The reason is that the wide spread of object technology has been accompanied by the loss of some of its more subtle but (in our opinion) critical principles, such as command-query separation (see "State Intervention" later in this chapter); this makes some brief reminders necessary. For the full rationale behind these ideas, see the cited references.

The Functional Examples

The overall goal of the article and presentation is to propose a convenient mechanism for describing and handling financial contracts, especially modern financial instruments that can be very complicated, as in this example from the presentation (in whose numerical values one can hear the nostalgic echo of a time when major currencies enjoyed a different relationship):

"Against the promise to pay USD 2.00 on December 27 (the price of the option), the holder has the right, on December 4, to choose between:

- Receiving USD 1.95 on December 29, or
- Having the right, on December 11, to choose between:
  - Receiving EUR 2.20 on December 28, or
  - Having the right, on December 18, to choose between:
    - Receiving GBP 1.20 on December 30, or
    - Paying immediately one more EUR and receiving EUR 3.20 on December 29"

(Throughout this section, extracts in quotes are direct citations from the presentation or the article. Elements not in quotes are our interpretations and comments.)

As a pedagogical device to illustrate the issues, the presentation starts with a toy example: *puddings* rather than contracts. From the precise description of a pudding, it should be possible

---

* Design by Contract is a trademark of Eiffel Software.

to "compute the sugar content," "estimate the time to make" the pudding, and obtain "instructions to make it." A "bad approach" would be to:

- "List all puddings (Trifle, lemon upside-down pudding, Dutch apple cake, Christmas pudding)
- For each pudding, write down sugar content, time to make, instructions, etc."

Although the presentation does not state why the approach is bad, we can easily surmise the reasons: as a collection of ad hoc descriptions, it has no reusability, since it does not take advantage of the property that different kinds of pudding may share the same basic parts; it has no extendibility, since any modification of a pudding part will require reworking all the puddings that rely on that part.

The pudding is a metaphor for the examples of real interest, contracts, but since it is easily understandable without a specialized knowledge domain, we continue with it. A "good approach" is to:

- "Define a small set of 'pudding combinators.'
- Define all puddings in terms of these combinators.
- Calculate sugar content from these combinators too."

A combinator is an operator that produces a composite object from similar objects. The tree shown in Figure 13-1, from the presentation, illustrates what the combinators may be in this example.

▶此设计利用了表达式树，设计的对象粒度小，易于重用与扩展，缺点是方案过于复杂。

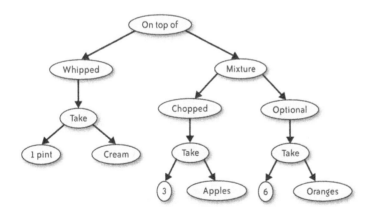

FIGURE 13-1. Ingredients and combinators describing a pudding recipe

> **NOTE**
> We share the reader's alarm at the unappetizing nature of the example, especially coming from a Paris-based author. The sympathetic explanation is that the presentation was directed to a foreign audience of which it assumed, along with unfamiliarity with the metric system, barbaric culinary habits. The present discussion relies on the assumption that bad taste in desserts is not a sufficient predictor of bad taste in language and architecture paradigms.

The nonleaf nodes of the tree represent combinators, applied to the subtrees. For example, "Take" is a combinator that assumes two arguments, a pudding part ("Cream" on the left, "Oranges" on the right) and a quantity ("1 pint" and "6"); the result of the application, represented by the tree node, is a pudding part or a pudding made of the given quantity of the given part.

It is also possible to write out such a structure textually, using a mini-"domain-specific language" (DSL) "for describing puddings" (boldface is added for operators):

```
"salad      = on_top_of topping main_part       -- Changed from
              "OnTopOf" for consistency
 topping    = whipped (take pint cream)
 main_part  = mixture apple_part orange_part
 apple_part = chopped (take 3 apples)
 orange_part = optional (take 6 oranges)"
```

This uses an anonymous but typical—the proper term might be "vanilla"—variant of functional programming notation, where function application is simply written as function args (for example, plus a b for the application of plus to a and b) and parentheses serve only for grouping.

With this basis, it becomes a piece of cake to define an operation such as sugar content (S) by case analysis on the combinators (similar to defining a mathematical function on recursively defined objects by using a definition that follows the same recursive structure):

```
"S  (on_top_of p1 p2)  = S (p1) + S (p2)
 S  (whipped p)        = S (p)
 S  (take q i)         = q * S(i)
 etc."
```

Not clear (to us) from the "etc." is how operators such as S deal with the optional combinator; there has to be some way of specifying whether a particular concoction has the optional part or not. This issue aside, the approach brings the benefits that the presentation claims for it:

- "When we define a new recipe, we can calculate its sugar content with no further work.
- Only if we add new combinators or new ingredients would we need to enhance S."

The real goal, of course, is not pudding but contracts. Here the presentation contains a sketch of the approach but the article is more detailed. It relies on the same ideas, applied to a more interesting set of elements, combinators, and operations.

The *elements* are financial contracts, dates, and observables (such as a certain exchange rate on a certain date). Examples of basic contracts include zero (can be acquired at any time, no rights, no obligations) and one (c) for a currency c (immediately pays the holder one unit of c).

Examples of *combinators* on contracts include: or, such that acquiring the contract (or c1 c2) means acquiring either of c1 and c2, and expiring when both have expired; anytime, such that (anytime c) can be acquired at any time before the expiration of c, and expiring whenever c expires; truncate, such that (truncate t c) is like c except that it expires at the earlier of t and the expiry of t; and get, so that acquiring (get c) means acquiring c at its expiry date. The paper lists about a dozen such basic combinators on contracts, and others on observables and dates. They make it possible to define advanced financial instruments such as a "European option" in a simple way:

```
european t u   = get (truncate t (or u zero))
```

Operations include the expiry date of a contract and—the most important practical benefit expected from all this modeling effort—its value process, a time-indexed sequence of expected values. As with the sugar content of a pudding, the functions are defined by case analysis on the basic constructors. Here are the cases involving the preceding basic elements and combinators for the operation H, which denotes the expiry date or "horizon":

```
H (zero)       = ∞    -- Where ∞ is a special value with the
       expected properties
H (or c1 c2)   = max (H (c1), H (c2))
H (anytime c)  = H (c)
H (truncate t c) = min (t, H (c))
H (get c)      = H (c)
```

The rules yielding value processes follow a similar structure, although the righthand sides are more sophisticated, involving financial and numerical computations. For more examples of applying combinators and functional programming ideas to financial applications, see Frankau (2008).

Assessing the Modularity of Functional Solutions

The preceding presentation, while leaving aside many contributions of the presentation and especially the article, suffices as a basis for discussing architectural features of the functional approach and comparing them with the OO view. We will freely alternate between the pudding example (which makes the ideas immediately understandable) and financial contracts (representative of real applications).

Extendibility Criteria

As pointed out by the presentation, the immediate architectural benefit is that it is easy to add a new combinator: "When we define a new recipe, we can calculate its sugar content with no further work." This property, however, is hardly a consequence of using a functional programming approach. The insight was to introduce the notion of a combinator, which creates pudding and pudding parts—or contracts—from components that can either be atomic or themselves result from applying combinators to more elementary components.

The article and presentation suggest that this is a new idea for financial contracts. If so, the insights should be beneficial to financial software. But as a general software design idea, they are not new. Transposed to the area of GUI design, the "bad approach" rejected at the beginning of the presentation (list all pudding types, for each of them compute sugar content, etc.) would mean devising every screen of an interactive application in its own specific way and writing the corresponding operations—display, move, resize, hide—separately in each case. No one ever does this. Any GUI design environment provides atomic elements, such as buttons and menu entries, and operations to combine them recursively into windows, menus, and other containers to make up a complete interface. Just as the pudding combinators define the sugar content and calorie count of a pudding from those of its ingredients, and contract combinators define the horizon and value sequence of a complex contract from those of its constituents, the display, move, resize, and hide operations on a composite figure apply these operations recursively on the components. The EiffelVision library (see the EiffelVision documentation at *http://eiffel.com*) is an example application of this compositional method, systematic but hardly unique. The article's contribution here is to apply the approach to a new application area, financial contracts. The approach itself, however, does not assume functional programming; any framework with a routine mechanism and recursion will do.

Interesting modularity issues arise not when existing combinators are applied to components of existing types, but when the combinators and component types change. The presentation indeed states: "Only if we add new combinators or new ingredients would we need to enhance S" (the sugar combinator). The interesting question is how disruptive such changes will be to the architecture.

The set of relevant changes is actually larger than suggested:

- Along with *atomic types* and *combinators*, we should consider changes in *operations*: adding a calorie count function for puddings, a delay operation for contracts, and a rotate operation for graphical objects.

- Besides such additions, we should include *changes* and *removal*, although for simplicity this discussion will continue to consider additions only.

Assessing the Functional Approach

The structure of the programs as given is simple—a set of definitions of the form:

$$O\ (a) \qquad = b_{a,O} \qquad [1]$$
$$O\ (c\ (x, y, \ldots)) = f_{c,O}\ (x, y, \ldots) \qquad [2]$$

for every operation O, atomic type a, and basic combinator c. The righthand sides involve appropriate constants b and functions f. Again for simplicity, we may view the atomic types such as a as 0-ary combinators, so that we only need to consider form [2]. With t basic combinators (on_top_of, hipped...) and f operations (sugar content, calories), we need t × f definitions.

Regardless of the approach, these t × f elements will have to be accommodated. The architectural problem is how we group them into modules to facilitate extension and reuse. This issue is not discussed in the article and presentation. Of course, the matter is not critical for small t and f; then all the definitions can be packed into a single module. This takes care of extendibility in a simple way:

- To add a basic combinator c, add f definitions of the above form, one for each existing operation.
- To add an operation O, add t definitions, one for each existing combinator.

This approach does not scale well; for larger developments, it will be necessary to divide the system into modules; the extendibility problem then becomes how to make sure that such modifications affect as few modules as possible.

Even with fairly small t and f, the one-module solution does not support reusability: if another program only needs a subset of the operations and combinators, it would suffer the usual dilemma of primitive modularization techniques:

Charybdis
Copy-paste the relevant parts, but then risk forgetting to update the derived modules when something changes in the original (possibly for such a prosaic reason as a bug fix).

Scylla
Use a module inclusion facility, as provided by many languages, to make the contents of an existing module available to a new one; but you end up loaded with a bigger baggage than necessary, which complicates updates and may cause conflicts (assuming the derived module defines a new combinator or function and a later version of the original module introduces a clashing definition).

▶前者体现了重复的坏处，即Martin Fowler所谓的"解决方案蔓延"坏味道；后者则说明了Robert Martin所谓的"包的复用原则"。前者体现了重用的好处，后者又说明了重用的代价。

These observations remind us in passing that reusability is closely connected to extendibility. An online critique of the OCaml functional language (Steingold 2007) takes a concrete example:[†]

> You cannot easily modify the behavior of a module outside of it. Suppose you use a Time module defining Time.date_of_string, which parses ISO8601 basic format ("YYYYMMDD"), but want to recognize ISO8601 extended format ("YYYY-MM-DD"). Tough luck: you have to get the module maintainer to edit the original function—you cannot redefine the function yourself in your module.

As software grows and changes, another aspect of reuse becomes critical: reuse of common properties. Along with European options, the article introduces "American options." Described as combinators, they have different signatures (Date → Contract → Contract and (Date, Date) → Contract → Contract). One suspects, however, that the two kinds of option have a number of

---

† This citation is slightly abridged. Inclusion of the citation does not imply endorsement of other criticism on that page.

第 13 章 软件架构：面向对象 vs. 面向函数

properties and operations in common, in the same way that puddings can be grouped into categories. Such groupings would help model and modularize the software, with the added benefit—if enough commonalities emerge—of reducing the number of required definitions. This requires, however, taking a new look at the problem domain: we must discover, beyond functions, the essential *types*.

Such a view will be at a higher level of abstraction. One can argue in particular with the fixation on functions and their signatures. According to the article (italics retained), "An *American option* offers more flexibility than a European option. Typically, an American option confers the right to acquire an underlying contract *at any time between two dates*, or not to do so at all." This suggests a definition by variation: either American options are a special case of European option, or they are both variants of a more general notion of option. Defining them as combinators immediately sets them apart from each other because of the extra Date in the signature. This is akin to defining a concept by its implementation—a mathematical rather than computer implementation, but still implying loss of abstraction and generality. Using types as the basic modularization mechanism, as in object-oriented design, will elevate the level of abstraction.

Levels of Modularity

Assessing functional programming against criteria of modularity is legitimate since better modularization is one of the main arguments for the approach. We have seen the presentation's comments on this issue, but here is a more general statement from one of the foundational papers of functional programming, by Hughes (1989), stating that with this approach:

> [Programs] can be modularized in new ways, and thereby greatly simplified. This is the key to functional programming's power—it allows greatly improved modularization. It is also the goal for which functional programmers must strive—smaller and simpler and more general modules, glued together with the new glues we shall describe.

The "new glues" described in Hughes's paper are the ones we have seen at work for the two examples covered—systematic use of stateless functions, including high-level functions (combinators) that act on other functions—plus the extensive use of lists and other recursively defined types, and the concept of lazy evaluation.

These are attractive techniques, but they address fine-grain modularization. Hughes develops a functional version of the Newton-Raphson computation of the square root of a number N with tolerance eps and initial approximation a0:

sqrt a0 eps N = **within** eps (**repeat** (next N) a0)

with appropriate combinators within, repeat, and next, and compares this version with a FORTRAN program involving goto instructions. Even ignoring the cheap shot (at the time of the paper's original publication, FORTRAN was already old hat and gotos despised), it is understandable why some people prefer such a solution, based on small functions glued through combinators, to the loop version. Then again, others prefer loops, and because we are talking about the fine-grain structure of programs rather than large-scale modularization, the issue hardly matters for software engineering; it is a question of style and taste. The more fundamental question of demonstrating correctness has essentially the same difficulty in both approaches; note, for example, that the definition of within in Hughes's paper, yielding the first element of a sequence that differs from the previous one by less than eps:

```
within eps ([a:b:rest])  =  if abs (a - b) <= eps then b
else within eps [b:rest]
```

seems to assume that the distances between adjacent elements are decreasing, and definitely assumes that one of these differences is no greater than eps.[‡] Stating this property would imply some Design by Contract-like mechanism to associate preconditions with functions (there is no such mechanism in common functional approaches); the proof that it guarantees termination of eps would be essentially the same as a proof of termination for the corresponding loop in the imperative style.

There seems to be no contribution to large-grain modularity or software architecture in this and earlier examples. In particular, the stateless nature of functional programming does not seem (positively or negatively) to affect the issue.

The Functional Advantage

There remains four significant advantages for the functional approach as illustrated in examples so far.

The first is notational. No doubt some of the attraction of functional programming languages comes from the terseness of definitions such as the above. This needs less syntactical baggage than routine declarations in common imperative languages. Several qualifications limit this advantage:

- In considering design issues, as in the present discussion, the notational issue is less critical. One could, for example, use a functional approach for design and then target an imperative language.
- Many modern functional languages such as Haskell and OCaml are strongly typed, implying the notation will be a little more verbose; for example, unless the designer wants to rely on type inference (not a good idea at the design stage), within needs the type declaration Double → [Double] → Double.
- Not everyone may be comfortable with the common practice of replacing multiargument functions by functions returning functions (known in the medical literature as RCS, for "Rabid Currying Syndrome," and illustrated by such signatures as (a → b → c) → Obs a → Obs b → Obs c in the financial article). This is a matter of style rather than a fundamental

---

[‡] In citing examples from Hughes's paper we have, with his agreement, used modern (Haskell) notation for lists, as in [a:b:rest], more readable than the original's cons notation, as in cons a (cons b rest).

property of the approach, which does not require it, but it is pervasive in these and many other publications.

Still, notation conciseness is a virtue even at the design and architecture level, and functional programming languages may have some lessons here for other design notations.

The second advantage (emphasized by Simon Peyton Jones and Diomidis Spinellis in comments on an earlier version of this chapter), also involving notation, is the elegance of combinator expressions for defining objects. In an imperative object-oriented language, the equivalent of a combinator expression, such as:

on_top_of topping main_part

would be a creation instruction:

create pudding .make_top (topping, main_part)

with a creation procedure (constructor) make_top that initializes attributes base and top from the given arguments. The combinator form is descriptive rather than imperative. In practice, however, it is easy and indeed common to use a variant of the combinator form in object-oriented programming, using "factory methods" rather than explicit creation instructions.

The other two advantages are of a more fundamental nature. One is the ability to manipulate operations as "first-order citizens"—the conventional phrase, although we can simply say "as objects of the program" or just "as data." Lisp first showed that this could be done effectively; a number of mainstream languages offered a way to pass routines as arguments to other routines, but this was not considered a fundamental design technique, and was in fact sometimes viewed with suspicion as reminiscent of self-modifying code with all the associated uncertainties. Modern functional languages showed the benefit of accepting higher-order functionals as regular program objects, and developed the associated type systems. This is the part of functional programming that has had the most direct effect on the development of mainstream approaches to programming; as will be seen below, the notion of agent, directly derived from these functional programming concepts, is a welcome addition to the original object-oriented framework.

▶由于函数式语言将函数（即所谓动词）视为一等公民，因而函数式语言的抽象性与原子性都比面向对象要强。Steve Yegge在《程序员的呐喊》第1章"编程语言里的宗教"中专门列出一节"名词王国里的执行"，以生动的方式对比了不同的编程范式。

The fourth significant attraction of functional programming is lazy evaluation: the ability, in some functional languages such as Haskell, to describe a computation that is potentially infinite, with the understanding that any concrete execution of that computation will be finite. The earlier definition of within assumes laziness; this is even more clear in the definition of repeat:

repeat f a = [a : **repeat** f (f a)]

which produces (in ordinary function application notation) the infinite sequence a, f (a), f (f (a)).... With next N x defined as (x + N / x) / 2, the definition of within as used by sqrt will stop evaluating that sequence after a finite number of elements.

This is an elegant idea. Its general application in software design calls for two observations.

First, there is the issue of correctness. The ease of writing potentially infinite programs may mask the difficulty of ensuring that they will always terminate. We have seen that within assumes a precondition, not stated in its presentation; this precondition, requiring that elements decrease to below eps, cannot be finitely evaluated on an infinite sequence (it is semi-decidable). These are tricky techniques for designers to use, as illustrated by the problem of how many lazy functional programmers it takes to change a light bulb. (It is hard to know in advance. If there are any lazy functional programmers left, ask one to change the bulb. If she fails, try the others.)

Second and last, lazy manipulation of infinite structures is possible in a nonfunctional design environment, without any special language support. The abstract data type approach (also known as object-oriented design) provides the appropriate solution. Finite sequences and lists in Eiffel libraries are available through an API relying on a notion of "cursor" (see Figure 13-2).

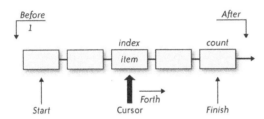

FIGURE 13-2. Cursors in Eiffel lists

Commands to move the cursor are start (go to first item), forth (move to next item), and finish. Boolean queries before and after tell if the cursor is before the first element or after the last. If neither holds, item returns the element at cursor position, and index its index.

It is easy to adapt this specification to cover infinite sequences: just remove finish and after (as well as count, the number of items). This is the specification of the deferred (abstract) class COUNTABLE in the Eiffel library. Some of its descendants include PRIMES, FIBONACCI, and RANDOM; each provides its implementations of start, forth, and item (plus, in the last case, a way to set the seed for the pseudo-random number generator). To obtain successive elements of one of these infinite sequences, it suffices to apply start and then query for item after any finite number of applications of forth.

Any infinite sequential structure requiring finite evaluation can be modeled in this style. Although this does not cover all applications of lazy evaluation, the advantage is to make the infinite structure explicit, so that it is easier to establish the correctness of a lazy computation.

State Intervention

The functional approach seeks to rely directly on the properties of mathematical functions by rejecting the assumption, present but implicit in imperative approaches, that computing operations can, in addition to delivering a result (as a mathematical function does), modify the *state* of the computation: either a global state or, in a more modular approach, some part of that state (for example, the contents of a specific object).

Although prominent in all presentations of functional programming, this property is not visible in the examples discussed here, perhaps because they follow from an initial problem analysis that already whisked the state away in favor of functional constructs. It is possible, for example, that a nonfunctional model of the notion of valuation of a financial contract would have used, instead of a function that yields a sequence (the value process), an operation that transforms the state to update the value.

It is nevertheless possible to make general comments on this fundamental decision of functional approaches. The notion of state is hard to avoid in any model of a system, computerized or not. One might even argue that it is the central notion of computation. (It has been argued [Peyton Jones 2007] that stateless programming helps address issues of concurrent programming, but there is not enough evidence yet to draw a general conclusion.) The world does not clone itself as a result of each significant event. Neither does the memory of our computers: it just overwrites its cells. It is always possible to model such state changes by positing a sequence of values instead, but this can be rather artificial (as suggested by the alternative answer to the earlier riddle: functional programmers never change a bulb, they buy a new lamp with a new cable, a new socket, and a new bulb).

▶ 不仅是monad，事实上函数式编程思想引入了许多与OO截然不同的概念，例如Functor、Applicative Functor、Semigroup、Monoid等，其中Semigroup、Identity等都是范畴学中群论的知识。FP还引入了Type-Lens的概念，它是一对函数：

get S -> T
putBack (S X T) -> S

putBack操作保证在发生状态变化时仍然保持了函数的不变性。这也是pure function的要求。

Recognizing the impossibility of ignoring the state for such operations as input and output, and the clumsiness of earlier attempts (Peyton Jones and Wadler 1993), modern functional languages, in particular Haskell, have introduced the notion of monad (Wadler 1995). Monads embed the original functions in higher-order functions with more complex signatures; the added signature components can serve to record state information, as well as any extra elements such as an error status (to model exception handling) or input-output results.

Using monads to integrate the state proceeds from the same general idea—used in the reverse direction—as the technique described in the last section for obtaining lazy behavior by modeling infinite sequences as an abstract data type: to emulate in a framework A a technique T that is *implicit* in a framework B, program in A an *explicit* version of T or of the key mechanism making T possible. T is infinite lists in the first case (the "key mechanism" is infinite lists evaluated finitely), and the state in the second case.

The concept of monad is elegant and obviously useful for semantic descriptions of programming languages (especially for the denotational semantics style). One may wonder, however, whether it is the appropriate solution as a mechanism to be used directly by programmers. Here we must be careful to consider the right arguments. The obvious objection to monads—that they are difficult to teach to ordinary programmers—is irrelevant; innovative ideas considered hard at the time of their introduction can fuse into the mainstream as educators develop ways to explain them. (Both recursion and object-oriented programming were once considered beyond the reach of "Joe the Programmer.") The important question is whether this is worth the trouble. Making the state available to functional programmers through monads is akin to telling your followers, after you convinced them to embrace chastity, that having children is actually good, if with you.

Is it really necessary to exclude the state in the first place? Two observations are enough to raise doubts:

- *Elementary* state-changing operations, such as assignment of simple values, have a clear mathematical model (Hoare rules, based on substitution). This diminishes the main benefit expected of stateless programming: to facilitate mathematical reasoning about programs.
- For the more *difficult* aspects of establishing the correctness of a design or implementation, the advantage of the functional approach is not so clear. For example, proving that a recursive definition has specific properties and terminates requires the equivalent of a loop invariant and variant. It is also unlikely that efficient functional programs can afford to renounce programmer-visible linked data structures, with all the resulting problems such as aliasing, which are challenging regardless of the underlying programming model.

If functional programming fails to bring a significant simplification to the task of establishing correctness, there remains a major practical argument: referential transparency. This is the notion of substitutivity of equals for equals: in mathematics, f (a) always means the same thing for given values of f and a. This is also true in a pure functional approach. In a programming language where functions can have side effects, f (a) can return different results in successive invocations. Renouncing such possibilities makes it much easier to understand program texts by retaining the usual modes of reasoning from mathematics; for example, we are all used to accepting that g + g and 2 × g have the same meaning, but this ceases to be guaranteed if g is a side effect-producing function. The difficulty here is not so much for automatic verification tools (which can detect that a function produces side effects) as for human readers.

Maintaining referential transparency in expressions is a highly desirable goal. It does not, however, justify removing the notion of state from the computational model. It is important to recall here the rule defined in the Eiffel method: *command-query separation principle* (Meyer 1997). In this approach the features (operations) of a class are clearly divided into two groups: commands, which can change the target objects and hence the state; and queries, which provide information about an object. Commands do not return a result; queries may not change the state—in other words, they satisfy referential transparency. In the above list example, commands are start, forth, and (in the finite case) finish; queries are item, index, count, before, and (finite case) after. This rule excludes the all-too-common scheme of calling a function to obtain a result *and* modify the state, which we guess is the real source of dissatisfaction with imperative programming, far more disturbing than the case of explicitly requesting a change through a command and then requesting information through a (side effect-free) query. The principle can also be stated as, "*Asking a question should not change the answer.*" It implies, for example, that a typical input operation will read:

```
io.read_character
Result := io.last_character
```

▶基于此原则，在架构层面上，又有人提出了CQRS原则，即 Command Query Responsibility Segregation。它通过为命令和查询提供不同的领域模型及消息处理方式来应对读远多于写的应用场景。

Here `read_character` is a command, consuming a character from the input; `last_character` is a query, returning the last character read (both features are from the basic I/O library). A contiguous sequence of calls to `last_character` would be guaranteed to return the same result repeatedly. For both theoretical and practical reasons detailed elsewhere (Meyer 1997), the command-query separation principle is a methodological rule, not a language feature, but all serious software developed in Eiffel observes it scrupulously, to the benefit of referential transparency. Although other schools of object-oriented programming do not apply it (continuing instead the C style of calling functions rather than procedures to achieve changes), it is in our view a key element of the object-oriented approach. It seems like a viable way to obtain the referential transparency goal of functional programming—since expressions, which only involve queries, will not change the state, and hence can be understood as in traditional mathematics or a functional language—while acknowledging, through the notion of command, the fundamental role of the concept of state in modeling systems and computations.

An Object-Oriented View

We now consider how to devise an object-oriented architecture for the designs discussed in the presentation and article.

Combinators Are Good, Types Are Better

So far we have dealt with operations and combinators. Operations will remain; the key step is to discard combinators and replace them with types (or classes—the distinction only arises with genericity as discussed below). This brings a considerable elevation of the level of abstraction:

- A combinator describes a specific way of building a new mechanism from existing ones. The combination is defined in a rigid way: a take combination (as in take 3 apples) associates one quantity element and one food element. As noted earlier, this is the mathematical equivalent of defining a structure by its implementation.

- A class defines a type of objects by listing the applicable features (operations). It provides abstraction in the sense of abstract data types: the rest of the world knows the corresponding objects solely through the applicable operations, not from how they were constructed. We may capture these principles of data abstraction and object-oriented design by noting that the approach means knowing objects not from what they *are* but through what they *have* (their public features and the associated contracts). This also opens the way to taxonomies of types, or *inheritance*, to keep the complexity of the model under control and take advantage of commonalities.

By moving from the first approach to the second one, we do not lose anything, since classes trivially include combinators as a special case. It suffices to provide features giving the constituents, and an associated creation procedure (constructor) to build the corresponding objects. In the take example:

▶ 我个人并不赞同这个观点。或许这会牵涉到OO与FP之间的论战，但我个人一贯坚持在软件设计领域，没有什么方法是最好，只能结合具体场景判断该方法是否适合。从抽象和重用的角度看，Function以及Combinator比OO中的类型更有优势，它的抽象粒度更细，重用粒度则更加地原子。例如fold函数对sum、product等函数的抽象；例如combinator提供的andThen或compose等便捷的组合方式，都可以很好地将问题域进行分解然后再组合。在OO中，我认为比较占优势的是类型对职责的封装，类型的抽象以及模块化设计的思想。多数情况下，我更倾向于二者在设计思想上的融合，而非厚此薄彼。

```
class REPETITION create
    make
feature
    base: FOOD
    quantity: REAL
    make (b: FOOD; q: REAL)
        -- Produce this food element from quantity units of base.
        ensure
            base = b
            quantity = q
        end
    ... Other features ...
end
```

This makes it possible to obtain an object of this type through **create** apple_salad.make (6.0, apple), equivalent to an expression using the combinator. It is possible, as mentioned, to bring the notation closer to combinators by using factory methods.

Using Software Contracts and Genericity

Since we are concentrating on design, the effect of make has been expressed in the form of a postcondition, but it really would not be a problem to include the implementation clause (do base := b; quantity := q). It is one of the consequences of well-understood OO design to abate the distance between implementation and design (and specification). In all this we are freely using state-changing assignment instructions and still have (we thank the reader for inquiring) most of our teeth and hair.

Unlike the combinator, however, the class is not limited to these features. For example, it may have other creation procedures. One can usually mix two repetitions of the same thing:

```
make (r1, r2: REPETITION)
    -- Produce this food element by combining r1 and r2.
    require
        r1.base = r2.base
    ensure
        base = r1.base
        quantity = q
```

The precondition expresses that the quantities being mixed are from the same basic food types. This requirement can also be made static through the type system; genericity (also available in typed functional languages, under the curious if impressive-sounding name of "parametric polymorphism") leads to defining the class as:

```
class REPETITION [FOOD] create
    ... As before ...
feature
    make (r1, r2: REPETITION [FOOD])
        ... No precondition necessary here ...
        ... The rest as before ...
end
```

Not only can classes have different creation procedures, they will generally have many more features. Specifically, the *operations* of our previous versions become features of the appropriate classes. (The reader may now have guessed that the variable name `t` stood for `type` and `f` for `feature`.) The pudding classes (including classes describing food variants such as `REPETITION`) have features such as `sugar` and `calorie_content`; the contract classes have features such as `horizon` and `value`. Two notes are in order:

- Since we started from a purely functional model, all the features mentioned so far are either creation procedures or queries. Although it is possible to keep this functional style in an object-oriented framework, the development might also introduce commands, for example, to change a contract in response to a certain event such as renegotiation. This issue—state, or not?—is largely irrelevant to the discussion of modularization.
- In the original, the `value` function yielded an infinite sequence. We can keep this signature by using a result of type `COUNTABLE`, permitting the equivalent of lazy computation; or we can give `value` an integer argument so that `value (i)` returns the `i`-th value.

The Modularization Policy

The modularization achieved so far illustrates the fundamental idea of object technology (at least the one we find fundamental [Meyer 1997]): *merging the concepts of type and module*. In its simplest expression, object-oriented analysis, design, and implementation means that we base every module of a system on a type of objects manipulated by the system. This is a more restrictive discipline than the modular facilities offered by other approaches: a module is no longer just an association of software elements—operations, types, variables—that the designer chooses to keep together based on any suitable criterion; it is the collection of properties and operations applicable to instances of a type.

The class is the result of this type-module merge. In OO languages such as Smalltalk, Eiffel, and C# (but not, for example, in C++ or Java), the merge is bidirectional: not only does a class define a type (or a type template if genericity is involved) but, the other way around, any type, including basic types such as integer, is formally defined as a class.

It is possible to retain classes in their type role only, separate from the modular structure. This is in particular the case with functional languages such as OCaml that offer both a traditional module structure and a type mechanism taken from object-oriented programming. (Haskell is similar, with a more restricted concept of class.) Conversely, it is possible to remove the requirement that all types be defined by classes, as with C++ and Java where basic types such as Int are not classes. The view of object technology taken here assumes a full merge, with the understanding that a higher-level of class grouping (packages as in Java or .NET, clusters in Eiffel) may be necessary, but as an organizational facility rather than a fundamental construct.

This approach implies the *primacy of types over functions* when it comes to defining the software architecture. Types provide the modularization criterion: every operation (function) gets attached to a class, not the other way around. Functions, however, take their revenge through the application of abstract data type principles: a class is defined, and known to the rest of the world, through an abstract interface (API) listing the applicable operations and their formal semantic properties (contracts: preconditions, postconditions, and, for the class as a whole, invariant).

The rationale for this modularization policy is that it yields better modularity, including extendibility, reusability, and (through the use of contracts) reliability. We must, however, examine these promises concretely on the examples at hand.

Inheritance

An essential contribution of the object-oriented method to modularity goals is inheritance. As we expect the reader to be familiar with this technique, we will only recall some basic ideas and sketch their possible application to the examples.

Inheritance organizes classes in taxonomies, roughly representing the "is-a" relation, to be contrasted with the other basic relation between classes, *client*, which represents usage of a class through its API (operations, signatures, contracts). Inheritance typically does not have to observe information hiding, as this is incompatible with the "is-a" view. While some authors restrict inheritance to pure subtyping, there is in fact nothing wrong with applying it to support a standard module inclusion mechanism. Eiffel actually has a "nonconforming inheritance" mechanism (Ecma International 2006), which disallows polymorphism but retains all other properties of inheritance. This dual role of inheritance is in line with the dual role of classes as types and modules.

In both capacities, inheritance captures commonalities. Elements of a tentative taxonomy for puddings might be as described by the inheritance graph shown in Figure 13-3.

▶要认识到继承的本质是"差异化编程"，于是推导出共性分析与可变性分析的方法。寻找共同特征，进而将其抽象为接口或父类，然后将差异下沉到子类中。继承不能离开多态，否则就很难应对需求的变化。此外，还需要多接触动态语言的一些语法特征，例如duck type, mixin等。

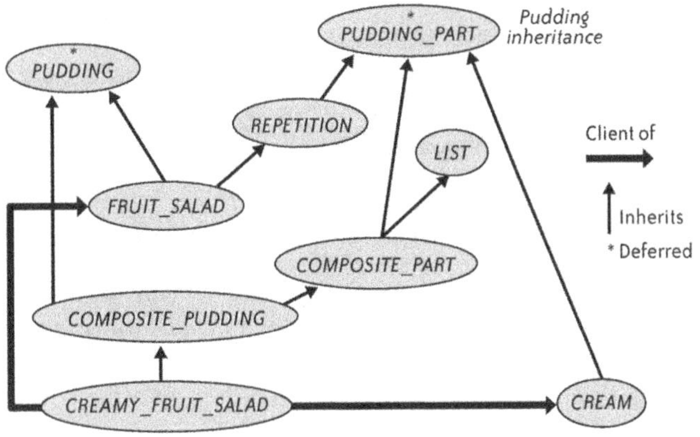

FIGURE 13-3. A class diagram of pudding ingredients

It is important to note the distribution of roles between inheritance and the client relation. A fruit salad is a pudding and is also a repetition in the earlier sense (we ignore generic parameters). A repetition is a special case not of pudding but of "pudding part," describing food ingredients. Some pudding parts (such as "composite puddings"), but not all, are also puddings. A fruit salad is a pudding and also a repetition (of fruit parts). A "creamy fruit salad," on the other hand, is *not* a fruit salad, if we take this notion to mean a pudding made of fruits only. It *has* a fruit salad and cream, as represented by the corresponding client links. It *is* a composite pudding, since this notion indeed represents concoctions that are made of several parts, like the more general notion of COMPOSITE_PART, and are also puddings. Here the parts, reflected in the client links, are a fruit salad and cream.

A similar approach can be applied to the contract example, based on a classification of contract types into such categories as "zero-coupon bonds," "options," and others to be obtained from careful analysis with the help of experts from that problem domain.

Multiple inheritance is essential to this object-oriented form of modeling. Note in particular the definition of a composite part, applying a common pattern for describing such composite structures (see Meyer 1997, 5.1, "Composite figures"):

```
class COMPOSITE_PART inherit
    PUDDING_PART
    LIST[PUDDING_PART]
feature
    ...
end
```

where square brackets introduce generic parameters. A composite part is both a pudding part, with all the applicable properties and operations (sugar content, etc.), and a list of pudding parts, again with all the applicable list operations: cursor movements such as `start` and `forth`, queries such as `item` and `index`, and commands to insert and remove elements. The elements of the list may be pudding parts of any of the available kinds, including—recursively—composite parts. This makes it possible to apply techniques of polymorphism and dynamic binding, as discussed next. Note the usefulness of having both genericity and inheritance; also, multiple inheritance should be the full mechanism for classes, not the form limited to interfaces (Java- and .NET-style) which would not work here.

Polymorphism, Polymorphic Containers, and Dynamic Binding

The contribution of inheritance and genericity to extendibility and extendibility comes in part from the techniques of polymorphism and dynamic binding, illustrated here by the version of `sugar_content` for class COMPOSITE_PART (see Figure 13-4):

```
sugar_content: REAL
    do
        from start until after loop
            Result := Result + item.sugar_content
            forth
```

            end
        end

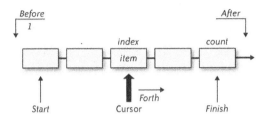

FIGURE 13-4. A polymorphic list with cursors

This applies the operations of class LIST directly to a COMPOSITE_PART, since the latter class inherits from the former. The result of item can be of any of the descendant types of PUDDING; since it may as a consequence denote objects of several types, it is known as a *polymorphic variable* (more precisely in this case, a polymorphic query). An entire COMPOSITE_PART structure, containing items of different types, is known as a *polymorphic container*. Polymorphic containers are made possible by the combination of polymorphism, itself resulting from inheritance, and genericity. (As these are two very different mechanisms, the functional programming term "parametric polymorphism" for genericity can cause confusion.)

The polymorphism of item implies that successive executions of the call item.sugar_content will typically apply to objects of different types; the corresponding classes may have different versions of the query sugar_content. *Dynamic binding* is here the guarantee that such calls will in each case apply the appropriate version, based on the type of the object actually attached to item. In the case of a part that is itself composite, this will be the above version, applied recursively; but it could be any other—for example, the version for CREAM.

Here as in most current approaches to OO design, polymorphism is controlled by the type system. The type of item's value is variable, but only within descendants of PUDDING as specified by the generic parameter of COMPOSITE_PART. This is part of a development that has affected both the functional programming world and the object-oriented world: using increasingly sophisticated type systems (still based on a small number of simple concepts such as inheritance, genericity, and polymorphism) to embody a growing part of the intelligence about a system's architecture into its type structure.

Deferred Classes and Features

Classes PUDDING and PUDDING_PART are marked as "deferred" (with an asterisk in the BON object-oriented modeling notation [Walden and Nerson 1994]) in the earlier diagram of the class structure. This means they are not completely implemented; another term is "abstract class." A deferred class will generally have deferred *features*, possessing a signature and (importantly) a contract, but no implementation. Implementations appear in nondeferred ("effective") descendant classes, adapted to the choice that each effective class has made for implementing the general concept defined by the deferred class. In the example, both classes `PUDDING` and `PUDDING_PART` have deferred features `sugar_content` and `calories`; descendants will "effect" (implement) it, for example, in `COMPOSITE_PART`, by defining the sugar content as the sum of the content of the parts, as shown earlier. In `COMPOSITE_PUDDING`, which inherits this version from `COMPOSITE_PART` and the deferred version from `PUDDING`, the effective version takes over, giving its implementation.

> **NOTE**
> The rule is that inheriting two features with the same name causes a name clash, which must be resolved through renaming, except if one of the features is deferred and the other effective, in which case they just yield a single feature with the available implementation. It is for this kind of sound application of the inheritance mechanism that name overloading brings intractable complexity, suggesting that this mechanism should not appear in object-oriented languages.

Deferred classes are more sophisticated than the Java and .NET notion of "interface" mentioned earlier, since they can be equipped with contracts that constrain future effectings, and also because they can contain effective features as well, offering the full spectrum between a fully deferred class, describing a pure implementation, and an effective one, defining a complete implementation. Being able to describe partial implementations is essential to the use of object-oriented techniques for architecture and design.

In the financial contract example, `CONTRACT` and `OPTION` would be natural deferred class candidates, although again they do not need to be *fully* deferred.

Assessing and Improving OO Modularity

The preceding section summarized the application of object-oriented architectural techniques to the examples at hand. We must now examine the sketched result in light of the modularity criteria stated at the beginning of this discussion. The contribution to reliability follows from the type system and contracts; we concentrate on reusability and extendibility.

Reusing Operations

One of the principal consequences of using inheritance is that common features can be moved to the highest applicable level; then descendants do not need to repeat them: they simply inherit them "as is." If they do need to change the implementation while retaining the functionality, they simply redefine (or "override") the inherited version. "Retaining the functionality" means here that, as noted, the original contracts still apply, whether the version being overridden was already effective or still deferred. This goes well with dynamic binding: a client can use the operation at the higher level—for example, `my_pudding.sugar_content`, or `my_contract.value`—without knowing what version of the routine is used, in what class, and whether it is specific to that class or inherited.

Thanks to commonalities captured by inheritance, the number of feature definitions may be significantly smaller than the maximum t × f. Any reduction here is valuable: it is a general rule of software design that repetition is always potentially harmful, as it implies future trouble in configuration management, maintenance, and debugging (if a fault found its way into the original, it must also be corrected in the copies). Copy-paste, as David Parnas has noted, is the software engineer's enemy.

The actual reduction clearly depends on the quality of the inheritance structure. We note here that abstract data type principles are the appropriate guidance here: since the key to defining types for object-oriented design is to analyze the applicable operations, a properly designed inheritance hierarchy will ensure that classes that collect features applicable to many variants appear toward the top.

There seems to be no equivalent to these techniques in a functional model. With combinators, it is necessary to define the variant of every operation for every combinator, repeating any common ones.

Extendibility: Adding Types

How well does the object-oriented form of architecture support extendibility? One of the most frequent forms of extension to a system will be the addition of new types: a new kind of pudding, pudding part, or financial contract. This is where object technology shines in its full glory. Just find the place in the inheritance structure where the new variant best fits—in the sense of having the most operations in common—and write a new class that inherits some features, redefines or effects those for which it provides its own variants, and add any new features and invariant clauses applicable to the new notion.

Dynamic binding is again essential here; the benefit of the OO approach is to remove the need for client classes to perform multibranch discriminations to perform operations, as in: "if this is a fruit salad, then compute in this way, else if it is a flan, then compute in that way, else …," which must be repeated for every operation and, worse, must be updated, for every single client and every single operation, any time a type is added or changed. Such structures, requiring client classes to maintain intricate knowledge of the variant structure of the supplier concepts on which they rely, are a prime source of architecture degradation and obsolescence in pre-OO techniques. Dynamic binding removes the issue; a client application can ask for `my_pudding.calories` or `my_contract.value` and let the built-in machinery select the appropriate version, not having to know what the variants are.

No other software architecture technique comes close to the beauty of this solution, combining the best of what the object-oriented approach has to offer.

▶这种可扩展性仍然存在缺陷，因为它较难支持行为的扩展。正如Kent Beck在《实现模式》中比较面向结构与面向对象之间的优缺点，如后文所说，通过visitor模式可以支持行为的扩展；但反过来它对类型扩展的支持又不够。这意味着类型与操作在扩展性方面在一定程度上是互斥的。相反，函数式编程语言中的Type Class反而较完美地解决了这一矛盾。它把结构与行为彻底解耦，使其对结构与行为都能各自扩展。对Type Class的理解，可以参考Martin Odersky等人的论文"Type Classes as Objects and Implicits"。

Extendibility: Adding Operations

The argument for object technology's support for extendibility comes in part (in addition to mechanisms such as information hiding and genericity, as well as the central role of contracts) from the assumption that the most significant changes in the life of a system are of the kind just discussed: introducing a type that shares some operations with existing types and may require new operations. Experience indeed suggests that this is the most frequent source of nontrivial change in practical systems, where object-oriented techniques show their advantage over others. But what of the other case: adding operations to existing types? Some client application relying on the notion of pudding might, for example, want to determine the cost of making various puddings, even though pudding classes do not have a cost feature.

Functional programming performs neither better nor worse for the addition of an operation than for the addition of a type: it's a matter of adding 1 to f rather than t. The object-oriented solution, however, does not enjoy this neutrality. The basic solution is to add a feature at the right level of the hierarchy. But this has two potential drawbacks:

- Because inheritance is a rather strong binding ("is-a") between classes, all existing descendants are affected. In general, adding a feature to a class at a high position in the inheritance structure can be a delicate matter.
- This solution is not available if the author of the client system is not permitted to modify the original classes, or simply does not have access to their text—a frequent case in practice since these classes may have been grouped into a library, for example, a financial contract library. It would make no sense to let authors of every application using the library modify it.

Basic object-oriented techniques (e.g., Meyer 1997) do not suffice here. The standard OO solution, widely used, is the *visitor pattern* (Gamma et al. 1994). The following sketch, although not quite the standard presentation, should suffice to summarize the idea. (It is summarized from Meyer's *Touch of Class: An Introduction to Programming Well* [2008], a first-semester introductory programming textbook—suggesting how fundamental these concepts have become.) Figure 13-5 lists the actors involved in the pattern.

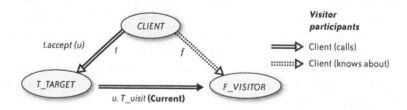

FIGURE 13-5. Actors of the Visitor pattern

The pattern turns the pas de deux between the application (classes such as CLIENT) and the existing types (such as T_TARGET for a particular type T, which could be PUDDING or CONTRACT in our examples) into a ménage à trois by introducing a visitor class F_VISITOR for every applicable operation F, for example, COST_VISITOR. Application classes such as CLIENT call an operation on the target, passing the appropriate visitor as an argument; for example:

    my_fruit_salad.accept (cost_visitor)

The command accept (v: VISITOR) performs the operation by calling on its *argument* v—cost_visitor in this example—a feature such as FRUIT_SALAD_visit, whose name identifies the target type. This feature is part of the class describing such a target class, here FRUIT_SALAD; it is applied to an object of the corresponding type (here a fruit salad object), which it passes as argument to the T_visit feature. *Current* is the Eiffel notation for the current object (also known as "this" or "self"). The *target* of the call, v on the figure, identifies the operation by using an object of the corresponding visitor type, such as COST_VISITOR.

The key question in software architecture when assessing extendibility is always distribution of knowledge; a method can only achieve extendibility by limiting the amount of knowledge that modules must possess about each other (so that one can add or change modules with minimum impact on the existing structure). To understand the delicate choreography of the visitor pattern, it is useful to see what each actor needs and does not need to know:

- The target class knows about a specific type, and also (since, for example, FRUIT_SALAD inherits from COMPOSITE_PUDDING and COMPOSITE_PUDDING from PUDDING) its context in a type hierarchy. It does *not* know about new operations requested from the outside, such as obtaining the cost of making a pudding.

- The visitor class knows all about a certain operation, such as cost, and provides the appropriate variants for a range of relevant types, denoting the corresponding objects through arguments: this is where we will find routines such as fruit_salad_cost, flan_cost, tart_cost, and such.

- The client class needs to apply a given operation to objects of specified types, so it must know these types (only their existence, not their other properties) and the operation (only its existence and applicability to the given types, not the specific algorithms in each case).

Some of the needed operations, such as accept and the T_visit features, must come from ancestors. Figure 13-6 is the overall diagram showing inheritance (FRUIT_SALAD abbreviated to SALAD).

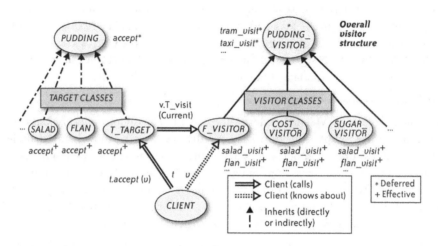

FIGURE 13-6. *Putting it all together: an architecture for constructing puddings*

Such an architecture is commonly used to provide new operations on an existing structure with many inheritance variants, without having to change that structure for every such operation. A common application is in language processing—for compilers and other tools in an Interactive Development Environment—where the underlying structure is an Abstract Syntax Tree (AST): it would be disastrous to have to update the AST class each time a new tool needs, for its own purposes, to perform a traversal operation on the tree, applying to each node an operation of the tool's choosing. (This is known as "visiting" the nodes, explaining the "visitor" terminology and the T_visit feature names.)

For this architecture to work at all, the clients must be able to perform t.accept (v) on any t of any target type. This assumes that all target types descend from a common class—here, PUDDING—where the feature accept will have to be declared, in deferred form. This is a delicate requirement since the goal of the whole exercise was precisely to avoid modifying existing target classes. Designers using the Visitor pattern generally consider the requirement to be acceptable, as it implies ensuring that the classes of interest have a common ancestor—which is often the case already if they represent variants of a common concept, such as PUDDING or CONTRACT—and adding just *one* deferred feature, accept, to that ancestor.

The Visitor pattern is widely used. The reader is the judge of how "beautiful" it is. In our view it is not the last word. Criticisms include:

- The need for a common ancestor with a special accept feature, in domain-specific classes that should not have to be encumbered with such concepts irrelevant to their application domain, whether puddings, financial contracts, or anything else.

- More worryingly, the class explosion, with numerous miniature F_VISITOR classes embodying a very specific kind of knowledge (a special operation on a set of special types). For the overall software architecture, this is just pollution.

Depollution requires adding a major new concept to the basic object-oriented framework: agents.

Agents: Wrapping Operations into Objects

The basic ideas of agents (added to the basic object-oriented framework of Eiffel in 1997; see also C# "delegates") can be expressed in words familiar in the functional programming literature: we treat operations (functions in functional programming, features in object-oriented programming) as "*first-class citizens.*" In the OO context, the only first-class citizens are, at runtime, objects, corresponding in the static structure to classes.

The Agent Mechanism

An agent is an object representing a feature of a certain class, ready to be called. A feature call x.f(u, ...) is entirely defined by the feature name f, the target object denoted by x, and the arguments u, ...; an agent expression specifies f, and may specify none, some, or all of the target and arguments, said to be *closed*. Any others, not provided in the agent's definition, are *open*. The expression denotes an object; the object represents the feature with the closed arguments set to the given values. One of the operations that can be performed on the agent object is call, representing a call to f; if the agent has any open arguments, the corresponding values must be passed as arguments to call (for the closed arguments, the values used are those specified in the agent's definition).

The simplest example of agent expression is **agent** f. Here all the arguments are open, but the target is closed. So if a is this agent expression—as a result of the assignment a := **agent** f, or of a call p (**agent** f) where the formal argument of p is a—then a call a.call ([u, v]) has the same effect as f (u, v). The difference, of course, is that f (u, v) directly names the feature (although dynamic binding means it could be a variant of a known feature), whereas in the form with agents, a is just a name, which may have been obtained from another program unit. So at this point of the program, nothing is known about the feature except for its signature and, on request, its contract. Because call is a general-purpose library routine, it needs a single kind of argument. The solution is to use a *tuple*, here the two-element tuple [u, v]. In this form, **agent** f, the target is closed (it is the current object) and both arguments are open.

A variant is **agent** x.f. Here, too, the arguments are open and the target is closed: that target is x rather than the current object. To make the target open, use **agent** {T}.f, where T is the type of x. Then a call needs a three-argument tuple: a.call ([x, u, v]). To keep some arguments open, you can use the same notation, as in **agent** x.f ({U}, v) (typical call a.call ([u])), but since the type U of u is clear from the context, you do not need to specify it explicitly; a question mark suffices, as in `agent x.f (?, v)`. This also indicates that the original forms with all arguments open, `agent f` and `agent x.f`, are abbreviations for `agent f (?, ?)` and `agent x.f (?, ?)`.

The `call` mechanism applies dynamic binding: the version of f to be applied will, as in non-agent calls, depend on the dynamic type of the target.

If f represents a query rather than a command, you can get from the corresponding agent the result of a call to f by using `item` instead of `call`, as in `a.item ([x, u, v])` (which performs a call and returns the value of its result); or you can call `call` and then access `a.last_result`, which, in accordance with the command-query separation principle, will return the same value, with no further `call`, in successive invocations.

For more advanced uses, rather than basing an agent on an existing feature f, it is also possible to write agents inline, as in `editor_window.set_mouse_enter_action (agent do text.highlight end)`, illustrating a typical use for graphical user interfaces, the basic style for event-driven programming in EiffelVision library. Inline agents provide the same mechanism as lambda expressions in functional languages: to write operations and make them directly available to the software as values to be manipulated like any other "first-class citizens."

More generally, agents enable the object-oriented framework to define higher-level functionals just as in functional languages, with the same power of expression.

Scope of Agents

Agents have turned out to be an essential and natural complement to the basic object-oriented mechanisms. They are widely used in particular for:

- Iteration: applying a variable operation, naturally represented as an agent, to all elements in a container structure.
- GUI programming, as just noted.
- Mathematical computations, as in the example of integrating a certain function, represented by an agent, over a certain interval.
- Reflection, where an agent provides properties of features (not just the ability to call them through `call` and `item`) and, beyond them, classes.

Agents have proved essential to our investigation of how to replace design patterns by reusable components (Arnout 2004; Arnout and Meyer 2006; Meyer 2004; Meyer and Arnout 2006). The incentive is that while the designer of any application needing a pattern must learn it in detail—including architecture and implementation—and build it from scratch into the application, a reusable component can be used directly through its API. Success stories include the Observer design pattern (Meyer 2004; Meyer 2008), which no one having seen the agent-based solution will ever be tempted to use again, Factory (Arnout and Meyer 2006), and Visitor, as will be discussed next.

An Agent-Based Library to Make the Visitor Pattern Unnecessary

The agent mechanism permits a much better solution to the problem addressed somewhat clumsily by the Visitor pattern: adding operations to existing types, without changing the supporting classes. The solution is detailed in Meyer and Arnout (2006) and available through an open source library available on the download site of the ETH Chair of Software Engineering (ETH Zurich, Chair of Software Engineering, at *http://se.ethz.ch*).

The resulting client interface is particularly simple. No change is necessary to the target classes (PUDDING, CONTRACT, and such): there is no more accept feature. One can reuse the classes exactly as they are, and accept their successive versions: there is no more explosion of visitor classes, but a single VISITOR library class, with only two features to learn for basic usage, register and visit. The client designer does not need to understand the internals of that class or to worry about implementing the Visitor pattern, but only needs to apply the basic scheme for using the API:

1. Declare a variable representing a visitor object, specifying the top target type through the generic parameter of VISITOR, and create the corresponding object:

   pudding_visitor: VISITOR [PUDDING]
   create pudding_visitor

2. For every operation to be executed on objects of a specific type in the target structure, register the corresponding agent with the visitor:

   pudding_visitor.register (**agent** fruit_salad_cost)

3. To perform the operation on a particular object—typically as part of a traversal—simply use the feature visit from the library class VISITOR, as in:

   pudding_visitor.visit (my_pudding)

That is all there is to the interface: a single visitor object, registration of applicable operations, and a single visit operation. Three properties explain this simplicity:

- The operations to be applied, such as fruit_salad_cost, would have to be written regardless of the architecture choice. Often they will already be available as routines, making the notation **agent** fruit_salad_cost possible; if not—especially if they are very simple operations—the client can avoid introducing a routine by using inline agents. In either case, there is no need for the spurious T_visit routines.

- It seems strange at first that a single VISITOR class, with a single register routine to add a visitor, should suffice. In the Visitor pattern solution the calls t.accept (v), the target t identified the target type (a particular kind of pudding), but here register does not specify any such information. How can the mechanism find the right operation variant to apply (the cost of a fruit salad, the cost of a flan)? The answer is a consequence of the reflective properties of the agent mechanism: an agent object embodies all the information about the associated feature, including its signature. So **agent** fruit_salad_cost includes the information that this is a routine applicable to fruit salads (from the signature `fruit_salad_cost (fs: FRUIT_SALAD)`, also available, in the case of an inline agent, from its text). This makes it possible to organize the internal data structures of `VISITOR` so that in a visiting call, such as `pudding_visitor.visit (my_pudding)`, the routine `visit` will find the right routine or routines to apply based on the dynamic type of the target, here `pudding_visitor: VISITOR [P]` for a specific pudding type P—also matching, as enforced statically by the type system, the type of the object dynamically associated with the argument, here the polymorphic `my_pudding`.

- This technique also enjoys the reuse benefits of inheritance and dynamic binding: if a routine is registered for a general pudding type (say, `COMPOSITE_PUDDING`) and no other has been registered for a more specific type (for example, the cost might be computed in the same way for all composite puddings), `visit` uses the best match.

The mechanism as described provides the complement to traditional OO techniques. When the problem is to add types providing variants of existing operations, inheritance and dynamic binding work like a charm. For the dual problem of adding operations to existing types without modifying these types, the solution described here will apply.

Applying the previous modularity criterion of distribution of knowledge—*who must know what?*—we see that in this approach:

- Target classes only know about fundamental operations, such as `sugar_content`, characterizing the corresponding types.
- An application only needs to know the interface of the target classes it uses, and the two essential features, `register` and `visit`, of the `VISITOR` library class. If it needs new operations on the target types, not foreseen in the design of the target classes, such as `cost` in our example, it need only provide the operation variants that it needs for the target types of interest, with the understanding that in the absence of overriding registration, the more general operations will be used for more specific types.
- The library class `VISITOR` does not know anything about specific target types or specific applications.

It seems impossible to go any further in minimizing the amount of knowledge required of the various parts of the system. The only question that remains open, in our opinion, is whether such a fundamental mechanism should remain available through a library or should somehow yield a language construct.

Assessment

The introduction of agents originally raised the concern that they might cause redundancy and hence confusion by offering alternative solutions in cases also amenable to standard OO mechanisms. (Such concerns are particularly strong in Eiffel, whose language design follows the principle of providing "*one good way to do anything*.") This has not happened: agents found right away their proper place in the object-oriented arsenal; designers have no trouble deciding when they are applicable and when not.

In practice, all nontrivial uses of agents—in particular, the cited pattern replacements—also rely on genericity, inheritance, polymorphism, dynamic binding, and other advanced OO mechanisms. This reinforces the conviction that the mechanism is a necessary component of successful object technology.

> **NOTE**
> For a differing opinion, see the Sun white paper explaining why Java does not need an agent- or delegate-like facility (Sun Microsystems 1997). It shows how to emulate the mechanism using Java's "inner classes." Although interesting and well-argued, it mostly succeeds, in our view, at demonstrating the contrary of its thesis. Inner classes do manage to do the job, but one can readily see, as in the elimination of the Visitor pattern with its proliferation of puny classes, the improvement in simplicity, elegance, and modularity brought by an agent-based solution.

Agents, it was noted above, allow object-oriented design to provide the same expressive power of functional programming through a general mechanism for defining higher-order functionals (operations that can use operations—themselves recursively enjoying the same property—as their inputs and outputs). Even lambda expressions find their counterpart in inline agents. These mechanisms were openly influenced by functional programming and should in principle attract the enthusiasm of its proponents, although one fears that some will view this debt acknowledgment as an homage that vice pays to virtue (La Rochefoucauld 1665).

Setting aside issues of syntax, the only major difference is that agents can wrap not only pure functions (queries without side effects) but commands. Ensuring full purity does not, however, seem particularly relevant to discussions of architecture, at least as long as we enforce the command-query separation principle, retaining the principal practical benefit of purity—referential transparency of expressions—without forcing a stateful model into the artificial stranglehold of stateless models.

Agents bring the final touch to object technology's contribution to modularity, but they are only one of its elements, together with those sketched in this discussion and a few more. The combination of these elements, going beyond what the functional approach can offer, makes object-oriented design the best available approach to ensure beautiful architecture.

Acknowledgments

I am grateful to several people who made important comments on drafts of this contribution; obviously, acknowledgment implies no hint of endorsement (as is obvious in the case of the functional programming grandees who were kind enough to share constructive reactions without, I fear, being entirely swayed yet by my argument). Particularly relevant were observations by Simon Peyton Jones, Erik Meijer, and Diomidis Spinellis. John Hughes's answer to my questions about his classic paper were detailed and illuminating. The Visitor library discussed in the last part of this chapter is the work of Karine Arnout (Karine Bezault). I thank Gloria Müller for further observations as part of an ETH master's thesis on implementing a library of Haskell-like facilities for Eiffel. I am especially grateful to the editors of this volume, Diomidis Spinellis and Georgios Gousios, for the opportunity to publish this discussion and for their extreme patience with my delays in finalizing it.

References

Arnout, Karine. 2004. "From patterns to components." Ph.D. thesis, ETH Zurich. Available at *http://se.inf.ethz.ch/people/arnout/patterns/*.

Arnout, Karine, and Bertrand Meyer. 2006. "Pattern componentization: the Factory example." *Innovations in Systems and Software Technology* (a NASA Journal). New York, NY: Springer-Verlag. Available at *http://www.springerlink.com/content/am08351v30460827/*.

Eber, Jean-Marc, based on joint theoretical work with Simon Peyton Jones and Pierre Weis. 2001. "Compositional description, valuation, and management of financial contracts: the MLFi language." Presentation available at *http://www.lexifi.com/Downloads/MLFiPresentation.ppt*.

Ecma International. 2006. *Eiffel: Analysis, Design and Programming Language*. ECMA-367. Available at *http://www.ecma-international.org/publications/standards/Ecma-367.htm*.

Frankau, Simon, Diomidis Spinellis, Nick Nassuphis, and Christoph Burgard. 2008. "Commercial uses: Going functional on exotic trades." *Journal of Functional Programming*, 19(1):2745, October.

Gamma, Erich, et al. 1994. *Design Patterns: Elements of Reusable Object-Oriented Software*. Boston, MA: Addison-Wesley.

Hughes, John. 1989. "Why functional programming matters." *Computer Journal*, vol. 32, no. 2: 98–107 (revision of a 1984 paper). Available at *http://www.cs.chalmers.se/~rjmh/Papers/whyfp.pdf*.

La Rochefoucauld, François de. 1665. *Réflexions ou sentences et maximes morales*.

Meyer, Bertrand, and Karine Arnout. 2006. "Componentization: the Visitor example." *Computer* (IEEE), vol. 39, no. 7: 23–30. Available at *http://se.ethz.ch/~meyer/publications/computer/visitor.pdf*.

Meyer, Bertrand. 1992. *Eiffel: The Language*. (Second printing.) Upper Saddle River, NJ: Prentice Hall.

Meyer, Bertrand. 1997. *Object-Oriented Software Construction*, Second Edition. Upper Saddle River, NJ: Prentice Hall. Available at *http://archive.eiffel.com/doc/oosc/*.

Meyer, Bertrand. 2004. "The power of abstraction, reuse and simplicity: An object-oriented library for event-driven design." *From Object-Orientation to Formal Methods: Essays in Memory of Ole-Johan Dahl*. Eds. Olaf Owe, Stein Krogdahl, and Tom Lyche. *Lecture Notes in Computer Science*, 2635, pp. 236–271. New York, NY: Springer-Verlag. Available at *http://se.ethz.ch/~meyer/publications/lncs/events.pdf*.

Meyer, Bertrand. 2008. *Touch of Class: An Introduction to Programming Well*. New York, NY: Springer-Verlag. See *http://touch.ethz.ch*.

Peyton Jones, Simon, Jean-Marc Eber, and Julian Seward. 2000. "Composing contracts: An adventure in financial engineering." Functional pearl, in *ACM SIGPLAN International Conference on Functional Programming* (ICFP '00), Montreal, Canada, September '00. ACM Press, pp. 280–292. Available at *http://citeseer.ist.psu.edu/jones00composing.html*.

Peyton Jones, Simon, and Philip Wadler. 1993. "Imperative functional programming." *Twentieth Annual ACM SIGPLAN-SIGACT Symposium on Principles of Programming Languages*. Charleston, South Carolina, pp. 71–84. Available at *http://citeseer.ist.psu.edu/peytonjones93imperative.html*.

Steingold, Sam. Online at *http://www.podval.org/~sds/ocaml-sucks.html*.

Sun Microsystems. 1997. "About Microsoft's 'Delegates.'" White paper by the Java Language Team at JavaSoft. Available at *http://java.sun.com/docs/white/delegates.html*.

Wadler, Philip. 1995. "Monads for functional programming." *Advanced Functional Programming*, Lecture Notes in Computer Science 925. Eds. J. Jeuring and E. Meijer. New York, NY: Springer-Verlag. Available at *http://homepages.inf.ed.ac.uk/wadler/papers/marktoberdorf/baastad.pdf*.

Walden, Kim, and Jean-Marc Nerson. 1994. *Seamless Object-Oriented Software Architecture*. Upper Saddle River, NJ: Prentice Hall. Available at *http://www.bon-method.com/index_normal.htm*.

CHAPTER FOURTEEN

Rereading the Classics

Panagiotis Louridas

IT SEEMS THAT IN ALL SCIENTIFIC FIELDS THERE ARE WORKS AND PEOPLE that one cannot avoid mentioning. The current living champion is probably Noam Chomsky. According to an April 1992 article in the MIT *Tech Talk*, Chomsky was one of the most cited individuals in works published in the previous 20 years. The full top 10 roster of the *Arts & Humanities Citation Index* included Marx, Lenin, Shakespeare, Aristotle, the Bible, Plato, Freud, Chomsky, Hegel, and Cicero. In the *Science Citation Index*, he was cited 1,619 times in the period from 1972 to 1992.

In software engineering, oak leaf clusters must probably go to *Design Patterns: Elements of Reusable Object-Oriented Software* (a.k.a. the "Gang of Four" book [Gamma et al. 1994]). A Google search on the exact book title returns about 173,000 results (in spring 2008). If we turn our attention to a more academic context, a search in the ACM Digital Library returns 1,572 results. The design patterns community has been one of the most vibrant communities in software engineering in the last 20 years. It is difficult to think of a software engineer plying her trade today who would not be familiar with this important body of work.

The present contribution happily pays its dues where it should by adding one to each of those citation counts.

The Gang of Four book can be credited with bringing design patterns to the masses. It can also be credited with not only providing a starting point for the design patterns movement, but a returning point as well: the material related to design patterns is extensive, yet most discussions on the subject revolve around the design patterns presented in that book. Without dismissing the others, it is the 32 patterns presented there that certainly deserve to be called classics.

A fascinating part of that book, however, is not the design patterns themselves but the introductory chapter, which provides the rationale behind many of them and a common thread for linking them together. There we find principles of reusable object-oriented design. The second such principle is *to favor object composition ("has") over class inheritance ("is-a")*. (The first principle is "to program to an interface, not an implementation," which should be clear to anybody who has seen any advice on encapsulation in the last 40 years.)

To programmers who have not followed the arrival of object-oriented programming to center stage in the 1980s and 1990s, this rule might not seem overly important. But if one recalls that time, one of the defining concepts in object-oriented programming was inheritance. Take, for instance, Bjarne Stroustrup's description of C++ in *The C++ Programming Language* (1985). We find there (p. 21) that:

> C++ is a general purpose programming language with a bias toward systems programming that:
> - Is a better C
> - Supports data abstraction
> - Supports object-oriented programming
> - Supports generic programming

If we then turn to see exactly what "supports object-oriented programming" entails (p. 39), we find:

> The programming paradigm is:
> - Decide which classes you want
> - Provide a full set of operations for each class
> - Make commonality explicit by using inheritance

Now consider by contrast yet another classic, Joshua Bloch's *Effective Java* (2008). We find there no less than three injunctions against inheritance:

- Favor Composition Over Inheritance
- Design and Document Inheritance or Else Prohibit It
- Prefer Interfaces to Abstract Classes

Is then inheritance to be avoided? That is not an academic point. Programming Microsoft Windows can be very frustrating, even when approached by way of a very pleasant book (Charles Petzold's *Programming Windows* [1999]). When the first frameworks for Windows programming came out (from Borland and Microsoft), they felt like a breath of fresh air. Before then, creating a single window was nowhere near simple: programmers were interested to learn that to program in Microsoft Windows they had to work with something called a window class, which had nothing to do with a C++ class. With the new frameworks, you needed only to create a subclass of a class provided by the framework, and that was that. We were happy to be suddenly freed from all the drudgery (or almost all), and also happy that we suddenly found such a neat application of object orientation.

Programming Microsoft Windows is just an example; the enthusiasm with object orientation and inheritance was pervasive. It is strange to learn now that we may have been wrong, but perhaps we were not that wrong. Inheritance may not be essentially bad. Like all technology, it can be put to good and bad uses, and the bad uses of inheritance have been documented in many places (the Gang of Four book being a good start). Here we will give an example of a beautifully crafted software system that has inheritance as its foundation. That system is Smalltalk.

Smalltalk is a pure object-oriented language, and although it never actually made it to the mainstream, it influenced language evolution in many ways. Perhaps the other language that exerted so much influence on later computer languages was Algol-60—another example that was more influential than actually used.

This is not a presentation of the Smalltalk programming language and its environment (these two really go together), but a presentation of basic architectural ideas and where they may lead us in our programming tasks. To borrow a term from the psychology of design, this is a discussion on basic design principles and the *affordances* they give the programmer. Donald Norman, in *The Psychology of Everyday Things* (1988), lucidly (and entertainingly) explains the notion of affordances. Simply put, an object by its appearance allows us, and even invites us, to do certain things. A hanging rope invites us to reach for it and pull it; a horizontal handle bar invites us to push it; and a door-handle invites us to reach for it and turn it. In the same way, the way a programming language looks to the programmer invites her to do certain things with it, and warns her against doing other things. A beautiful crafted language has a beautiful architecture, and that will show through the programs we make with it.

A strong expression of this is the Shapir-Whorf Hypothesis (SWH), which states that language determines thought. The hypothesis has excited linguists and language designers for many years. The preface to the first edition of *The C++ Programming Language* starts with SWH, and the 1980 Turing Award lecture by K. E. Iverson (of APL fame) was devoted to the importance of notation for expressing our thoughts. The SWH is controversial—after all, everybody has had the experience of not being able to find the words for something, so we are able to think of more than we can say—but in computer code the link between languages and programs is clear. We know that computer languages are Turing complete, but we also know that for some things some languages fit better than others.

But apart from influencing program architecture, a language's architecture is interesting in its own right. We will take a glimpse at Smalltalk's own architecture—implementation choices, design concepts, and patterns. We see many of them today in more recent programming languages; some of those that we do not see give us reason to pause and reflect on the reasons why.

We do not assume any prior knowledge of Smalltalk here, although we will cover a significant part of Smallktalk by the time we reach the end of this chapter. We will highlight design principles by using small code snippets to illustrate them. One of the strengths of strong design principles is that there are few things to learn, and once you grasp them, the whole infrastructure flows from them. The Smalltalk system we will be referring to is Squeak (*http://www.squeak.org*), an open source implementation. Some code examples may be difficult to understand at first reading, since we introduce concepts somewhat informally, but they are illustrated in subsequent examples, so it is prudent to make an effort to go through to the end and then return to any offending parts. At the same time, we do not insult the reader's intelligence.

Exploring Smalltalk will show language features that are not necessarily available in your preferred language. That should not be a problem. It is a time-tested maxim in software development that it is not necessary for the language you are using to natively support a feature in order to program with it; with some diligence, you can find an elegant way to find an alternative for it in your language of choice. According to Steve McConnell's *Code Complete* (2004), this is called programming *into* a language (p. 69):

> Understanding the distinction between programming in a language and programming into one is critical.... Most of the important programming principles depend not on specific languages but on the way you use them. If your language lacks constructs that you want to use or is prone to other kinds of problems, try to compensate for them. Invent your own coding conventions, standards, class libraries, and other augmentations.

Indeed, in a reversal of the SWH that is an homage to programmer creativity (or stubbornness), the author remembers, when object orientation had become *de rigueur* in 1990, coming upon a book at the local technical bookstore treating the subject of object-oriented assembly language. More recently, Randall Hide's High Level Assembler (HLA) has combined assembly with classes, inheritance, and other trappings.

We will approach a programming language as we would approach a classic book. Before we go on in earnest, let us begin with a few suggested definitions from Italo Calvino's essay "Why Read the Classics" (1986):

> The classics are the books of which we usually hear people say, "I am rereading..." and never "I am reading...."

> We use the word "classics" for books that are treasured by those who have read and loved them; but they are treasured no less by those who have the luck to read them for the first time in the best conditions to enjoy them.

The classics are books that exert a peculiar influence, both when they refuse to be eradicated from the mind and when they conceal themselves in the folds of memory, camouflaging themselves as the collective or individual unconscious.

Every rereading of a classic is as much a voyage of discovery as the first reading.

A classic is a book that has never finished saying what it has to say.

The classics are the books that come down to us bearing the traces of readings previous to ours, and bringing in their wake the traces they themselves have left on the culture or cultures they have passed through (or, more simply, on language and customs).

A classic does not necessarily teach us anything we did not know before. In a classic we sometimes discover something we have always known (or thought we knew), but without knowing that this author said it first, or at least is associated with it in a special way. And this, too, is a surprise that gives much pleasure, such as we always gain from the discovery of an origin, a relationship, an affinity.

The classics are books which, upon reading, we find even fresher, more unexpected, and more marvelous than we had thought from hearing about them.

A classic is a book that comes before other classics; but anyone who has read the others first, and then reads this one, instantly recognizes its place in the family tree.

Books are not computer languages, of course, and yet these definitions may also apply to our task.

Everything Is an Object

Today's popular object-oriented computer languages (C++, Java, and C#) are not purely object-oriented. Not everything is an object. Some types are *primitive*. Hence, for example, we cannot subclass an integer. Arithmetic is performed the usual way on pure numbers, not by invoking methods on objects. This brings performance benefits, and it may make things easier for people coming to object orientation from a procedural language background.

But if we do decide to have everything as an object, then the situation changes drastically. In Smalltalk, integers up to 31 bits long are instances of the SmallInteger class (in fact, there is an abstract Integer class and its subclasses SmallInteger, LargePositiveInteger, and LargeNegativeInteger, with conversions carried out automatically by the system as necessary). We can perform ordinary arithmetic on them, but we can also do more. The SmallInteger class offers no less than 670 methods (or *selectors*, in Smalltalk parlance), as we can easily find out with the following code snippet:

```
SmallInteger allSelectors size
```

It is instructive to examine how this code works. `allSelectors` is a class selector that does what its name suggests. It returns all the selectors of a class in a `Set` (in fact, an `IdentitySet`, but this does not make any difference to us here). The `Set` is itself a first-class object with its own selectors; one of them, `size`, tells us the number of elements it contains.

Among the `SmallInteger` selectors we find the expected arithmetic operators. We also find trigonometric and logarithmic functions, functions to compute factorials, greatest common divisor, and least common multiple. There are functions for bit manipulation, and many others.

What we meet as integer primitives in other languages are actually `SmallInteger` instances in Smalltalk. This explains why:

```
2 raisedTo: 5
```

works, and more intriguingly, why the following works as well:

```
(7 + 3) timesRepeat: [ Transcript show: 'Hello, World'; cr ]
```

Selectors that take arguments have a colon (:) appended to them before each argument. Selectors standing for arithmetic and logical operators, such as + just shown, are exempt from that rule. `Transcript` is a class representing something similar to the system console. `cr` stands for carriage return, and a semicolon (;) cascades messages, so `cr` is sent to `Transcript`. We can evaluate this code in an interpreter window (usually called a *workspace* in Smalltalk) and see directly what happens.

Of course, not all 670 `SmallInteger` methods are defined in `SmallInteger`. `SmallInteger` is part of an inheritance hierarchy, shown in Figure 14-1, where we also see the number of selectors for each of the `SmallInteger` ancestors. Most of the selectors are inherited by `Object`, and understanding what `Object` offers explains a great deal of Smalltalk architecture (in Squeak, the actual root of the hierarchy is `ProtoObject`, but this is a minor detail).

Instances of `Object` have at their disposal comparison selectors (both equality, denoted by =, and identity, denoted by ==); selectors for making copies (both deep, by invoking `deepCopy`, and shallow, by invoking `shallowCopy`); and selectors for printing on streams, error handling, debugging, message handing, and others. Only a few of the hundreds of `Object` methods are of use for everyday programming. Methods in Smalltalk are organized in groups called *protocols*, and checking the protocol descriptions makes finding methods easier.

Methods themselves are first-class objects in Smalltalk. To see what this means in terms of the overall architecture, take the code:

```
aRectangle intersects: anotherRectangle
```

where aRectangle and anotherRectangle are instances of class `Rectangle`. When aRectangle, the *receiver* (as objects that receive a message are called in Smalltalk) receives the `intersects:` message, the Smalltalk interpreter will do the following (Conroy and Pelegri-Llopart 1983):

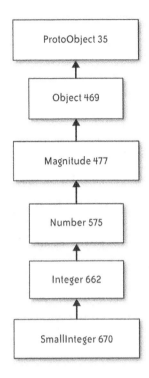

FIGURE 14-1. SmallInteger hierarchy

1. Determine the class of the receiver.
2. Search for the message selector in the class and the class ancestors.
3. Retrieve the method associated with the message selector at the class where it was found.

Not only are things like numbers objects in Smalltalk, classes are objects in Smalltalk too. Hence, `SmallInteger`, `Object`, `Rectangle`, etc. are all objects. When the interpreter searches for the message selector in a class (step 2 in the list), it searches for it in the contents of the corresponding class object. To be more precise, it looks it up in its method dictionary. An instance of the `Dictionary` class associates key with values; a method dictionary associates each selector with an instance of the corresponding `CompiledMethod`.

As an aside, `intersects:` can be implemented elegantly in Smalltalk as:

    (origin max: aRectangle origin) < (corner min: aRectangle corner)

To see why this works, you need to know that the `origin` selector returns the upper-left point of a rectangle (an instance of class `Point`), the `corner` selector returns the bottom-right point of a rectangle, `max:` returns the lower-right corner of the rectangle uniquely defined by the receiver and the argument, and `min:` returns the upper-left corner of the rectangle uniquely defined by the receiver and the argument. Although pithy, this is not the most optimal solution; Squeak provides an alternative one:

```
intersects: aRectangle
    "Answer whether aRectangle intersects the receiver anywhere."
    "Optimized; old code answered:
        (origin max: aRectangle origin) < (corner min: aRectangle corner)"

    | rOrigin rCorner |
    rOrigin := aRectangle origin.
    rCorner := aRectangle corner.
    rCorner x <= origin x ifTrue: [^ false].
    rCorner y <= origin y ifTrue: [^ false].
    rOrigin x >= corner x ifTrue: [^ false].
    rOrigin y >= corner y ifTrue: [^ false].
    ^ true
```

Faster, but less beautiful. It gives us the opportunity, though, to introduce some Smalltalk syntax. Variables local to a method are declared inside | |. The assignment operator is :=, ^ is the equivalent of return in C++ and Java, and the period (.) separates statements. The code inside square brackets ([ ]) is called a *block*, a key concept in the Smalltalk architecture. A block is a *closure*—that is, a piece of code that can access the variables defined in its surrounding scope. Blocks in Smalltalk are represented by the class `BlockContext`. The contents of a block are executed when the block object receives the message `value`, and in most cases (like here), the message is sent implicitly. Comments in Smalltalk are inside double quotes; single quotes are used for strings.

A `BlockContext` shares the receiver, the arguments, the temporary variables, and the sender of the context that creates it. There is a similar class, `MethodContext`, representing all the dynamic state associated with the execution of a method (which, as we saw, is represented by `CompiledMethod`, an array of bytecodes). As we would expect from an object-oriented language, both `BlockContext` and `MethodContext` are subclasses of class `ContextPart`. The `ContextPart` class adds execution semantics to its superclass, `InstructionStream`. On its part, `InstructionStream` is the class whose instances can interpret Smalltalk code. The superclass of `InstructionStream` is `Object`, where the chain ends.

Apart from the `value` selector, blocks also have the `fork` selector, which implements concurrency in the language. As everything in Smalltalk is an object, processes are instances of the `Process` class. The `Delay` class allows us to suspend an execution of a process for a specified period of time; a `Delay` object will suspend the current executing process when it is sent the `wait` message. Combining all this together, a trivial clock can be implemented with the following code (Goldberg and Robson 1989, p. 266):

```
[[true] whileTrue:
    [Time now printString displayAt: 100@100.
    (Delay forSeconds: 1) wait]] fork
```

The `whileTrue:` selector will execute its block argument as long as its own receiver block is true. The @ character is a selector of class `Number` that constructs instances of class `Point`.

Blocks also give us a basic error-handling functionality, the idea being that we specify blocks to be executed when something goes wrong. For instance, in a Collection object (a container of objects) the method remove: will try to remove the specified element from the collection. The method remove:ifAbsent: will try to remove the specified element from the collection, and if the element does not exist, it will execute the block passed as argument in ifAbsent:. In a nice example of minimizing code, the first is implemented in terms of the second:

```
remove: oldObject
    "Remove oldObject from the receiver's elements. Answer oldObject
    unless no element is equal to oldObject, in which case, raise an error.
    ArrayedCollections cannot respond to this message."

    ^ self remove: oldObject ifAbsent: [self errorNotFound: oldObject]

remove: oldObject ifAbsent: anExceptionBlock
    "Remove oldObject from the receiver's elements. If several of the
    elements are equal to oldObject, only one is removed. If no element is
    equal to oldObject, answer the result of evaluating anExceptionBlock.
    Otherwise, answer the argument, oldObject. ArrayedCollections cannot
    respond to this message."

    self subclassResponsibility
```

self is a reference to the current object (equivalent to this in C++ and Java); it is a reserved name, a *pseudovariable* with fixed semantics. There are some more pseudovariables: super is a reference to the superclass (equivalent to super in Java); nil, true, and false have the expected meanings; and finally there is also thisContext, which we see in action in the definition of the subclassResponsibility method. This selector is defined in Object and is simply an indicator that the subclass must override it:

```
subclassResponsibility
    "This message sets up a framework for the behavior of the class's subclasses.
    Announce that the subclass should have implemented this message."

    self error: 'My subclass should have overridden ', thisContext
sender selector printString
```

The thisContext pseudovariable is a reference to the current executing context—that is, the current executing method or block—so it is an instance of the current executing instance of MethodContext or BlockContext. The sender selector returns the context that sent the message; selector gives the selector of the method. *All these are objects.*

Handling code as objects is not new; a strength of Lisp is its similar handling of code and data. It allows us to program using reflection—that is, to do metaprogramming.

Metaprogramming is an idea whose importance has been increasing with time. In statically typed compiled languages such as C or C++, support for metaprogramming is meager. An effort to put metaprogramming to work in C++ (Forman and Danforth 1999), seems to have been more influential in Python than in C++ itself (see PEP 253, "Subtyping Built-in Types," at *http://www.python.org/dev/peps/pep-0253/)*. Template metaprogramming in C++ is a rather different approach: we use the fact that the C++ compiler will generate template code at compile time to carry out computation right then (Abrahams and Gurtovoy 2005). It is a technique with exciting possibilities that calls, however, for esoteric programming skills. In Java, metaprogramming, via reflection, is an integral part of the language, although Java code that uses reflection tends to be cumbersome.

A related issue is highlighted when we are faced with the problem of constructing a menu at runtime. A menu associates items with handlers that are invoked when the user selects the associated label. If we are able to refer to handlers by name, then we can construct a menu dynamically using code such as this:

```
CustomMenu new addList: #(
    #('red' #redHandler)
    #('green' #greenHandler)
    #('blue' #blueHandler)); startUpWithCaption:'Colors'.
```

In Smalltalk the delimiters #( ) enclose arrays. We create a new menu with its items and handlers given by a list that contains label-handler pairs. The handlers are selectors (in real code, we would need to give the implementation for them). The handlers are prefixed by the character #, which denotes symbols in Smalltalk. We can think of symbols like strings. Symbols are typically used for class and method names.

Reflection allows us to implement the Abstract Factory design pattern in a succinct way (Alpert et al. 1998, pp. 43–44). If we want a factory class that instantiates objects of classes specified at runtime by the user, we can do the following:

```
makeCar: manufacturersName
    "manufacturersName is a Symbol, such as #Ford, #Toyota, or #Porsche."
    | carClass |
    carClass := Smalltalk
                    at: (manufacturersName, #Car) asSymbol
                    ifAbsent: [^ nil].
    ^ carClass new
```

After the user has given a manufacturer's name, we create the name of the class by appending the word `Car` to it. For `#Ford`, the class name will be `#FordCar`, for `#Toyota` it will be `#ToyotaCar`, and so on. Concatenation in Smalltalk is denoted by comma (,), and we want the resulting concatenated string to be made a symbol again, so we call its `asSymbol` method. All class names in Smalltalk are stored in the `Smallktalk` dictionary, the unique instance of the `SystemDictionary` class. Once we find the required class name, we return an instance of the class.

We have seen several examples of Smalltalk code, but not a class definition. In Smalltalk, classes are constructed in the same way as anything else: by sending the necessary message to the appropriate receiver. We start by completing the following template in the Smalltalk environment:

```
NameOfSuperclass subclass: #NameOfSubclass
    instanceVariableNames: ''
    classVariableNames: ''
```

```
        poolDictionaries: ''
        category: 'Unknown'
```

We substitute the actual names for `NameOfSuperclass` and `NameOfSubclass`. In `instanceVariableNames`, we list the instance variables, in `classVariableNames` the class variables, and in `category`, we mark the category our class falls under (classes are grouped in categories in Smalltalk, similar to the namespaces or packages in other languages). The `poolDictionaries` slot lists the dictionaries that we are sharing with other classes; this is a mechanism for sharing variables in Smalltalk. When the details of the template are filled in, they are passed to the `subclass` selector of class `Class`:

```
subclass: t instanceVariableNames: f classVariableNames: d
poolDictionaries: s category: cat
    "This is the standard initialization message for creating a new class as a
    subclass of an existing class (the receiver)."
    ^(ClassBuilder new)
        superclass: self
        subclass: t
        instanceVariableNames: f
        classVariableNames: d
        poolDictionaries: s
        category: cat
```

The `subclass` selector creates an instance of class `ClassBuilder`, which creates new classes or modifies existing ones. We send the required information to the `ClassBuilder` instance so that the new class can be created according to what we have entered in the class template.

Doing everything on objects by sending messages gives us an economy of concepts that we have to grasp. It also allows us to limit the number of syntactic constructs in the language. Minimalism in programming languages goes a long way back. In the first paper on Lisp (McCarthy 1960), we find that Lisp comprised two classes of expressions: s-expressions (or syntactic expressions), which were expressions built from lists, and m-expressions (or meta-expressions), which were expressions using s-expressions as data. In the end, programmers opted for using s-expressions all the way, and hence Lisp became what we know today: a language with almost no syntax, as everything, program and data, is a list. Depending on where you stand on Lisp, this is proof that one single idea is enough for expressing the most complicated constructs (or that humans can be coerced into accepting anything).

Smalltalk does not limit itself to one syntactic element, but still, Smalltalk programs are composed of six building blocks:

1. Keywords, or pseudovariables, of which there are only six (`self`, `super`, `nil`, `true`, `false`, and `thisContext`)
2. Constants
3. Variable declarations
4. Assignments

5. Blocks
6. Messages

What we do not see included in this list is perhaps more interesting than what is included: we do not see any elements for expressing control flow, no conditionals or loops. They are not needed, as they are implemented in terms of messages, objects, and blocks (which are objects). The following method implements the factorial function in class Integer:

```
factorial
    "Answer the factorial of the receiver."

    self = 0 ifTrue: [^ 1].
    self > 0 ifTrue: [^ self * (self - 1) factorial].
    self error: 'Not valid for negative integers'
```

Operators = and > are message selectors returning objects of abstract class Boolean, which has two subclasses, True and False. If the receiver of selector ifTrue: is an instance of True, then its argument is executed. The argument is the block delimited by [ ]. There is a symmetrical selector ifFalse: with the opposite semantics. In general it is a good idea to use loops instead of recursion, so here is a loop implementation of the factorial function:

```
factorial
    "Implement factorial function using a loop"

    | returnVal |
    returnVal := 1.
    self >= 0
        ifTrue: [2
                    to: self
                    do: [:n | returnVal := returnVal * n]]
        ifFalse: [self error: 'Not valid for negative integers'].
    ^ returnVal
```

The bulk of the job is carried out inside two blocks. The first block is executed for positive values starting at 2 until the value of the receiver. Each iterated value is passed to the inner block, where we calculate the result. Block arguments are prefixed by a colon (:) and separated from the body of the block with a vertical bar (|). There may be more than one block argument, as in the following definition of factorial (Black et al. 2007, p. 212):

```
factorial := [:n | (1 to: n) inject: 1 into: [:product :each | product * each ] ].
```

The to: selector returns an instance of class Interval, which effectively enumerates the values from 1 to 10. For a number n, the factorial block will do the following. First, it will set the product argument of the inner block to 1. Then, it will call the inner block for each value from 1 to n, calculating the product of each iterated number and the current product, and storing the result to product. To evaluate the factorial of 10, we need to write factorial value: 10. To borrow Herbert Simon's quotation of early Dutch physicist Simon Stevin in *The Sciences of the Artificial* (1996):

Wonder, en is gheen wonder. That is to say: "Wonderful, but not incomprehensible."

Types Are Defined Implicitly

Although everything in Smalltalk, even classes, is an object, classes do not correspond to types in the way they do in languages such as C++ and Java. Types are defined implicitly by what they do, and by their interfaces. This is described by names such as *latent typing* or *duck typing*.

Latent typing is the only typing mechanism in Smalltalk (and also in some other dynamically typed languages), but that does not mean it is of no importance to strongly typed languages. In C++, for instance, latent typing is the basis of generic programming via templates. It makes sense to see it first in that language. Take a look at the following introductory example of C++ templates (Vandervoorde and Josuttis 2002, 2.4):

```
// maximum of two int values
inline int const& max (int const& a, int const& b)
{
    return  a < b ? b : a;
}

// maximum of two values of any type
template <typename T>
inline T const& max (T const& a, T const& b)
{
    return  a < b ? b : a;
}

// maximum of three values of any type
template <typename T>
inline T const& max (T const& a, T const& b, T const& c)
{
    return ::max (::max(a,b), c);
}

int main()
{
    ::max(7, 42, 68);       // calls the template for three arguments
    ::max(7.0, 42.0);       // calls max<double> (by argument deduction)
    ::max('a', 'b');        // calls max<char> (by argument deduction)
    ::max(7, 42);           // calls the nontemplate for two ints
    ::max<>(7, 42);         // calls max<int> (by argument deduction)
    ::max<double>(7, 42);   // calls max<double> (no argument deduction)
    ::max('a', 42.7);       // calls the nontemplate for two ints
}
```

As we see in main(), the function ::max will work for any type that implements the comparison operator. In C++, that type may be a primitive, or it may be user-defined. There are no restrictions that it should inherit from a specific class. It can be of any type, as long as it obeys the basic requirement on comparison. The implicit type definition is: anything for which operator < makes sense.

In school we learned that there are two ways to define a set. One is explicit, by enumerating its elements. This is called *extensional definition*. The set of natural numbers we know is

{1, 2, 3, ...}. The other way to define a set is by describing what the members of the set have in common. This is called *intentional definition*. The intensional definition of the set of even natural numbers is "whole numbers greater than zero." When we declare an object in C++, we are effectively using an extensional definition: we say that the object belongs to the given type. When we use templates, however, we are effectively using an intensional definition: we say that the set of objects for which this code is applicable is the set of objects that have the given properties (that is, they offer the required operations).

The situation is unfortunately muddled in Java. Java offers generics, but they bear only a surface syntactic resemblance to C++ templates. An eloquent explanation of the problem has been given by Bruce Eckel (see *http://www.mindview.net/WebLog/log-0050*). In a language such as Python, which supports latent typing, we can do this:

```python
class Dog:
    def talk(self):  print "Arf!"
    def reproduce(self): pass

class Robot:
    def talk(self): print "Click!"
    def oilChange(self): pass

a = Dog()
b = Robot()
speak(a)
speak(b)
```

The two invocations of `speak()` will work without caring for the type of its argument. We can do the same with C++:

```cpp
class Dog {
public:
    void talk() { }
    void reproduce() { }
};

class Robot {
public:
    void talk() { }
    void oilChange() { }
};

template<class T> void speak(T speaker) {
    speaker.talk();
}

int main() {
    Dog d;
    Robot r;
    speak(d);
    speak(r);
}
```

We *cannot* do the same in Java, as the following will not compile:

```
public class Communicate {
    public <T> void speak(T speaker) {
        speaker.talk();
    }
}
```

Confusingly, this *does* compile, as under the hood, generic types are converted to instances of Object in Java (this is called erasure):

```
public class Communicate {
    public <T> void speak(T speaker) {
      speaker.toString(); // Object methods work!
    }
}
```

So, we have to do something like this:

```
interface Speaks { void speak(); }

public class Communicate {
    public <T extends Speaks> void speak(T speaker) {
      speaker.speak();
    }
}
```

But this pretty much defeats the advantages of generics, as we are defining a type via the Speak interface. The lack of generality also shows in that Java primitive types cannot be used with the generics mechanism. As a workaround, Java offers wrapper classes, true object classes that correspond to primitive types. Converting between primitives and wrappers used to be a chore in Java programming. In recent versions of the language it is less so, thanks to the autoboxing feature that performs automatic conversions in certain circumstances. Be that as it may, we can write List<Integer>, but not List<int>.

Latent typing has been popularized recently thanks to its widespread adoption in the Ruby programming language. The term "duck typing" is a tongue-in-cheek reference to inductive reasoning, attributed to James Whitcomb Riley, which goes:

> If it walks like a duck and quacks like a duck, I would call it a duck.

To see the importance of duck typing, take an essential feature of object-oriented programming, *polymorphism*. Polymorphism stands for the use of different types in the same context. One way to achieve polymorphism is through inheritance. A subclass can be used (more precisely, should be used, because programmers can be careless) wherever a superclass can be used. Duck typing offers an additional way to achieve polymorphism: a type can be used anywhere it offers methods fitting the context. In the pet and robot example shown earlier in Python and C++, Dog and Robot do *not* share a superclass.

Of course it is possible to program your way around duck typing only the inheritance type of polymorphism. A programmer, however, is wealthier if she has more tools at her disposal for solving the problem she faces. As long as the plurality of tools does not get into the way, she can choose among them as best fits the situation. This has been expressed very elegantly by Bjarne Stroustrup in *The Design and Evolution of C++* (1994, p. 23):

> My interest in computers and programming languages is fundamentally pragmatic.
>
> I feel most at home with the empiricists rather than with the idealists.... That is, I tend to prefer Aristotle to Plato, Hume to Descartes, and shake my head sadly over Pascal. I find comprehensive "systems" like those of Plato and Kant fascinating, yet fundamentally unsatisfying in that they appear to me dangerously remote from everyday experiences and the essential peculiarities of individuals.
>
> I find Kierkegaard's almost fanatical concern for the individual and keen psychological insights much more appealing than the grandiose schemes and concern for humanity in the abstract of Hegel or Marx. Respect for groups that doesn't include respect for individuals of those groups isn't respect at all. Many C++ design decisions have their roots in my dislike for forcing people to do things in some particular way. In history, some of the worst disasters have been caused by idealists trying to force people into "doing what is good for them." Such idealism not only leads to suffering among its innocent victims, but also to delusion and corruption of the idealists applying the force. I also find idealists prone to ignore experience and experiment that inconveniently clashes with dogma or theory. Where ideals clash and sometimes even when pundits seem to agree, I prefer to provide support that gives the programmer a choice.

Going back to Smalltalk, consider the problem of going over a collection of objects, applying a function on each one of them, and collecting the results. This is implemented as follows:

```
collect: aBlock
    "Evaluate aBlock with each of the receiver's elements as the argument.
    Collect the resulting values into a collection like the receiver. Answer
    the new collection."

    | newCollection |
    newCollection := self species new.
    self do: [:each | newCollection add: (aBlock value: each)].
    ^ newCollection
```

To understand this method, it is enough to know that the `species` method returns either the class of the receiver or a class similar to it—the difference is too subtle to make a difference to us here. What is interesting is that to construct the new collection we only need something that has a selector called `value`, which we call. Blocks do have a selector called `value`, so any block can be used. But the fact that we are talking about blocks is incidental: anything that implements `value` will do.

Anything that returns a value has turned out to be important enough in programming that it has earned a name: it is now called a *function object*, and is a fundamental constituent of C++ STL algorithms. Traditionally, it is a staple of functional programming, usually called a *map* function. It is of course available in Lisp. It is also offered in Python, allowing us to do things like:

```
def negate(x): return -x
map(negate, range(1, 10))
```

whereas in Perl we would write:

```
map { -$_ } (1..10)
```

and the C++ STL allows us to do the equivalent (Josuttis 1999, 9.6.2):

```
vector<int> coll1;
list<int> coll2;

// initialize coll1

//negate all elements in coll1
transform(coll1.begin(), coll1.end(),        //source range
          back_inserter(coll2),              //destination range
          negate<int>());                    //operation
```

It is perhaps a tragedy of our times that many C++ programmers will write the code just shown using loops over arrays.

Duck typing can create controversy—and it indeed it has. In statically typed languages, such as C++, the compiler will check that the object used in an expression involving a latent type does offer the required interface. In a dynamically typed language such as Smalltalk, this will be discovered at runtime when it produces an error.

This cannot be dismissed out of hand. Strongly typed languages prevent programmers from oversight; they are particularly helpful in big projects, where a good structure aids maintenance. The most important change from traditional Unix C to ANSI C in the 1980s was the introduction of a stronger type system: C got proper function prototypes and function arguments would from now on be checked at compile time. We now frown upon cavalier conversions between types. In short, we exchanged some freedom for some discipline—or, alternatively, chaos for some order.

In Smalltalk, it is generally assumed that chaos should not result, as we should be writing small code fragments while testing them at the same time. Testing is easy in Smalltalk. Because there is no distinct compile and build cycle, we can write small code fragments in a workspace and see directly how our code behaves. Unit testing is also easy; we can write unit tests before our code without worrying about compiler messages referring to undeclared types or methods. It may not be accidental that the community that brought us JUnit shares a lot with the Smalltalk community. We can achieve much of the same in Java by using Java scripting, with BeanShell or Groovy for instance. In fact, strongly typed languages can give us a false sense of security:

> If a program compiles in a strong, statically typed language, it just means that it has passed some tests. It means that the syntax is guaranteed to be correct.... But there's no guarantee of correctness just because the compiler passes your code. If your code seems to run, that's also no guarantee of correctness.

> The only guarantee of correctness, regardless of whether your language is strongly or weakly typed, is whether it passes all the tests that *define the correctness of your program.* (*http://www.mindview.net/WebLog/log-0025*)

Sometimes it pays to be cautious, though. We may be adding objects to some collection and expecting these objects to obey a certain interface; we would be loath to discover at runtime that some of these objects in fact do not. We can guard against such mishaps by using Smalltalk metaprogramming facilities.

Metaprogramming complements latent typing by allowing us to check that the specified interface is really offered, and giving us the opportunity to react if it is not—for instance, by delegating nonimplemented calls to delegates that we know they implement them, or by simply handling failure gracefully and avoiding crashes. So in the pickaxe book (Thomas et al. 2005, pp. 370–371), we find the following Ruby example, where we try to add song information to a string:

```
def append_song(result, song)
  # test we're given the right parameters
  unless result.kind_of?(String)
    fail TypeError.new("String expected")
  end
  unless song.kind_of?(Song)
    fail TypeError.new("Song expected")
  end
  result << song.title << " (" << song.artist << ")"
end
```

This is what we would do if we adopted a Java or C# programming style. In Ruby-style duck typing, it would simply be:

```
def append_song(result, song)
  result << song.title << " (" << song.artist << ")"
end
```

This code will work with any object that appends using <<; for those objects that do not, we will get an exception. If we really want to be defensive, we can do so by checking the object's capabilities and not its type:

```
def append_song(result, song)
  # test we're given the right parameters
  unless result.respond_to?(:<<)
    fail TypeError.new("'result' needs `<<' capability")
  end
  unless song.respond_to?(:artist) && song.respond_to?(:title)
    fail TypeError.new("'song' needs 'artist' and 'title'")
  end
  result << song.title << " (" << song.artist << ")"
end
```

Smalltalk offers the `respondsTo:` method, defined in `Object`, which we can use to see in runtime whether a given receiver has a given selector.

```
respondsTo: aSymbol
    "Answer whether the method dictionary of the receiver's class contains
    aSymbol as a message selector."

    ^self class canUnderstand: aSymbol
```

The implementation is trivial, delegating the check to selector `canUnderstand:` defined in `Behavior`:

```
canUnderstand: selector
    "Answer whether the receiver can respond to the message whose selector
    is the argument. The selector can be in the method dictionary of the
    receiver's class or any of its superclasses."

    (self includesSelector: selector) ifTrue: [^true].
    superclass == nil ifTrue: [^false].
    ^superclass canUnderstand: selector
```

Finally, `includesSelector:` is also defined in `Behavior`, where it boils down to checking the method dictionary of the class:

```
includesSelector: aSymbol
    "Answer whether the message whose selector is the argument is in the
    method dictionary of the receiver's class."

    ^ self methodDict includesKey: aSymbol
```

When a receiver gets a message it does not understand, its standard response is sending the `doesNotUnderstand:` message to the system. If we would rather try to remedy the situation ourselves, we only need to override the message, doing something like the following:

```
doesNotUnderstand: aMessage
    "Handles messages not being understood by attempting to
    proxy to a target"
    target perform: aMessage selector withArguments: aMessage arguments].
```

We assume that `target` refers to the proxy object that we hope will be able to handle the misdirected message.

Latent typing is not an excuse for sloppy programming. We duck types, not responsibilities.

Problems

> Public inheritance means "is-a." That requires careful thinking from the part of the programmer to come up with class hierarchies that really fit this pattern. If you have a class with a method and a subclass where you realize that method makes no sense, it is beyond the scope of public inheritance, and it should be a sign of bad design. Languages, however, prove accommodating.

▶ 通过Liskov替换原则（LSP）可以验证。

In C++, you can get away with it by making the nonsensical method either return an error or throw an exception. The prototypical example concerns birds (Meyers 2005, item 32):

```
class Bird {
public:
    virtual void fly();    // birds can fly
    // ...
};

class Penguin: public Bird {    // penguins are birds
public:
    virtual void fly() { error("Attempt to make a penguin fly!"); }
    // ...
};
```

C++ programmers may alternatively hide the offending method:

```
class Base {
public:
    virtual void f()=0;
};

class Derived: public Base {
private:
    virtual void f() {
    }
};
```

So, `Derived` is no longer an abstract class, but it still does not have an `f()` function that can be of any use. Such shenanigans are to be avoided.

In Java, you can get away with it again by returning an error or throwing an exception; you can also making the nonsensical method abstract in the subclass, thus making the class hierarchy abstract from that point downward until you remake the method concrete. Again, such shenanigans are to be avoided.

The usual way to design around the problem of inapplicable or irrelevant methods in a class hierarchy is to redesign that hierarchy. A class hierarchy that would better reflect the peculiarities of the avian world would introduce a `FlyingBird` subclass of `Birds`, for those birds who do fly, and make `Penguin` a direct subclass of `Bird`, and not of a `FlyingBird`.

In Squeak we find a mere 45 methods that send the `shouldNotImplement` message, which is used when a method inherited by a superclass is not applicable in the current class. This is a very small proportion of the total number of methods and objects in Smalltalk, so the language is not fraught with badly designed class hierarchies. However, even the `shouldNotImplement` message is actually an implementation. This hints at a deeper issue in Smalltalk, which is that we do not have real abstract classes or methods. Methods are abstract by convention; there are no methods that have no implementation at all.

Instead of using the `shouldNotImplement` message, we could specify that a given method is the responsibility of a subclass, which we have seen is what the `subclassResponsibility` message is for. The class from which we send `subclassResponsibility` is then by convention abstract. For instance, the `Collection` class gives a generic interface for adding and removing objects, but does not provide implementation, as the implementation varies depending on the subclass we are dealing with (it could be a dictionary, an array, a list...). Method `add:` is to be implemented in the subclasses:

```
add: newObject
    "Include newObject as one of the receiver's elements. Answer newObject.
    ArrayedCollections cannot respond to this message."

    self subclassResponsibility
```

This "abstract" definition of `add:` even allows us to define in the Collection methods that use it, such as `add:withOccurrences:`, which is:

```
add: newObject withOccurrences: anInteger
    "Add newObject anInteger times to the receiver. Answer newObject."

    anInteger timesRepeat: [self add: newObject].
    ^ newObject
```

We can even do without defining `add:` at all; `add:withOccurrences:` would still be defined as just shown, and Smallktalk will not balk as long as at runtime the receiving object has `add:` defined. (By the way, `add:WithOccurrences:` is a nice little implementation of the Strategy pattern.) At the same time, the comment in `add:` points out that some subclasses of Collection, those rooted at its ArrayedCollection subclass, should not implement the message at all. This, again, is enforced only at runtime, by using `shouldNotImplement`:

```
add: newObject
    self shouldNotImplement
```

There is nothing inherently wrong in using conventions in programming; part of the art is mastering conventions. What can be problematic, however, is depending solely on collections to obtain the required result. Smallktalk will not warn us if we forget not to implement `add:` in ArrayedCollection. We will only fail miserably at runtime.

We saw earlier how easy it is to implement a proxy class in Smalltalk. The truth is, though, that if we actually want a proxy class that stands as a proxy for only a small number of methods, things get more complicated. The reason has to do with the absence of real abstract classes. A proxy class may be a subclass of the class we want to proxy, in which case it will inherit all the methods of the proxied class, and not just those methods that we want to proxy. Alternatively, we may employ latent typing and define in the proxy class only the proxied methods; the problem is that, since everything is an object, the proxy class will inherit all the methods of the Object class. Ideally, we would like a class that proxies two methods to have these two methods only, but it is not obvious how we can achieve that. The class will have all the methods inherited by its ancestors. We can resort to tricks to minimize the number of inherited methods; for example, in some Smalltalk dialects it is possible to make a class a subclass of `nil` instead of `Object`. In this way nothing is inherited, but we need to copy and paste some necessary methods from Object (Alpert et al. 1998, p. 215). In Squeak, we can make it a subclass of `ProtoObject` instead of `Object`.

▶这可以认为是继承体系中父类对子类的污染。尤其对于支持多重继承或混入方式的语言，这种情况更加严重。

> ▶ Ruby的meta programming即借鉴于此。

When we start looking into what really gets inherited from where, things become intricate. As everything in Smalltalk is an object, including classes, we may wonder what kind of instances classes are. It turns out that classes are instances of metaclasses; a class's metaclass is named after the class name with the word "class" after it. For example, the metaclass of `Object` is `Object class`. This is a sensible convention, but it does not really answer the question, as we may wonder what kind of instances metaclasses are. It turns out that metaclasses are instances of class `Metaclass` (no class suffix here). And `Metaclass` is an instance of another class, called `Metaclass class` (class suffix included), which again must be an instance of something, so it was made an instance of `Metaclass`. (Readers should be forgiven here for thinking they are reading part of the script of Terry Gilliam's *Twelve Monkeys*.) Things get even more intricate if we take into account the inheritance relationships as well (we have been talking only about *instances* until now). There we find that `Object class` is a subclass of `Class`, which is an instance of `Class class`; `Class` is a subclass of `ClassDescription`, and that of `Behavior`, and that of `Object`, and in Squeak we also have `ProtoObject` at the apex of the hierarchy. The situation is presented graphically in Figure 14-2. Note that we do not include any class below `Object` here. The reader can try and combine Figure 14-1 and Figure 14-2 (don't forget to add the metaclasses, starting from `SmallInteger class` and working your way up). It may be that we cannot have too much of a good thing; it is nice to have everything an object, but following the axiom to all its consequences does not necessarily lead to an easily accessible structure. Fortunately, most Smalltalk programmers do not need to be concerned with such matters.

There are other consequences of the "everything is an object" and "everything is carried out by messages" dicta that do concern everyday programmers. Any programmer with a minimum of programming experience would expect that:

```
3 + 5 * 7 == 38
```

would be true, but alas, in Smalltalk the following is true:

```
3 + 5 * 7 == 56
```

The reason is that the binary arithmetic operators are in fact nothing but selectors of the number classes. They are therefore parsed as any other message, and there is nothing dictating that one binary message should have precedence over another. True, it makes sense once you are familiar with the rules of the game. But it still makes one wonder. It is easy to check arithmetic calculations in a workspace in a Smalltalk environment, but it might be better if the programmer never really had to think about such issues.

The Smalltalk environment itself is one of the great innovations of Smallktalk. Back in the 1980s, when graphical displays were a rarity and most programming was done on monochrome text terminals, Smalltalk implemented a graphical user interface (GUI) for a language built on top of a virtual machine. That might have been ahead of its time. Virtual machines had to wait another decade for Java to bring them into mainstream programming. GUIs won the battle, but only when hardware prices came down. In the meantime, an environment with these requirements was deemed (and perhaps it was) unsuitable for most purposes at the time. But there might also be something subtler going on.

The Smallktalk environment is really an ecosystem of classes with which you not only *can work*, but with which you *have to work*. The only reasonable way to create a new class in Smalltalk is to use the appropriate browser to find the class you want, from which you will subclass it. You may be used to your favorite editor and your well-honed skills with build command-line tools in several languages, but Smalltalk is different. You may love the class libraries that Smalltalk designers thoughtfully provided for you, but if you think otherwise, there is not much you can do. It is nice when you are inside, but you do have to move inside to do anything with it. This may not be what programmers are prepared to do. It is possible that, after reading the first few pages here, you rushed to download and install a Smalltalk installation to see for yourself, and that you threw your hands up in despair when you realized how different and unfamiliar everything is.

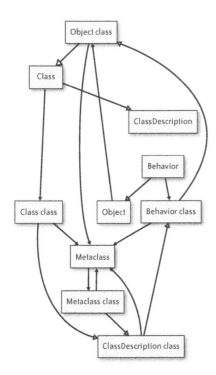

FIGURE 14-2. *Spaghetti à la Smalltalk*

Which hints at why Smallktalk never really made it to the mainstream. Smalltalk is an adamant language; it did not make compromises. It defined a new programming model, used concepts that were later adopted by many other languages, set in many ways an example, and was imitated. This is not very different from the world of architecture.

Brick and Mortar Architecture

Among architectural wonders, the most celebrated house in America must be Fallingwater, designed by Frank Lloyd Wright in 1935, "probably the most frequently illustrated house of the twentieth century" (Nuttgens 1997, p. 264). Fallingwater is built above a waterfall in a ravine called Bear Run. The house was designed for Edgar Kaufmann, Sr., a millionaire Pittsburgh businessman. It was used as the Kaufmanns' weekend home from 1937 to 1963, when it was donated to the Western Pennsylvania Conservancy, and opened to the public as a museum in 1964.

Even photographs of the building (see Figure 14-3) evoke serenity and partake of the beauty of the house and its surroundings. Wright strove to integrate nature and architecture, to make art and nature reflect each other; in Fallingwater nature is drawn inside the building, and the building becomes a part of it. This is a masterpiece of modern architecture that still invites us to contemplate the meanings that the architect wanted to put into his art.

But a house is not only an object to behold, but something to live in. We cannot live in Fallingwater—at best we can only visit it—but perhaps we can imagine how living in it would be.

It is very likely that we would be wrong. In Steward Brand's *How Buildings Learn* (1995, p. 58), we learn that:

> Wright's late-in-life triumph, Fallingwater in Pennsylvania, celebrated by that AIA poll as the "best all-time work of American architecture," lives up to its name with a plague of leaks; they have marred the windows and stone walls and deteriorated the structural concrete. To its original owner, Fallingwater was known as "Rising Mildew," a "seven-bucket building." It is indeed a gorgeous and influential house, but unlivable. (Quoted from Judith Donahue, "Fixing Fallingwater's Flaws," *Architecture*, Nov. 1989, p. 100.)

This judgment may have been too harsh. Edgar Kaufmann, Jr., who did live in Fallingwater, expressed the situation a bit differently:

> Mistakes have plagued Fallingwater, yet the extraordinary beauty of the house and the delight it brought to the life of its inhabitants form the context in which its construction should be evaluated. Life at Fallingwater did include flaws and the efforts to overcome them. (Kaufmann 1986, p. 49)

This judgment should again be taken with a pinch of salt. Edgar Kaufmann, Jr. was not a disinterested party. He entered Wright's Taliesin Fellowship in 1934. (Twenty-three apprentices came to live and learn in 1932 at Taliesin, in Spring Green, Wisconsin, thus starting what still exists as "The Frank Lloyd Wright School of Architecture"; the school now has two campuses, Taliesin and Taliesin West in Arizona.) He introduced the architect to his parents, and he was often the intermediary between Wright and his parents during the building's design and construction.

FIGURE 14-3. Fallingwater

Another architectural wonder, perhaps the most influential house of the 20th century (influential in the sense that its style defined the morphology of modern cities the world over), is the Villa Savoye, at Poissy, outside Paris, designed by the Swiss architect Charles-Édouard Jeanneret-Gris, a.k.a. Le Corbusier (see Figure 14-4). Villa Savoye, like Fallingwater, was designed as a weekend house. It was constructed between 1928 and 1931.

FIGURE 14-4. The Villa Savoye

Wright had strong views on the role of architecture; Le Corbusier was also strong-willed, but had a rather different set of ideals. Whereas Wright worked on the relationship between art and nature:

> Le Corbusier invented a proportional system—the "Modulor," which drew together the Golden Section, a six-foot human figure, and harmonic proportions in an elaborate Corbusian theorem supposedly reconciling mechanization and "natural order." (Curtis 1996, p. 412)

The Villa Savoye has provoked awe on architects:

> Like any work of a high order, the Villa Savoye evades facile categorization. It is simple and complex, cerebral and sensuous. Laden with ideas, it still expresses these directly through shapes, volumes, and spaces "in a certain relationship." A "classic" moment of modern architecture, it also has affinities with the architecture of the past. It was a central concern of Le Corbusier's philosophy that a vision of contemporary life be given expression in architectural forms of perennial value, and in the Villa Savoye one recognizes echoes of old classical themes: repose, proportion, clarity, simple trabeation. (Curtis 1996, p. 284)

People may disagree on matters of state, and for some the Villa Savoye may look like a white box ready for lift-off; after all, *de gustibus et coloris non est diputandum*. It is on a different set of considerations, that of its practical value as house, that we may pass more unequivocal judgement. Here Le Corbusier's clients took a rather different view:

> In spite of initial protests from the Savoyes, Le Corbusier insisted—supposedly on technical and economic grounds alone—that a flat roof would be preferable to a pitched one. It would, he assured his clients, be cheaper to construct, easier to maintain and cooler in summer, and Madame Savoye would be able to do her gymnastic exercises on it without being bothered by damp vapours emanating from the ground floor. But only a week after the family moved in, the roof sprang a leak over Roger's [the Savoyes' son] bedroom, letting in so much water that the boy contracted a chest infection, which turned into pneumonia, which eventually required him to spend a year recuperating in a sanatorium in Chamonix. (De Botton 2006, p. 65)

This reads like a practical joke from the architect on his clients, but it seems that the architect had a narrow escape himself:

> In September 1936, six years after the villa's official completion, Madame Savoye compressed her feelings about the performance of the flat roof into a (rain-splattered) letter: 'It's raining in the hall, it's raining on the ramp, and the wall of the garage is absolutely soaked. What's more, it's still raining in my bathroom, which floods in bad weather, as the water comes in through the skylight.' Le Corbusier promised that the problem would be fixed straightaway, then took the opportunity to remind his client of how enthusiastically his flat-roofed design had been received by architectural critics worldwide: 'You should place a book on the table in the downstairs hall and ask all your visitors to inscribe their names and addresses in it. You'll see how many fine autographs you will collect'. But this invitation to philography was of little comfort to the rheumatic Savoye family. 'After innumerable demands on my part, you have finally accepted that this house which you built in 1929 is uninhabitable,' admonished Madame Savoye in the autumn of 1937. 'Your responsibility is at stake and I have no need to foot the bill. Please render it habitable immediately. I sincerely hope that I will not have to take recourse to legal action.' Only the outbreak of the Second World War and the Savoye family's consequent flight from Paris saved Le Corbusier from having to answer in a courtroom for the design of his largely uninhabitable, if extraordinarily beautiful, machine-for-living. (De Botton 2006, pp. 65–66)

Another figurehead of modern architecture, Ludwig Mies van der Rohe, used a minimalistic formal system based on I-beams. He "kept full scale I-beam details by his desk to get the proportions just so. He thought [the I-beam] was the modern equivalent of the Doric Column" (Jencks 2006, p. 13).

Mies van der Rohe is more famous for his use of "less is more," by which he wanted to bring architecture back to the main essentials: no ornament, decoration, or superfluous elements, unless they served a functional purpose. I-beams are part of the essentials of a building.

Or so it would seem to be. One of Mies van der Rohe's major works is the Seagram building in New York City, completed in 1958 (see Figure 14-5). There the architect faced a conundrum: much as he wanted the I-beams to show through the construction, that was impossible under American construction regulations, which demanded that metal supports should be enclosed in fireproof material, for example, concrete. But strangely, a close look at the Seagram building shows I-beams on its surface. These are not the real supporting I-beams. Mies van der Rohe added fake I-beams on the surface so that they would "reveal" the structure underneath. Moreover, so that the visual impression of the building would not be jeopardized, the window shades could stand in only three positions: open, closed, and half-way—probably not the best arrangement to protect against sun glare (Wolfe 1982, pp. 75–76).

FIGURE 14-5. Seagram building

The leading modern architect Louis Sullivan (one of the creators of the high-rise building, the skyscraper, and mentor to Wright, among other things) famously wrote:

> Whether it be the sweeping eagle in his flight, or the open apple-blossom, the toiling work-horse, the blithe swan, the branching oak, the winding stream at its base, the drifting clouds, over all the coursing sun, *form ever follows function*, and this is the law. Where function does not change

form does not change. The granite rocks, the ever-brooding hills, remain for ages; the lightning lives, comes into shape, and dies in a twinkling.

It is the pervading law of all things organic and inorganic, of all things physical and metaphysical, of all things human and all things superhuman, of all true manifestations of the head, of the heart, of the soul, that the life is recognizable in its expression, that form ever follows function. *This is the law*. (Sullivan 1896)

Mies van der Rohe then seems to be putting "form follows function" on its head. Or it may be that such aphorisms are more useful to provoke, rather than to describe what is actually happening. Paul Rand was arguably the leading graphic American designer. He was responsible for the IBM, ABC, and the original UPS logos; he collaborated with Steve Jobs in NeXT Computer; he wrote influential books exposing his theory of design. He commented:

That the separation of form and function, of concept and execution, is not likely to produce objects of aesthetic value has been repeatedly demonstrated. Similarly any system that sees aesthetics as irrelevant, that separates the artist from his product, that fragments the artist from his product, that fragments the work of the individual, or create by committee, or makes mincemeat of the creative process will in the long run diminish not only the product but the maker as well. (Rand 1985, p. 3)

It is easy to pour opprobrium on modern architecture, especially since the work of Mies van der Rohe and Le Corbusier has been imitated badly the world over, and held responsible for many ugly neighborhoods, crime-infested worker housing, and soulless business centers. It is more interesting to look deeper into the criticism of the masters themselves. There we see that Wright, Le Corbusier, and Mies van der Rohe are all criticized because they were unyielding; they were criticized for being adamant, for their lack of compromise. Their strident views gave us beautiful buildings, but not buildings that would provide us with material comfort.

Being adamant is not necessarily a flaw. In an interview in Doctor Dobb's Journal in April 1996, Donald Knuth was asked his opinion on Edsger Dijkstra. "His great strength," he replied, "is that he is uncompromising. It would make him physically ill to think of programming in C++." Uncompromising to the point of not touching a computer for years, and writing "his really terrific essay on the *Humble Programmer* discussing this." Dijkstra was one of the most influential computer scientists; his writings are still alive with valuable advice, and they make a salutary reading when attacked by the latest fashion or silver bullet in programming. His adamant stance made his work all the more precious to those programmers who do have to write programs that run on computers in the real world.

This may be the key to understand the role that Smalltalk, like Algol before it, has to play in our professional lives. There are some architects who blaze new paths and create monuments for the generations to come, and it may be the nature of these buildings to be more manifestos than actual houses or offices. Nobody disputes Fallingwater's power to move the visitor and to inspire young architects, even if we may argue about it being a home, or we may clamor after "form follows function" and let ornament in through the back door. There are software systems that, similarly, have been more successful at influencing code than writing it.

If we want to write code then, for business or for pleasure, we need inspiration from beautiful architecture, but we may not be able to work with it. Our work will have to exemplify beautiful architecture, but it has to be practical. The purest and most beautiful of intellectual edifices is the world of pure mathematics; we have a lot to learn from it, but we cannot program in it. We have to make something work, and this is where things start getting messy. Sometimes it is easy to get lost in design methodology, forgetting that our aim is different. Christopher Alexander, the architect father of design patterns had this to say:

> A whole academic field has grown up around the idea of "design methods"—and I have been hailed as one of the leading exponents of these so-called design methods. I am very sorry that this has happened, and want to state, publicly, that I reject the whole idea of design methods as a subject of study, since I think it is absurd to separate the study of designing from the practice of design. In fact, people who study design methods without also practicing design are almost always frustrated designers who have no sap in them, who have lost, or never had, the urge to shape things. (Alexander 1971)

As programmers, we must make things that work, not simply things that look beautiful. These two are not necessarily incompatible. Figure 14-6 shows the Salginatobel Bridge designed by Robert Maillart and completed in 1930. Maillart, a Swiss, studied engineering, but his works, his bridges especially, are exemplars of architectural beauty. Crucially, they are not just beautiful. Maillart built his bridges by winning contracts in competitive tenders, and for Salginatobel Bridge, he won the contract by outbidding 19 other competing designs. Construction of the bridge and the road cost only 700,000 Swiss Francs at that time, less than $4 million today. The bridge is no minnow. It has a span of 90 meters and it vaults 80 meters above the ravine of the Salgina brook (Billington 2000). The slenderness and lightness of the structure is exactly what makes it economical. This bridge is economical thanks to its elegance.

It could be that Maillart's chief virtue was his pragmatism. He arrived at his designs by a form of creative intuition. He eschewed decoration and ornaments, as well as imitation of traditional architectural styles. The structures he designed could not be analyzed with the mathematical tools of the time (and the lack of computers), so they could not be proven to be sound. He evaluated the feasibility of his designs using a simplified graphical analysis. If Maillart had to wait for rigorous validations of his designs, none of them would be constructed (he died in 1940). Maillart "found that innovation, especially in bridge design, came not from laboratory work and mathematical theories, but from design offices and construction sites. Numbers play an essential role in engineering. But innovation in bridge design was the product of visual-geometric imagination, not the outcome of abstract numerical studies or deduction from general theories" (Billington 1997, pp. 1–2).

FIGURE 14-6. Salginatobel Bridge

Programming, like architecture, is a story of practice. We had better avoid being dogmatic, and instead focus on what works:

> Architecture is a hazardous mixture of omnipotence and impotence. Ostensibly involved in "shaping" the world, for their thoughts to be mobilized architects depend on the provocations of others—clients, individual or institutional. Therefore, incoherence, or more precisely, randomness, is the underlying structure of all architects' careers: they are confronted with an arbitrary sequence of demands, with parameters they did not establish, in countries they hardly know, about issues they are only dimly aware of, expected to deal with problems that have proved intractable to brains vastly superior to their own. Architecture is by definition a *chaotic adventure*. (Koolhas et al. 1998, p. xix)

Architecture is a chaotic adventure because beautiful architecture alone is not enough; not only beauty, but also usefulness, is the law for architecture and programming alike.

References

Abrahams, David, and Aleskey Gurtovoy. 2005, *C++ Template Metaprogramming: Concepts, Tools, and Techniques from Boost and Beyond*. Boston, MA: Addison-Wesley.

Alexander, Christopher. 1971. *Notes on the Synthesis of Form*, Preface to the paperback edition. Cambridge, MA: Harvard University Press.

Alpert, Sherman R., Kyle Brown, and Bobby Woolf. 1998. *The Design Patterns Smalltalk Companion*. Boston, MA: Addison-Wesley.

Billington, David P. 2000. "The Revolutionary Bridges of Robert Maillart." *Scientific American*. July, pp. 85–91.

Billington, David P. 1997. *Robert Maillart: Builder, Designer, and Artist*. New York, NY: Cambridge University Press.

Black, Andrew P., et al. 2007. *Squeak By Example*. Square Bracket Publishing.

Bloch, Joshua. 2008. *Effective Java*, Second Edition. Boston, MA: Addison-Wesley.

Brand, Stewart. 1997. *How Buildings Learn: What Happens After They're Built*, Revised Edition. London, UK: Phoenix Illustrated.

Calvino, Italo. 1986. "Why Read the Classics?" *The Uses of Literature*. Translated by Patrick Creagh. New York, NY: Harcourt Brace Jovanovich.

Conroy, Thomas J., and Eduardo Pelegri-Llopart. 1983. "An Assessment of Method-Lookup Caches for SmallTalk-80 Implementations." *Smalltalk-80: Bits of History, Words of Advice*. Ed. Glenn Krasner. Boston, MA: Addison-Wesley.

Curtis, William J. R. 1996. *Modern Architecture Since 1900*, Third Edition. New York, NY: Phaidon Press.

De Botton, Alain. 2006. *The Architecture of Happiness*. London, UK: Hamish Hamilton.

Gamma, Erich, et al. 1994. *Design Patterns: Elements of Reusable Object-Oriented Software*. Boston, MA: Addison-Wesley.

Goldberg, Adele, and David Robson. 1989. *Smalltalk-80: The Language*. Boston, MA: Addison-Wesley.

Forman, Ira R., and Scott H. Danforth. 1999. *Putting Metaclasses to Work: A New Dimension in Object Oriented Programming*. Boston, MA: Addison-Wesley.

Jencks, Charles. 2006. *The New Paradigm in Architecture: The Language of Post-Modernism*. New Haven, CT: Yale University Press.

Josuttis, Nicolai M. 1999. *The C++ Library: A Tutorial and Reference*. Boston, MA: Addison-Wesley.

Kaufmann, Edgar Jr. 1986. *Fallingwater: A Frank Lloyd Wright Country House*. New York, NY: Abbeville Press.

Koolhas, Rem, et al. 1998. *S, M, L, XL*, Second Edition. New York, NY: The Monacelli Press.

McCarthy, John. 1960. "Recursive Functions of Symbolic Expressions and Their Computation by Machine, Part I." *Communications of the ACM*, April 1960.

McConnell, Steve. 2004. *Code Complete*, Second Edition. Redmond, WA: Microsoft Press.

Meyers, Scott. 2005. *Effective C++: 55 Ways to Improve your Programs and Designs*. Boston, MA: Addison-Wesley.

Norman, Donald. 1988. *The Psychology of Everyday Things*. New York, NY: Basic Books.

Nuttgens, Patrick. 1997. *The Story of Architecture*, Second Edition. New York, NY: Phaidon Press.

Petzold, Charles. 1999. *Programming Windows*, Fifth Edition. Redmond, WA: Microsoft Press.

Rand, Paul. 1985. *A Designer's Art*. New Haven, CT: Yale University Press.

Simon, Herbert. 1996. *The Sciences of the Artificial*. Cambridge, MA: MIT Press.

Stroustrup, Bjarne. 1985. *The C++ Programming Language*. Boston, MA: Addison-Wesley.

Stroustrup, Bjarne. 1994. *The Design and Evolution of C++*. Boston, MA: Addison-Wesley.

Sullivan, Louis H. 1896. "The tall office building artistically considered." *Lippincott's Magazine*, March 1896.

Thomas, David, et al. 2005. *Programming Ruby: The Pragmatic Programmers Guide*, Second Edition. Raleigh, NC, and Dallas, TX: The Pragmatic Bookshelf.

Vandervoorde, David, and Nicolai M. Josuttis. 2002. *C++ Templates: The Complete Guide*. Boston, MA: Addison-Wesley.

Wolfe, Tom. 1982. *From Bauhaus to Our House*. London, UK: Jonathan Cape.

Afterword

Building Beautifully

William J. Mitchell

LOOSE ANALOGIES ARE OFTEN MADE BETWEEN SOFTWARE SYSTEMS and works of architecture. But the structural similarity between these different sorts of systems is actually closer, and more rigorously specifiable, than it might seem.

The code comprising software systems consists of one-dimensional strings of symbols, put together from well-defined vocabularies in accordance with precise syntactic rules, and intended to produce useful results when run on appropriate machines.

Works of architecture obviously aren't one-dimensional, but otherwise they are very much like code. They are three-dimensional assemblies of discrete physical components, put together from fairly well-defined component vocabularies in accordance with reasonably rigorous syntactic rules, and intended to serve useful purposes. (Architects do, in practice, have a little more latitude with vocabulary and syntax than programmers.)

In both cases we can write formal grammars to establish the rules of the game. In general, formal grammars tell you how to put things together. More precisely, according to the standard definition used in linguistics and computer science, a formal grammar consists of: a finite set N of nonterminal things; a finite set T of terminal things; a finite set R of replacement rules; and an initial thing S. A replacement rule has an assembly of things on its left side, an arrow in the middle, and another assembly on its right side. It specifies that you can replace the assembly on the left side with the assembly on the right side. Complete, correctly structured assemblies are derived by recursively applying the replacement rules *R* to the initial thing *S*. The language specified by the grammar consists of all the assemblies of terminal things that can be derived in this fashion.

Formal grammars commonly apply to vocabularies of words and tell you how to put together complete and correct sentences, but they are not limited to this. They can apply to many different sorts of things and tell you how to put together useful assemblies of them.

In the case of a grammar for a programming language, the things to be put together are recognizable symbols, the replacement rules have one-dimensional strings of symbols on their left and right sides, and these rules derive complete and correct expressions in that language. In the case of a two-dimensional shape grammar, the things to be put together are two-dimensional shapes, the replacement rules have two-dimensional assemblies of shapes on their left and right sides, and they derive complete and correct graphic designs. In the case of three-dimensional architectural grammars, the nonterminal things to be put together are construction lines and the like, the terminal things to be put together within skeletons of nonterminals are actual architectural components, and the replacement rules derive complete and correct compositions of these components—in other words, designs in the architectural language that the grammar specifies.

Decades ago, for example, George Stiny and I published a formal grammar, in this format, for the famously beautiful villas of the great Italian Renaissance architect Andrea Palladio.* It derives all of the known villa designs by Palladio, together with large numbers of convincing fake Palladian villas. (Or, you might say, it derives all the villas that Palladio might have designed if he had lived longer and had more clients.) Furthermore, it provides a cogent explanation of the underlying principles of Palladio's villa architecture. Since then, numerous architectural grammars have been written for other bodies of design work.

One of the most important functions of an architectural grammar is to capture the principles or modularity and hierarchical organization that characterize works in some particular architectural style. In the precisely defined and widely used language of classical architecture, for example, a column has a base, a shaft, and a capital. The capital decomposes further into a hierarchy of components—different for Doric, Ionic, and Corinthian—and so on. Moving up the hierarchy of subassemblies, regularly spaced columns form colonnades. Then columns, entablatures, and pediments form porticos. Eventually, all the components and subassemblies fit nicely together to constitute complete, grammatical, classical compositions. These compositions can, like sentences, be parsed into named parts.

From a geometric or CAD system perspective, the components and subassemblies of a building are discrete shapes that can be transformed and assembled to produce larger spatial compositions. From a supply chain and construction perspective, they are material elements

---

* George Stiny and William J. Mitchell, "The Palladian Grammar," *Environment and Planning B*, vol. 5, no. 1, 5–18 (1978).

that are fabricated (often in a lengthy sequence of steps, each one of which adds value), acquired, transported to the required location, and finally fixed into place within an assembly. From a building maintenance perspective, they are replaceable parts. And from a functional perspective, they are modules that perform identifiable tasks within a building, and that can be composed with other modules to create subsystems that perform higher-level tasks. When modules are composed in this way, they not only have spatial relationships that may delight the eye, they also transfer something—structural loads, for example—across their interfaces by virtue of these relationships.

Similarly, programming languages provide ways of breaking code down into modules, and hierarchically assembling modules to produce higher-level modules and eventually complete software systems. As all programmers know, good code isn't a mess; it has a clear, logical structure of modules and hierarchies, implemented using the language's abstraction and organization constructs. The organizational clarity of classical architecture provides an excellent model for this.

With occasional exceptions, works of architecture follow additional principles of internal order as well. Columns lined up in a row, for example, are normally regularly spaced. If you want to write code to generate a CAD model of a colonnade, then you don't specify the location of each column individually. You employ iteration, with a location parameter that is incremented at each step. In other words, you express the principles of the architecture in a more concise and elegant way, and one that provides the reader of the code with more insight.

What if you want to generate a regular column grid? You employ nested iteration. First you iterate a column to generate a regularly spaced row of columns, and then you iterate the row of columns as many times as you want to create the grid.

What if you want to make the corner columns different from the internal columns? (Architects often do this in response to the different structural and other conditions that exist at the corners of buildings.) You use a conditional: *if* it's a corner condition, *then* substitute the alternative column design. If you want to vary the spacing of columns at the center to mark the importance of the central axis, distinguish external columns from internal columns, and so on, you just introduce additional conditionals.

Modularity, hierarchy, and regular repetition are by no means the only ordering principles commonly employed by architects. If you analyze architectural compositions carefully, you can often find regularities in dimensions and proportions, symmetries (and artfully broken symmetries), nested self-similar shapes as in fractals, and parametrically varied motifs. I leave it as an exercise for the reader to imagine the expression of these principles in code.

Sometimes, though, internal order seems to be lacking. An architect might, for some reason, scatter columns randomly. What then? It turns out that the most concise way to describe this sort of configuration is to specify the location of each column individually. There is no shorter, more economical description.

As these simple examples suggest, good architects don't construct their designs by ad-hoc, brute force means, and they certainly try to avoid clumsy hacks. In response to the complex demands of site and context, climate, the activities to be accommodated, supply chains of materials and components, construction processes, and budgets, they design commensurately varied and complex buildings. But they try to do this with conceptual elegance—following principles of economy of means, and rigorously applying their own version of Occam's Razor. Underlying the apparent variety and complexity of beautiful works of architecture, then, you can usually discover some simple, elegant principles of functional organization and formal order. Discovering these principles takes intellectual engagement—which is a crucial part of the experience and pleasure of architecture.

If you *can* figure out these principles, you can construct models of these works with a few equally elegant lines of code in some standard programming language, or (in a less traditional programming environment) a few shape rewriting rules. You might even be able to generalize, and write code that produces designs, following the same principles, that respond appropriately to ranges of varied conditions and requirements. But if you can't figure out these principles, then you are condemned to write lengthier, less insightful code.

The evident complexity of a building derives from the complexity of the requirements to which the architect has responded, and it is measured by the length of its brute force, point-by-point description. The principles that the architect followed to produce that response can usually be captured in a much shorter sequence of code that executes to generate all that complexity. To a pretty good approximation, the lower the ratio of this short description to the long description, the more beautiful the building.

Architects, then, admire the beauty of buildings that employ a few simple, elegant principles to accomplish many complex things. Similarly, software architects and programmers admire the beauty of code that cleanly and concisely (without sacrificing readability or maintainability) performs many complex tasks. And scientists admire the beauty and explanatory power of simple laws that describe widely varied phenomena. These are all particular cases of a beautiful general principle.

Contributors

TILL ADAM spent his youth studying philosophy, comparative literature, American studies, and musicology while making a living as a musician. Having failed to become rich and famous that way, he went for a M.Sc. in maths, computer science, and business. Some years of contributing to Free Software, in particular KDE, taught him to program, which led to his employment at Klarälvdalens Datakonsult AB, where he now coordinates the KDE- and Free Software-related activities of the company, among other things. Till lives with his wife and daughter in Berlin, Germany.

JIM BLANDY maintained GNU Emacs for the Free Software Foundation from 1990 to 1993, and released Emacs version 19 with Richard Stallman. He is one of the original designers of the Subversion version control system. He has also contributed to the CVS version control system, the GNU Debugger (GDB), the Guile extension language library, and an Emacs made for editing gene sequences. He now works for the Mozilla Corporation on SpiderMonkey, Mozilla's implementation of the JavaScript programming language. Jim lives in Portland, Oregon with his wife and two daughters.

MIRKO BOEHM has been a KDE developer since 1997, and was a member of the KDE e.V. board from 1999 to 2006. He is a business graduate from Helmut Schmidt University in Hamburg, Germany. In his private life he tries to stay away from computers by reading books printed on real paper and spending time with his family. He currently works at Klarälvdalens Datakonsult AB in Berlin, Germany, where he specializes in cross-platform and embedded software development.

CHRISTOPHER DENNIS has been the lead developer on the JPC project since its outset in 2005. Chris adopted Java while working on his doctorate at the University of Oxford. Prior to that, he worked with all manner of programming languages, from Z80 machine code written on hex keypads to PHP and JavaScript. A fan of corner cases, coding tricks, and the occasional ugly hack, Chris enjoys writing compact, elegant code in any language.

DAVE FETTERMAN is an engineering manager at Facebook, where he founded the Facebook Platform project. Prior to joining Facebook in 2006, he worked as a software engineer on projects in Microsoft's developer division, including the .NET Common Language Runtime (CLR). He loves creating software for other developers and talking at great length about it to anyone who will listen. He received a bachelor's degree in applied mathematics and a master's degree in computer science in 2003 from Harvard University.

KEIR FRASER is a founder of XenSource, now part of Citrix Systems, and lead architect of the Xen hypervisor. Keir implemented the original version of Xen in 2002 as a diversion from his Ph.D. studies in the Cambridge Computer Laboratory, and continued to lead development as the project grew into a large community effort. He was awarded a Ph.D. for his work on lock-free concurrency control in 2004, and became a member of the faculty the same year.

PETE GOODLIFFE is a programmer, columnist, speaker, and author who never stays at the same place in the software food chain. Pete's popular book, *Code Craft* (No Starch Press), is a practical and entertaining investigation of the entire programming pursuit—all in about 600 pages, which is no mean feat! He has a passion for curry and doesn't wear shoes.

GEORGIOS GOUSIOS is a researcher by profession, a software engineer by education, and a software enthusiast by passion. Currently, he is working on his Ph.D. thesis at the Athens University of Economics and Business in Greece. His research interests include software engineering, software quality, virtual machines, and operating systems, and he holds an M.Sc. with distinction from the University of Manchester, UK. Gousios has contributed code to various OSS projects and also has worked in various research and development projects in both academic and commercial settings. He was the project manager, design authority, and core development team member of the SQO-OSS project, researching novel ways for evaluating software quality. In his academic life, Gousios has published 10 technical papers in referred conferences and journals. Gousios is a member of the ACM, the IEEE, the Usenix Association, and the Technical Chamber of Greece.

DAVE GROVE is a research staff member in the Dynamic Optimization Group at IBM's T.J. Watson Research Center. His primary research interests include the analysis and optimization of object-oriented languages, virtual machine design and implementation, JIT compilation, online feedback directed optimization, and garbage collection. He joined the Jalapeño project in 1998 and was a key contributor to the initial implementation of the optimizing compiler and the adaptive optimization system. Since Jalapeño went open source as Jikes RVM in 2001, he has been active as a member of the Jikes RVM core team and steering committee.

John Klein is a senior member of technical staff at the Software Engineering Institute (SEI), where he works on architecture methods for systems-of-systems and helps individuals, teams, and organizations improve their software architecture competence. Before joining SEI, John was a chief architect at Avaya, Inc. There his responsibilities included development of multimodal agents, architectures for communication analytics, and the creation and enhancement of architectures in the Customer Interaction product portfolio. Prior to that, John was a software architect at Quintus, where he designed the first commercially successful multichannel integrated contact center product and led the technology integration of the product portfolio as Quintus acquired two other companies. Before joining Quintus, John worked for several companies in the video conferencing and video networking industry. He began his professional career at Raytheon, where he developed hardware and software solutions for radar signal processing, multispectral image processing, and parallel processing architectures and algorithms. John holds a B.E. degree from the Stevens Institute of Technology and an M.E. degree from Northeastern University. He is a member of the ACM and IEEE Computer Society.

Greg Lehey spent his long career in Germany and Australia, working for the German space research agency, for computer manufacturers such as Univac, Tandem, Siemens-Nixdorf, and IBM, for nameless software houses as a large user, and for himself as a consultant. His activities ranged from kernel development to product management, from systems programming to systems administration, from processing satellite data to programming petrol pumps, and from the production of CD-ROMs of ported free software to DSP instruction set design. He was a member of the FreeBSD Core Team and president of the Australian Unix User Group. He is a developer in the FreeBSD and NetBSD projects and the author of *Porting Unix Software* and *The Complete FreeBSD*, Fourth Edition (both O'Reilly). He has also been known to write commercial applications software. Greg retired in 2007 and has spent the ensuing time trying to find a life. His leisure activities now take up most of his time, which still isn't enough, and include classical woodwind music, cooking, brewing beer (he has developed a computer-controlled fermentation system), gardening, horse riding, and photography. He is also interested in a number of historical topics, including ancient and obscure European languages. Browse his home page at *http://www.lemis.com/grog/*.

Panagiotis Louridas got involved with computers in the 1980s with a Sinclair ZX Spectrum; he has been coding, and enjoying, the language of machines ever since. He received his diploma in computer science from the Department of Informatics of the University of Athens, and his M.Sc. and Ph.D. in computation from the University of Manchester. Over the years he has developed software for the private sector and is currently working for the Greek Research and Education Network (GRNET). He is also a member of the Athens University of Economics and Business (AUEB) Software Engineering and Security (SENSE) research group. He has published in subjects ranging from anthropology to cryptography, and from instrument representation to software engineering. He especially enjoys seeking connections from the world of computing to other fields.

Stephen J. Mellor is an internationally recognized pioneer in creating effective, engineering approaches to software development. In 1985, he published the widely read Ward-Mellor trilogy, *Structured Development for Real-Time Systems* (Prentice Hall), and in 1988, the first books defining object-oriented analysis. Stephen also published *Executable UML: A Foundation for Model-Driven Architecture* (Addison-Wesley Professional) in 2002. His latest book, *MDA Distilled: Principles of Model-Driven Architecture* (Addison-Wesley Professional), was published in 2004. He is active in the Object Management Group, chairing the consortium that added executable actions to the UML, and he recently completed on a standard for executable UML. He is a signatory to the Agile Manifesto. He was a two-term member of the OMG Architecture Board, chair of the *IEEE Software* Advisory Board, and, until recently, chief scientist of the Embedded Software Division at Mentor Graphics.

Bertrand Meyer is professor of software engineering at ETH Zurich and chief architect of Eiffel Software, where he led the design of the EiffelStudio environment and numerous libraries. He is the author of several best-selling books, including *Object-Oriented Software Construction* (Prentice Hall), winner of the Jolt Award. He also received the ACM Software System Award and the Dahl-Nygaard Award for his work on object technology and Eiffel, and an honorary doctorate from the State Technical University of St. Petersburg. His research interests cover object technology, programming languages, and software verification, including test, concurrency, and formal methods. He is also an active consultant and lecturer.

William J. Mitchell is Alexander Dreyfoos Professor of Architecture and Media Arts and Sciences at MIT, where he directs the Smart Cities group in the MIT Media Laboratory and the MIT Design Laboratory. He previously served as Dean of the School of Architecture and Planning at MIT. His recent books include *World's Greatest Architect* and *Imagining MIT* (both MIT Press).

Derek Murray is a Ph.D. student at the University of Cambridge's Computer Laboratory. He joined the Xen project in 2006 and worked on improving Xen security by rearchitecting the control stack. His research now looks at improving failure tolerance in large-scale distributed systems, but he still visits ring zero occasionally. Derek was awarded an M.Sc. in high performance computing from the University of Edinburgh in 2006, and a B.Sc. in computing science from the University of Glasgow in 2005.

Rhys Newman adopted Java while completing his doctorate at Oxford University over a decade ago, when Java was only a couple of years old. In his early research he demonstrated how high-performance real-time vision processing could be done, even with these early JITed JVMs, within a pure Java environment. Since then he has worked in both academia and industry, proving time and again how flexible, productive, and fast the Java platform really is. He has won several industry awards for technical excellence over a 20-year software engineering career, and most recently returned to Oxford to undertake groundbreaking research in the field of Grid Computing. JPC is one part of this latest research effort.

MICHAEL NYGARD strives to raise the bar and ease the pain for developers across the country. He shares his passion and energy for improvement with everyone he meets, sometimes even with their permission. Michael has spent the better part of 20 years learning what it means to be a professional programmer who cares about art, quality, and craft. He's always ready to spend time with other developers who are fully engaged and devoted to their work—the "wide awake" developers. On the flip side, he cannot abide apathy or wasted potential. Michael has been a professional programmer and architect for nearly 20 years. During that time, he has delivered running systems to the U.S. government, the military, banking, finance, agriculture, and retail industries. More often than not, Michael has lived with the systems he built. This experience with the real world of operations changed his views about software architecture and development forever. He worked through the birth and infancy of a Tier 1 retail site and has often served as "roving troubleshooter" for other online businesses. These experiences give him a unique perspective on building software for high performance and high reliability in the face of an actively hostile environment. Most recently, Michael wrote *Release It! Design and Deploy Production-Ready Software* (Pragmatic Programmers), which won a Jolt Productivity award in 2008. His other writings can be found at *http://www.michaelnygard.com/blog*.

IAN ROGERS is a Research Fellow in the University of Manchester's Advanced Processor Technology research group. His Ph.D. research work in the Dynamite binary translator was exploited commercially and now forms part of many binary translator products, including Apple's Rosetta. His recent academic work has been in programming language design, runtime, and virtual machine environments—in particular how to allow them to automatically create and efficiently exploit parallelism. He is a leading contributor to the Jikes Research Virtual Machine, where he serves as a core team member.

BRIAN SLETTEN is a liberal arts-educated software engineer with a focus on forward-learning technologies. He has worked as a system architect, a developer, a mentor, and a trainer. He has spoken at conferences around the world and writes about web-oriented technologies for several online publications. His experience has spanned the defense, financial, and commercial domains. He has designed and built network matrix switch control systems, online games, 3D simulation/visualization environments, Internet-distributed computing platforms, P2P, and Semantic Web-based systems. He has a B.S. in computer science from the College of William and Mary and currently lives in Fairfax, Virginia. He is the president of Bosatsu Consulting, Inc., a professional services company focused on web architecture, resource-oriented computing, the Semantic Web, advanced user interfaces, scalable systems, security consulting, and other technologies of the late 20th and early 21st centuries.

DIOMIDIS SPINELLIS is an Associate Professor in the Department of Management Science and Technology at the Athens University of Economics and Business in Greece. His research interests include software engineering, computer security, and programming languages. He has written the two "Open Source Perspective" books published by Addison-Wesley: *Code Reading* (which won the Software Development Productivity Award in 2004), and *Code Quality* (which won the Software Development Productivity Award in 2007). He has also written dozens of scientific papers. He is a member of the *IEEE Software* editorial board, authoring the regular "Tools of the Trade" column. Diomidis is a FreeBSD committer and the developer of UMLGraph and other open source software packages, libraries, and tools. He holds an MEng in Software Engineering and a Ph.D. in computer science, both from Imperial College London. Diomidis is a senior member of the ACM, and a member of the IEEE and the Usenix Association.

JIM WALDO is a distinguished engineer with Sun Microsystems Laboratories, where he investigates next-generation large-scale distributed systems. He is currently the technical lead of Project Darkstar, a multithreaded, distributed infrastructure for massive multiplayer online games and virtual worlds. Prior to his current assignment with Sub Labs, he was the lead architect for Jini, a distributed programming system based on Java. Jim edited the book *The Evolution of C++: Language Design in the Marketplace of Ideas* (MIT Press), and was one of the authors of *The Jini Specification* (Addison-Wesley). He was the cochairman of the National Academies board that produced the book *Engaging Privacy and Information Technology in a Digital Age*, which he edited. Jim is also an adjunct faculty member of Harvard University, where he teaches distributed computing and topics in the intersection of policy and technology in the department of computer science. Jim received his Ph.D. in philosophy from the University of Massachusetts (Amherst).

DAVID WEISS has a B.S. in math from Union College, and an M.S. and Ph.D. in computer science from the University of Maryland. He is currently the head of the Software Technology Research Department at Avaya Laboratories, and he is looking into the problem of how to improve the effectiveness of software development in general and of Avaya's software development processes in particular. In this latter capacity, he heads the Avaya Resource Center for Software Technology. Previously he was the director of the Software Production Research Department at Lucent Technologies Bell Laboratories, which conducted research on how to improve the effectiveness of software development. Before joining Bell Labs, he was director of the Reuse and Measurement Department of the Software Productivity Consortium (SPC), a consortium of 14 large U.S. aerospace companies. Prior to joining SPC, Dr. Weiss spent a year at the Office of Technology Assessment, where he was coauthor of a technology assessment of the Strategic Defense Initiative. During the 1985–1986 academic year he was a visiting scholar at the Wang Institute, and for many years he was a researcher at the Computer Science and Systems Branch of the Naval Research Laboratory (NRL) in Washington D.C. He has also worked as a programmer and as a mathematician. Dave's principal research interests are in the area of software engineering, particularly in software development processes and methodologies, software design, and software measurement. He is best known for his invention of the goal-question-metric approach to software measurement, his work on the modular structure of software systems, and his work in software product-line engineering as a coinventor of the Synthesis process and its successor, the FAST process. He is coauthor and coeditor of two books: *Software Product-Line Engineering* and *Software Fundamentals: Collected Papers of David L. Parnas* (both Addison-Wesley Professional).